Time Reversibility, Computer Simulation, Algorithms, Chaos

2nd Edition

ADVANCED SERIES IN NONLINEAR DYNAMICS*

Editor-in-Chief: R. S. MacKay *(Univ. Warwick)*

Published

- Vol. 11 Rayleigh–Bénard Convection
  A. V. Getling

- Vol. 12 Localization and Solitary Waves in Solid Mechanics
  A. R. Champneys, G. W. Hunt & J. M. T. Thompson

- Vol. 13 Time Reversibility, Computer Simulation, Algorithms, Chaos (2nd Edition)
  W. G. Hoover & C. G. Hoover

- Vol. 14 Topics in Nonlinear Time Series Analysis – With Implications for EEG Analysis
  A. Galka

- Vol. 15 Methods in Equivariant Bifurcations and Dynamical Systems
  P. Chossat & R. Lauterbach

- Vol. 16 Positive Transfer Operators and Decay of Correlations
  V. Baladi

- Vol. 17 Smooth Dynamical Systems
  M. C. Irwin

- Vol. 18 Symplectic Twist Maps
  C. Gole

- Vol. 19 Integrability and Nonintegrability of Dynamical Systems
  A. Goriely

- Vol. 20 The Mathematical Theory of Permanent Progressive Water-Waves
  H. Okamoto & M. Shoji

- Vol. 21 Spatio-Temporal Chaos & Vacuum Fluctuations of Quantized Fields
  C. Beck

- Vol. 22 Energy Localisation and Transfer
  eds. T. Dauxois, A. Litvak-Hinenzon, R. MacKay & A. Spanoudaki

- Vol. 23 Geometrical Theory of Dynamical Systems and Fluid Flows (Revised Edition)
  T. Kambe

- Vol. 24 Microscopic Chaos, Fractals and Transport in Nonequilibrium Statistical Mechanics
  R. Klages

- Vol. 25 Smooth Particle Applied Mechanics – The State of the Art
  W. G. Hoover

- Vol. 26 Geometry of Nonholonomically Constrained Systems
  by R. H. Cushman, J. Śniatycki & H. Duistermaat

*For the complete list of titles in this series, please visit
http://www.worldscibooks.com/series/asnd_series.shtml

ADVANCED SERIES IN NONLINEAR DYNAMICS
VOLUME 13

Time Reversibility, Computer Simulation, Algorithms, Chaos

2nd Edition

William Graham Hoover
Carol Griswold Hoover

University of California, Davis, USA

World Scientific

NEW JERSEY · LONDON · SINGAPORE · BEIJING · SHANGHAI · HONG KONG · TAIPEI · CHENNAI

Published by

World Scientific Publishing Co. Pte. Ltd.
5 Toh Tuck Link, Singapore 596224
USA office: 27 Warren Street, Suite 401-402, Hackensack, NJ 07601
UK office: 57 Shelton Street, Covent Garden, London WC2H 9HE

British Library Cataloguing-in-Publication Data
A catalogue record for this book is available from the British Library.

Advanced Series in Nonlinear Dynamics — Vol. 13
TIME REVERSIBILITY, COMPUTER SIMULATION, ALGORITHMS, CHAOS
2nd Edition

Copyright © 2012 by World Scientific Publishing Co. Pte. Ltd.

All rights reserved. This book, or parts thereof, may not be reproduced in any form or by any means, electronic or mechanical, including photocopying, recording or any information storage and retrieval system now known or to be invented, without written permission from the Publisher.

For photocopying of material in this volume, please pay a copying fee through the Copyright Clearance Center, Inc., 222 Rosewood Drive, Danvers, MA 01923, USA. In this case permission to photocopy is not required from the publisher.

ISBN-13 978-981-4383-16-5
ISBN-10 981-4383-16-3

Printed in Singapore.

For Beau, Denali, and Elias
and all those others
who love learning

Preface

A decade after publishing *Time Reversibility, Computer Simulation, and Chaos*, World Scientific requested a Second Edition. We accepted. Carol took up my suggestion that we do this work together. In the decade since the original publication in 1999, and reprinting in 2001, questions from students, progress by researchers, plus the inevitable improvements in our understanding suggested this new book, "Time Reversibility, Computer Simulation, Algorithms, and Chaos". *Time-Reversible Computer Simulation* remains the fundamental basis of this work. *Algorithms* has been added to the title. They are the *"sine qua non"* of computer simulation. The recent literature has impressed us with the need for simple descriptions of the basics of computer simulation, the construction and implementation of algorithms. Our ability to formulate physical problems for computational solution and analysis has led us to an understanding of *Chaos* which has in turn enriched our understanding of irreversibility within the confines of a *Time-Reversible* description. We have found this understanding satisfying, rewarding, and continually challenging. We feel an obligation to pass it on to present and future readers.

The plan of this book is reflected in our new title. In discussing Computer Simulation we emphasize the basics, *Models* describing selected aspects of the Real World and *Algorithms* suited to determining the models' properties and behavior. Models based on classical mechanics, both atomistic and continuous, support our work. These mechanical models can help us to formulate and generate both reversible and irreversible system trajectories. Irreversible behavior can be explicit. More often it emerges from formally reversible microscopic models. Traditional macroscopic treatments of processes follow Thermodynamics or Fluid Mechanics. The one-way increase of entropy found in these macroscopic treatments contrasts with the two-way time-reversible fluctuating nature of classical Hamiltonian Mechanics.

The purpose of this book is to make the unity of all these disparate descriptions of physical phenomena more transparent. It would not be possible without the algorithms for our computers, without the introduction of nonequilibrium constraints into mechanics, simulating irreversible processes, and without the graphical displays to bolster and stimulate our understanding. The powerful combination of all these computational tools provides us with a clearer understanding of the development of irreversibility from a time-reversible basis.

One might expect that a soundly-based theoretical approach to nonequilibrium problems would provide a reliable path to "understanding". In fact, the judicious computer simulation of illustrative models is more productive and reliable than is theory. The fractal phase-space objects, the variety of computational thermostats, some of which "work" and some of which don't, the asymmetric Lyapunov spectra away from equilibrium, are all significant examples which are easily "understood" in retrospect but hard to predict.

Throughout the book we illustrate the main points with worked-out computational examples. We sketch, or even display, the algorithms necessary to reproduce the examples. We believe that there is no understanding in the absence of such examples—the most efficient path to understanding necessarily builds on example problems. All the examples given here are appropriate to modern laptop or desktop serial computers. We prefer the simplicity and clarity of Fortran ("Formula Translation") as a computer language. It is closest in form to the underlying mathematical formulation.

We have intentionally made the notation as spare as possible, resisting the use of bold face, arrows, overbars, and *sans serif* type to distinguish scalars, vectors, and tensors. *All* of them are written in italics. The components of a vector are usually identified in either of two ways. Vector components are listed in parentheses, for example, $r = (x, y, z)$. *Sets* of them are often enclosed in braces, $\{x, y, z\}$. Tensors are written as matrices of elements or components. Mostly, the tensors found in continuum mechanics are second-rank tensors, like stress, strain, and strain rate. With a bit of practice we trust the reader to make these distinctions for us. Likewise, in the computer codes and fragments, we have omitted the six blanks that precede a standard line of Fortran code. We have *not* included special exercises or "points to ponder". An excellent way forward is to reproduce numerical results given in the text and figures. In emergencies we'll be happy to respond to emailed suggestions or requests for clarification.

It is appropriate to highlight those topics newly selected for inclusion after an additional decade of reflection and research. The numerical solution of the ordinary differential equations required for computer simulation is discussed in more detail. We emphasize the need for judgment and common sense in formulating and interpreting algorithms and their results. We discuss two newly-fashionable ideas, "covariant" Lyapunov vectors and "Fluctuation Theorems". We also expand our description of strong fluid shockwaves to include results of our recent investigations. We recently returned to that topic, believing that it represents the ultimate challenge to a theoretical understanding of materials in motion. As to new books which have appeared in the last decade we have found Clint Sprott's *Strange Attractors; Creating Patterns in Chaos* and James Gleick's *Information* particularly clear and pertinent to our interests. These join Gleick's *Chaos*, Lorenz' *The Essence of Chaos*, and Bass' *Eudæmonic Pie* on our shelf of favorite readings. Rainer Klages' 2007 book *Microscopic Chaos, Fractals, and Transport in Nonequilibrium Statistical Mechanics* has a fairly comprehensive bibliography.

Improvements in electronics and the internet have changed our ways profoundly. Now we get and give same day responses to queries, free electronic copies of papers new and old, all with portable highly-reliable computation, with terabytes of data, on our laptops. During our work on this book two of our desktop Dell computers gave up their ghosts nearly simultaneously. Their portable replacements, MacBook Pro laptops, typify change, and progress. Our lost software was replaced through a combination of writing anew, scavenging old emails, and the help of colleagues. Frequent low-cost backups are now our standard procedure.

Just as before, it has been one of life's greatest pleasures for us to share this work with each other and with our many colleagues now so tightly linked together through email and the internet. The quick response of the internet plus the sheer volume of interesting and relevant research to be found there have made this work a very pleasant luxury. It is a joy to know that some of the pleasure and excitement of creating this book is shared with readers around this World. This book is very much a labor of love. Our pleasure in working together to serve interested students, present and future, is our reward.

Wg Hoover   *Carol G Hoover*

Ruby Valley, Elko County, Nevada, the United States of America, 2012

Preface to the First Edition

Today a small army of physicists, chemists, mathematicians, and engineers has joined forces for a renewed attack on a classic problem, the "reversibility paradox". The paradox is simply stated: "How can the irreversible Second Law of Thermodynamics be compatible with, and result from, an underlying time-reversible mechanics?" Building on the ideas of van der Waals and Maxwell, Boltzmann provided the classic nineteenth-century resolution of the paradox by using a probabilistic analysis of dilute-gas collisions. Here I bring Boltzmann's classic analysis up to date by adopting modern tools. This approach augments and generalizes Boltzmann's statistical understanding. The new interpretive tools are (i) Linear-response theory, a consequence of Gibbs' statistical mechanics, (ii) Chaos theory and (iii) the Fractal geometry to which it leads, and (iv) Computers, which make possible the simulations and analyses which were not available to Boltzmann.

As the available tools change, so do the targets and the points of view. Philosophers interested in the reversibility paradox have provided some insight too. We will seek correlations with their work. The present book describes both the scientific and the philosophical work from the perspective of computer simulation, emphasizing my own thermomechanical approach to resolving the reversibility paradox by analyzing the consequences of time-reversible thermostats. Computer simulation has made it possible to probe and characterize time reversibility from a variety of directions. "Chaos theory" or "nonlinear dynamics" has supplied a vocabulary detailing a useful set of concepts, which allows for a fuller explanation of irreversibility than was available to Boltzmann or even that provided by the linear response theory of Green, Kubo, and Onsager.

Throughout my own research career I have spent countless hours rereading the fruits of others' work. This rereading has been made necessary by the lack of clear example problems illustrating the meanings of the concepts used in these works. While this lack was understandable in the precomputer era of Boltzmann, Krylov, and Zubarev, it is inexcusable today. Throughout the book I emphasize the clear illustration of fundamental concepts with simple example problems, suited to desktop computation. I have also clarified the concepts by including a glossary of technical terms from the specialized fields which are combined here to focus on a common theme.

I am personally quite satisfied with the modern resolution of the reversibility paradox, as presented here. I see thermodynamic irreversibility as an inevitable, understandable outcome of an underlying time-reversible dynamics. This understanding is predicated on accepting that no dynamics is a *perfect* replica of nature. I cannot imagine any completely comprehensive "unified theory", able to include all of nature, together with its observers. My goal, in this book, is more modest. I wholeheartedly embrace classical mechanics as the most useful basis for an understanding of the physical world on the length and time scales relevant to us humans. By generalizing classical mechanics to include temperature and thermostated "thermal boundaries", we obtain "thermomechanics". This discipline will be our main model for the exploration and explanation of the links between time-reversible micromechanics, macroscopic irreversible thermodynamics, computer simulation, modern chaos theory, and fractal geometry.

The book begins with a discussion contrasting the idealized deterministic reversibility of basic physics with the pragmatic unpredictable irreversibility of what we call "real life" or "nature". The chaotic complexity discovered by Maxwell, Boltzmann, and Poincaré suggests that the unpredictability of life is intrinsic. This view is quite consistent with Gödel's undecidability proof, as well as with our quite evident ability to affect the future by exercising "free will". Computational models and simple thermomechanical simulations based on them are discussed and illustrated throughout the book. The simulations provide a reliable means to assimilate complex concepts through worked out examples. Such analyses, from the point of view of dynamical systems, are applied to simple two-dimensional maps and higher-dimensional dynamical systems, as well as to many-body examples from nonequilibrium molecular dynamics and to chaotic irreversible flows from finite-difference, finite-element, and particle-based continuum simulations. Two necessary concepts from dynamical systems theory—fractal distributions and Lyapunov instability—are fundamental to interpreting the results of the computational approach.

Undergraduate-level physics, calculus, ordinary differential equations, and a taste for computation are sufficient background for a full appreciation of the book. For nearly twenty years the Academy of Applied Science (Concord, New Hampshire) has sponsored the summer work of bright high-school seniors in the University of California's Davis Campus' Department of Applied Science at Livermore. The example problems worked out in the book are representative of the summer projects to which these students have contributed. The book is intended to appeal to advanced undergraduate as well as to graduate students, and to research workers. I fervently hope that the generous assortment of examples that I have worked out in the text will stimulate readers to explore and enjoy the rich and fruitful field of study which links fundamental reversible laws of physics to the irreversibility which surrounds us all. I have chosen mainly one- and two-dimensional examples in order to permit me to convey ideas with simple pictures. I stress here that the ideas so illustrated are not essentially different in three space dimensions.

To summarize the view I have reached, as the result of a decade of research, the Second Law of Thermodynamics is most simply described as a ubiquitous time-symmetry breaking which invariably accompanies the dynamics of a sufficiently chaotic system connected to its environment. Now it is certainly true that the "chaos" and symmetry breaking found in computer simulations are idealizations of the chaos and irreversibility of "nature". Our simulations are classical and nonrelativistic. They have a finite and digital representation. Nevertheless it is well-established by now that computational "pseudochaos" provides results which show no important differences from the idealizations of nature in the minds of mathematicians and the real-world observations of experimentalists.

Some of the popular books dealing with chaos and irreversibility seek an understanding of the macroscopic irreversibility of nature in terms of a comprehensive quantum mechanical and cosmological explanation, by linking the present state of the Universe to its "initial conditions". To me it is completely implausible that particular initial conditions, cosmological or not, are at all relevant to understanding the irreversibility present in everyday diffusive, viscous, or conducting flows. None of these ambitious books takes seriously the need for including boundary conditions and constraints in dynamics, which seems to me a crucial ingredient to obtaining irreversible behavior from time-reversible laws. It is clear that computer simulation has been the catalyst for our new understanding of irreversible flows.

There are *many* books addressing irreversibility. Most of them at least mention computer simulation. But there are really only two, Evans and Morriss' *Statistical Mechanics of Nonequilibrium Liquids* and my own *Computational Statistical Mechanics*, which emphasize the primary importance of simulation to a proper understanding of the "reversibility paradox" and the involvement of fractal distributions in resolving it. More mathematical expositions of some of the underlying ideas can be found in Dorfman and Gaspard's books. More philosophical expositions include Coveney and Highfield's, Dudeney's, Hawking's, Penrose's, Price's, and Sklar's. For me, Sklar's is the most interesting of all of them. He emphasizes the classical aspects of the reversibility paradox while simultaneously exploring and expounding a wide variety of alternative points of view.

My own approach is, I think, much the simplest, and proceeds by way of defining nonequilibrium states, in order to reach a compelling understanding of *irreversibility* in terms of the straightforward, but subtle, consequences of *time-reversible* chaotic differential equations. Generating the nonequilibrium states *via* computer simulation is illustrated here in the many example problems. It is my fond hope that the reader will find this approach palatable.

I would like to thank Richard Lim and Robert MacKay for stimulating and encouraging the effort needed to write this book. I also owe a longstanding continuing debt of thanks to my colleagues, friends, and students, for helping me to understand, and to the Livermore Laboratory, the University of California, and Universities in Australia, Austria, Germany, Japan, Korea, and Poland, for having provided me with the resources and havens necessary to teaching, research, and good fellowship.

Hideout Canyon Ranch—Red Skunk Ranch, 1998-1999
Stanislaus and Santa Clara Counties, California USA

Contents

Preface vii

Preface to the First Edition xi

Glossary of Technical Terms xxi

1. Time Reversibility, Computer Simulation, Algorithms, Chaos — 1
   - 1.1 Microscopic Reversibility; Macroscopic Irreversibility — 1
   - 1.2 Time Reversibility of Irreversible Processes — 6
   - 1.3 Classical Microscopic and Macroscopic Simulation — 8
   - 1.4 Continuity, Information, and Bit Reversibility — 10
   - 1.5 Instability and Chaos — 11
   - 1.6 Simple Explanations of Complex Phenomena — 13
   - 1.7 The Paradox: Irreversibility from Reversible Dynamics — 15
   - 1.8 Algorithm: Fourth-Order Runge-Kutta Integrator — 16
   - 1.9 Example Problems — 20
     - 1.9.1 Equilibrium Baker Map — 21
     - 1.9.2 Equilibrium Galton Board — 25
     - 1.9.3 Equilibrium Hookean Pendulum — 29
     - 1.9.4 Nosé-Hoover Oscillator with a Temperature Gradient — 32
   - 1.10 Summary and Notes — 36
     - 1.10.1 Notes and References — 37

2. Time-Reversibility in Physics and Computation — 39
   - 2.1 Introduction — 39
   - 2.2 Time Reversibility — 41

	2.3	Levesque and Verlet's Bit-Reversible Algorithm	44
	2.4	Lagrangian and Hamiltonian Mechanics	46
	2.5	Liouville's Incompressible Theorem	49
	2.6	What *Is* Macroscopic Thermodynamics?	50
	2.7	First and Second Laws of Thermodynamics	52
	2.8	Temperature, Zeroth Law, Reservoirs, Thermostats	54
	2.9	Irreversibility from Stochastic Irreversible Equations	58
	2.10	Irreversibility from Time-Reversible Equations?	60
	2.11	An Algorithm Implementing Bit-Reversible Dynamics	61
	2.12	Example Problems	67
		2.12.1 Time-Reversible Dissipative Map	68
		2.12.2 A Smooth-Potential Galton Board	73
	2.13	Summary	77
		2.13.1 Notes and References	78

3. Gibbs' Statistical Mechanics — 81

	3.1	Scope and History	81
	3.2	Formal Structure of Gibbs' Statistical Mechanics	83
	3.3	Initial Conditions, Boundary Conditions, Ergodicity	86
	3.4	From Hamiltonian Dynamics to Gibbs' Probability	89
	3.5	From Gibbs' Probability to Thermodynamics	90
	3.6	Pressure and Energy from Gibbs' Canonical Ensemble	92
	3.7	Gibbs' Entropy *versus* Boltzmann's Entropy	93
	3.8	Number-Dependence and Thermodynamic Fluctuations	96
	3.9	Green and Kubo's Linear-Response Theory	97
	3.10	An Algorithm for Local Smooth-Particle Averages	99
	3.11	Example Problems	103
		3.11.1 Quasiharmonic Thermodynamics	104
		3.11.2 Hard-Disk and Hard-Sphere Thermodynamics	106
		3.11.3 Time-Reversible Confined Free Expansion	108
	3.12	Summary	111
		3.12.1 Notes and References	112

4. Irreversibility in Real Life — 113

	4.1	Introduction	113
	4.2	Phenomenology — the Linear Dissipative Laws	116
	4.3	Microscopic Basis of the Irreversible Linear Laws	117
	4.4	Solving the Linear Macroscopic Equations	119

	4.5	Nonequilibrium Entropy Changes	120
	4.6	Fluctuations and Nonequilibrium States	123
	4.7	Deviations from the Phenomenological Linear Laws	124
	4.8	Causes of Irreversibility à la Boltzmann and Lyapunov	126
	4.9	Rayleigh-Bénard Algorithm with Atomistic Flow	128
	4.10	Rayleigh-Bénard Algorithm for a Continuum	135
	4.11	Three Rayleigh-Bénard Example Problems	140
		4.11.1 Rayleigh-Bénard Flow *via* Lorenz' Attractor	142
		4.11.2 Rayleigh-Bénard Flow with Continuum Mechanics	144
		4.11.3 Rayleigh-Bénard Flow with Molecular Dynamics	154
	4.12	Summary	159
		4.12.1 Notes and References	160
5.	Microscopic Computer Simulation	163
	5.1	Introduction	163
	5.2	Integrating the Motion Equations	164
	5.3	Interpretation of Results	165
	5.4	Control of a Falling Particle	168
	5.5	Second Law of Thermodynamics	176
	5.6	Simulating Shear Flow and Heat Flow	177
	5.7	Shockwaves	181
	5.8	Algorithm for Periodic Shear Flow with Doll's Tensor	184
	5.9	Example Problems	188
		5.9.1 Isokinetic Nonequilibrium Galton Board	189
		5.9.2 Heat-Conducting One-Dimensional Oscillator	192
		5.9.3 Many-Body Heat Flow	195
	5.10	Summary	196
		5.10.1 Notes and References	197
6.	Shockwaves Revisited	199
	6.1	Introduction	199
	6.2	Equation of State Information from Shockwaves	201
	6.3	Shockwave Conditions for Molecular Dynamics	203
	6.4	Shockwave Stability	206
	6.5	Thermodynamic Variables	214
	6.6	Shockwave Profiles from Continuum Mechanics	215
		6.6.1 Shockwave Profile with Shear Viscosity	217

		6.6.2	Shockwave Profile with Viscosity and Conductivity	220

        6.6.2 Shockwave Profile with Viscosity and Conductivity 220
        6.6.3 Shockwave Profiles with Tensor Temperatures . . . 222
        6.6.4 Flow Algorithm with Maxwell-Cattaneo Time Delays 223
   6.7 Comparing Model Profiles with Molecular Dynamics . . . 229
   6.8 Lyapunov Instability in Strong Shockwaves 232
   6.9 Summary 238
        6.9.1 Notes and References 238

7. Macroscopic Computer Simulation   **241**

   7.1 Introduction 241
   7.2 Continuity and Coordinate Systems 243
   7.3 Macroscopic Flow Variables 245
   7.4 Finite-Difference Methods 246
   7.5 Finite-Element Methods 248
   7.6 Smooth Particle Applied Mechanics [SPAM] 251
   7.7 A SPAM Algorithm for Rayleigh-Bénard Convection . . . 255
        7.7.1 Initial Conditions 255
        7.7.2 SPAM Evaluation of the Particle Densities 257
        7.7.3 SPAM Evaluation of $\{\nabla u\}$ and $\{\nabla T\}$ 258
        7.7.4 SPAM Evaluation of the Constitutive Relations . 260
   7.8 Applications of SPAM to Rayleigh-Bénard Flows 262
        7.8.1 SPAM with and without a Core Potential 266
        7.8.2 SPAM and Kinetic-Energy Fluctuations 268
   7.9 Summary 271
        7.9.1 Notes and References 271

8. Chaos, Lyapunov Instability, Fractals   **273**

   8.1 Introduction 273
   8.2 Continuum Mathematics 277
   8.3 Chaos 278
   8.4 The Spectrum of Lyapunov Exponents 279
   8.5 Fractal Dimensions 284
   8.6 A Simple Ergodic Fractal 288
   8.7 Fractal Attractor-Repeller Pairs 290
   8.8 A Global Second Law from Reversible Chaos 292
   8.9 Coarse-Grained and Fine-Grained Entropy 297

	8.10	Oscillators, Lyapunov Algorithms, Fractal Dimensions	298
		8.10.1 A Thought-Provoking Oscillator Exercise	298
		8.10.2 Doubly-Thermostated Oscillator; Lyapunov Spectra	300
		8.10.3 Lyapunov Spectra for a Chaotic Double Pendulum	310
		8.10.4 Coarse-Grained Galton Board Entropy	312
		8.10.5 Color Conductivity	313
	8.11	Summary	316
		8.11.1 Notes and References	317
9.	Resolving the Reversibility Paradox		319
	9.1	Introduction	319
	9.2	Irreversibility from Boltzmann's Kinetic Theory	320
	9.3	Boltzmann's Equation Today	325
	9.4	Gibbs' Statistical Mechanics	327
	9.5	Jaynes' Information Theory	330
	9.6	Green and Kubo's Linear Response Theory	332
	9.7	Thermomechanics	334
	9.8	The Delay Times Separating Causes from their Effects	336
	9.9	A Fluctuation Theorem	337
	9.10	Are Initial Conditions Relevant?	340
	9.11	Constrained Hamiltonian Ensembles	343
	9.12	Anosov Systems and Sinai-Ruelle-Bowen Measures	344
	9.13	Trajectories *versus* Distribution Functions	347
	9.14	Are Maps Relevant?	348
	9.15	Irreversibility ⟵ Time-Reversible Motion Equations	351
	9.16	Boltzmann-Equation Shockwave-Structure Algorithm	353
	9.17	Summary	359
		9.17.1 Notes and References	361
10.	Afterword—a Research Perspective		363
	10.1	Introduction	363
	10.2	What do We Know?	364
	10.3	Why Reversibility is Still a Problem	366
	10.4	Change and Innovation	369
	10.5	Role of Examples	372
	10.6	Role of Chaos and Fractals	374

10.7	Role of Mathematics	374
10.8	Remaining Puzzles	376
10.9	Summary	379
10.10	Acknowledgments	383

Bibliography   387

Index   397

Glossary of Technical Terms

Algorithm : computational recipe suitable for computer evaluation

Attractor : long-time-averaged contracting *flow* solution, a fractal sink

Bit Reversibility : exact time reversibility, "to the very last bit"

Boltzmann's Entropy : $-Nk\langle \ln f_1 \rangle$; f_1 : one-body distribution function

Boundary Conditions : specified $u(r,t)$, $T(r,t), \ldots$ on system boundary

Canonical Distribution : probability density proportional to $e^{-\mathcal{H}(q,p)/kT}$

Central Limit Theorem : $\mathrm{prob}[\langle x \rangle_N] \longrightarrow$ Gaussian with $\sigma \propto \sqrt{1/N}$

Chaos : confined Lyapunov instability

Closed System : isolated system, with constant mass and energy

Coarse Graining : division of solution space into small cells

Conjugate Momentum : $p \equiv (\partial \mathcal{L}/\partial \dot{q})$, where q is a generalized coordinate

Conservative System : isolated system, with constant mass and energy

Constitutive Equation : state- (and sometimes rate-) dependence of fluxes

Constraint : specified value or history of thermomechanical variable

Continuity Equation : $(\partial \rho/\partial t) = -\nabla \cdot (\rho u) \longrightarrow (d \ln \rho/dt) = -\nabla \cdot u$

Continuum : system with continuously varying ρ, e, P, $Q, \ldots$

Convergence : reaching a limit

Deterministic : time rate-of-change depends on current state and history

Diffusion Equation : $(\partial \rho/\partial t) = D \nabla^2 \rho$

Dissipation : conversion of mechanical energy into heat

Dynamical Systems : coupled sets of ordinary differential equations

Energy : E, the power to do work or create heat

Energy Equation : $\rho \dot{e} = \sigma : \nabla u - \nabla \cdot Q = -\nabla P : \nabla u - \nabla \cdot Q$

Ensemble : many similar systems with common specified state variables

Enthalpy : $H = E + PV$, thermodynamic potential for fixed pressure

Entropy : $\Delta S = \int (1/T) dQ_{\text{reversible}}$ from reference to current state

Equation of Motion : differential equation for acceleration

Equation of State : relation linking pressure, energy, volume, temperature

Ergodicity : reaching all states consistent with macroscopic variables

Eulerian Coordinates : coordinates with respect to a fixed frame

Flux : quantity of flow per unit area and time

Fourier's Law : $Q = -\kappa \nabla T$

Fractal : power-law dependence of density distribution on (small) distance

Friction Coefficient : ζ, an inverse time; frictional force is $-\zeta p$

Gauss' Mechanics : $\{\dot{p} = F - \zeta p\}$; $\zeta = \sum [(F \cdot p)/(2K)]$

Generalized Coordinates : any variables sufficient to describe microstate

Gibbs' Entropy : $S = -k \langle \ln f_N \rangle$; f_N : N-body distribution function

Gibbs' Free Energy : $G = E + PV - TS$, potential for P and T

Hamiltonian : $\mathcal{H}(q, p) \to (\dot{q}, \dot{p})$; usually an energy sum, $\mathcal{H} = K + \Phi$

Hausdorff Dimension : small-ϵ limit of $\langle \ln \# \rangle / \ln(1/\epsilon)$; (avoid this term)

Heat : energy in the form of velocity fluctuations

Heat Flux : Q : comoving energy flow, per unit area and time

Helmholtz' Free Energy : $A = E - TS$, potential for V and T

H Theorem : $H \equiv k \langle \ln f_1 \rangle \to \langle \dot{H} \rangle \leq 0$ [Boltzmann's Idea!]

Ideal Gas : dilute gas with energy $E(T)$ and $PV \equiv NkT$

Information Dimension : small-ϵ limit of $\langle \ln \text{prob} \rangle / \ln \epsilon$

Glossary of Technical Terms

Invertibility : ability to recover past from future and *vice versa*

Irreversibility : lack of time reversibility, usually dissipative

Lagrangian : $\mathcal{L}(q,\dot{q})$; usually an energy difference, $\mathcal{L} \equiv K - \Phi$

Lagrangian Coordinates : coordinates comoving with a moving material

Linear Response : ensemble-averaged response to small perturbations

Liouville Equation : phase-space continuity equation for $\dot{f}_N(q,p)$

Liouville's Theorems : $\dot{f} = -f \nabla \cdot v$, where usually $v \equiv \{\dot{q}, \dot{p}\}$ or $\{\dot{q}, \dot{p}, \dot{\zeta}\}$

Lyapunov Exponents : long-time-averaged comoving growth/decay rates

Lyapunov Instability : exponential growth of infinitesimal perturbations

Map : recipe for generating next configuration from present one

Mass Flux : mass flow, per unit area and time, ρv

Maxwell-Cattaneo Relaxation : $\sigma + \tau \dot{\sigma} = \eta \dot{\epsilon}$; $Q + \tau \dot{Q} = -\kappa \nabla T$

Microcanonical Distribution : all accessible states of fixed energy

Mixing : long-time loss of correlation with initial condition

Multifractal : fractal distribution with spatially varying power law

Newtonian Viscosity : $\sigma = \eta[\nabla u + \nabla u^t] + \lambda \nabla \cdot u I$

Nosé-Hoover Mechanics : $\{\dot{p} = F - \zeta p\}$; $\dot{\zeta} = [(K/K_0) - 1]$; $K_0 \propto T$

Open System : system interacting with surroundings

Ordinary Differential Equations : expressions for *time* derivatives

Pair Potential : interaction between two bodies

Partial Differential Equations : expressions linking *space-time* derivatives

Periodic Boundary : opposite sides of system connected

Phase Space : $\{q, p\}$ space, with points giving complete state description

Poincaré Section : surface, or plane, intersecting phase-space trajectories

Pressure Tensor : $P = -\sigma$: comoving momentum flux (or force/area)

Probability : likelihood, with $\sum$ prob $\equiv 1$

Random Numbers : uncorrelated, usually uniform, with $0 < R < 1$

Repeller : time-reversed attractor, source of flow in remote past

Reservoir : external source of mass, momentum, or energy for system

Reversibility : same equations of motion in either time direction

Second Law : increase of entropy

Self Similar : same topological structure on smaller and smaller scales

Shock : adiabatic irreversible transformation with (P, E, T, S) increasing

Simulation : computer-generated solution of physical differential equations

Stability : indifference to small perturbations

Stationary State : boundary conditions fixed, see steady state

Steady State : boundary conditions fixed, see stationary state

Stochastic Boundary : boundary interaction includes random numbers

Strain : relative deformation : $\epsilon_{xx} = (\partial \delta x/\partial x); \epsilon_{xy} = (\partial \delta x/\partial y) + (\partial \delta y/\partial x)$

Stress : tensile force per unit area, $\{\sigma_{ij}\} \equiv -\{P_{ij}\}$

Temperature : comoving $T_{xx} = \langle mv_x^2/k \rangle = \langle p_x^2/mk \rangle$

Thermodynamics : systematic study of state changes *via* heat and work

Thermomechanics : mechanics including heat reservoirs

Thermostat : algorithmic mechanism for temperature control

Virial Series : series expansion of pressure in powers of density

Wave Equation : $(\partial^2 u/\partial t^2) = c^2 \nabla^2 u$

Work : energy change due to varying macroscopic coordinates

Chapter 1

Time Reversibility, Computer Simulation, Algorithms, Chaos

The moving finger writes, and having writ,
moves on; nor all thy piety nor wit
shall lure it back to cancel half a line;
nor all thy tears wash out a word of it.

Omar Khayyám, as translated by Edward FitzGerald

1.1 Microscopic Reversibility; Macroscopic Irreversibility

The "moving finger", the "passing scene", the "unfolding of events". These familiar concepts all convey the notion of motion, linking the present to the past and to the future. Science dissects and idealizes our universal experience of time's passing according to a variety of "physical theories". Any useful physical theory, though approximate and idealized, with limited validity, must be deterministic. A useful theory needs to describe and foretell motion, predicting the unfolding of events with the passage of time.

Fundamental physical theories invariably replace the physical description of nature by a mathematical representation, a set of coupled differential equations. The differential equations give explicit time derivatives for all the interesting variables in terms of present values, allowing the "theory" to predict the future or to recover the past by integration. For the usual nonlinear equations the integration is necessarily numerical, and relies on fast computers. In classical physics the variables are macroscopic observables, such as coordinates, velocities, temperatures, or electric fields. Schrödinger's version of quantum physics provides the time derivative of an intermediate more abstract, and complex, "wave function" $\Psi(r,t)$, which can be used to compute observables through a spatial averaging, with local probability $\Psi(r,t)\Psi^*(r,t)dr$. As mentioned in the Preface we don't use a special notation to distinguish scalars from vectors from tensors. All of them are written in *italics*. Here r indicates a set of spatial coordinates, such as $\{x_i, y_i\}$ in two dimensions or $\{x_i, y_i, z_i\}$ in three. As usual, t is *time*.

Feynman's equivalent formulation, though much closer to classical mechanics, seems to us also to be much more complex. Feynman relates the evolving quantum probability to a weighted sum over *all* possible paths $\{r(t)\}$, among which only an infinitesimal fraction could correspond to possible classical evolutions. Because time is a continuous variable, numerical solutions of either the classical or the quantum evolution equations require a sufficiently small timestep Δt. The histories which result are *approximate* descriptions, with the interesting variables given at a discrete number of closely-spaced times, $\{n\Delta t\}$.

Though some philosophers choose to formulate, address, and debate definitions of "time" more detailed than "what a clock measures" we will not. Here we take the concept of time as basic and fundamental. Our view of the basic mechanical quantities, time, space, mass, and force, is that these concepts are best grasped intuitively, through examples, and that it is futile and unproductive to seek a deeper philosophical understanding of them. Newton put it thus: "I do not define time, space, place, and motion, since they are well known to all."

Newton's best-known physical theory, classical point mechanics, describes the accelerations $\{a = \dot{v} = dv/dt = \ddot{r} = d^2r/dt^2\}$ governing the time-evolution of particle coordinates $\{r\}$, as determined by "forces" $\{F\}$ depending on these coordinates (and sometimes on their velocities $\{v\}$). Notice that this description must implicitly *assume* that directions of increasing spatial coordinates and time are given.

All physical theories also *assume*, in their formulation, a "subjective" or "psychological" "arrow of time" in order to *define* time derivatives. Because Newton's particles are points, the forces accelerating them represent "action at a distance". Newtonian mechanics is simply a set of differential equations, with variables, constraints, boundaries, and initial conditions which correspond, more or less roughly, to laboratory or astronomical observables, or to idealized models.

In parallel to Newton's classical seventeenth-century theory for moving particles, Maxwell's theory for the propagation of electromagnetic waves, and Schrödinger's 1926 theory for the evolution of quantum probabilities, likewise depend upon time through the underlying differential equations which define the corresponding physical theories. And none of these differential equations—Newton's, Maxwell's, or Schrödinger's—displays any explicit distinction between the future and the past. These microscopic equations are all "time-reversible". For instance, this means that any coordinate sequence $\{r(n\Delta t)\}$ obtained by solving Newton's motion equations corresponds also to a solution of the same equations (but with different initial conditions) in the *opposite* time direction.

This ubiquitous quality of *time reversibility*, shared by all the fundamental physical theories, is evidently dictated, or at least strongly suggested, to us by nature. This conclusion is based on more than three centuries of careful observation and controlled experimentation. Apart from relativistic corrections, still time-reversible, heavenly bodies move according to Newton's time-symmetric differential equations, $\{m\ddot{r} = \Sigma F\}$, where the sum of forces includes each mass' gravitational interaction with its fellows. The same motion equations describe simple mechanical observations on the scale of laboratory experiments.

But the time-symmetric formal reversibility common to both large-scale observation and simple earthbound laboratory experiments seems to lack relevance to our intuitive sense of passing time. The apparent incompatibility of "Time's Arrow" with time-reversible physics was nicely captured by the September 1984 cover of Discover Magazine. The cover shows, traffic-sign-fashion, a *two*-headed arrow bearing the label "One Way". The paradoxical label expresses the one-way "arrow of time" which sums up our psychological experience of passing time. Our everyday macroscopic experiences are filled with irreversible processes. We see living things age and die; chinaware cracks, chips, and shatters; viscous flows and heat flows generate entropy—they "dissipate" or "degrade" energy into heat. All these dissipative processes can be summarized by phenomenological physical theories included in "irreversible thermodynamics".

Dissipative processes, like aging, fracture, friction, and conduction, are *never* time-reversible, but are nevertheless quite real. In each case the symmetry of time reversibility is broken through interactions which link a system to its surroundings. Though complex, these links can often be modeled by a simple selection of boundary, constraint, and driving forces. Indeed, in computer simulations, just as in nature, no steady dissipative state can be achieved in the absence of such links. They *must* be present, both to govern the flow and to remove the generated heat. Boundary conditions and initial conditions are prerequisites to any solution of dynamical equations.

Besides transmitting mass, momentum, and energy between a system and its surroundings, boundaries, those necessary links to the surrounding outside world, also transmit *information*—the detailed description of the mass, momentum, and energy transfers. If any of this transmitted information is discarded, or lost, and degraded into heat, the ability to recapture the past is also lost.

The irreversibility of all macroscopic phenomena leads inexorably to a future where the *order* of the past has been replaced by relative *disorder*. This irreversible growth of disorder can equivalently be described as a loss of information. From this latter point of view it is natural to view macroscopic thermodynamic irreversibility as resulting from a loss of the information required to describe or recover the past. The lost information makes it impossible to reverse the flow. Information is a legitimate focus for any physical study, for our information is strictly limited, both by our attention span, and by our capacity for storage. Our best petaflop and petabyte tools for solving the differential equations of physics and storing the solutions are quite limited in their capacities for processing and retaining information. And in a real sense, we cannot hope ever to make more reliable predictions from our physical theories than those produced by computer simulations.

None of the foregoing microscopic physical theories is well-equipped for discussing *irreversible* processes. At first glance it is tempting to declare that no time-reversible theory can ever provide irreversible consequences. Though we will see that this is wrong, the declaration is almost correct where classical mechanics is concerned, as has been emphasized by a host of scientists. But see Chapter 6 for some contrary evidence. Loschmidt and Zermélo pointed out that the isolated systems of classical mechanics cannot provide a consistent description of irreversible processes. Feynman, Gibbs, Krylov, Ruelle, and Smale are among those who have reïterated this basic point. Gibbs was too cautious to give a precise formulation of the coupling between system and surroundings. But he suggested that the entropy (or the available phase-space volume) of a two-part compound system would somehow seek out its maximum value.

Gibbs' reticence to discuss detailed mechanisms for maximizing entropy resulted in Krylov's criticism of his contributions. Krylov devoted considerable thought to the "foundations" of statistical mechanics, emphasizing that Liouville's incompressible theorem for the evolution of the N-body phase-space probability density,

$$\dot{f}_N \equiv (\partial f_N/\partial t) + \sum \dot{q}(\partial f_N/\partial q) + \sum \dot{p}(\partial f_N/\partial p) \equiv (df_N/dt) \equiv 0 \, ,$$

makes Gibbs' entropy-maximizing arguments invalid, or at least incomplete. Only Krylov's critique of earlier work was published. Although incomplete, and lacking examples, his critical descriptions of Gibbs' work make it clear enough that he had in mind distributions something like the fractal distributions which we now know characterize nonequilibrium states. But he lacked not only the *computers* needed to solve and characterize precise examples but also the *time* to develop and express his own interpretation of irreversibility. From a modern perspective, the ideas which Krylov was able to express appear not only incomplete but also unconvincing.

Loschmidt pointed out the incompatibility separating time-reversible mechanics from irreversible thermodynamics. He lacked our modern understanding of chaos' role in the loss of information responsible for irreversibility. It is a corollary of Liouville's incompressible theorem, that an idealized many-body trajectory confined to a bounded region in phase space will eventually return arbitrarily close to its starting point. Zermélo emphasized that this "Poincaré recurrence" implies that isolated systems cannot display irreversible behavior for too long. Both Loschmidt's and Zermélo's arguments depended upon the concept of precisely-defined trajectories. The existence of chaos shows that this trajectory concept is flawed. In practice, continuous numerical trajectories can neither be computed nor reversed with infinite precision.

Additionally, the importance of the boundaries and the shielding required to contain an isolated system are generally not discussed. Ruelle and Smale suggested that mechanics needed to be generalized to include boundary effects. A study of models for boundaries was a natural part of the first "realistic" nonequilibrium simulations. The important role of boundaries is particularly apparent in the example of a confined free expansion, treated in Section 3.11.3, page 108.

There is a widespread peculiar view of the importance of *initial conditions* which needs comment. Some philosophers state that Newton's, Maxwell's, and Schrödinger's theories can only be made to provide irreversible solutions by selecting just the right initial conditions. Sklar, for example, repeatedly emphasizes the point of view that choosing the *right* initial conditions is essential to an "understanding" of irreversibility. With time-reversible solutions it is hard to argue for an irreversible future. Any solution which does provide "irreversible behavior" (growing disorder, entropy increase) in the future could be played backward (gaining order, losing entropy). This objection to the simple "explanation" in terms of initial conditions was Loschmidt's. The answer Loschmidt requires became clearer a century later, when chaos theory showed the vital role of Lyapunov instability in making the initial conditions rapidly become irrelevant.

The misguided emphasis on initial conditions likewise ignores the empirical basis of physics. Experiments are useful aids to understanding precisely because microscopic initial conditions are quite irrelevant to the macroscopic outcome. If initial conditions *were* crucial there could be no bodies of knowledge, like physics and chemistry, which describe the evolution of macroscopic behavior independently of the microscopic details.

In addition to this justifiable complaint, choosing the "right" initial conditions appears hopeless whenever, as is usual, the dynamics is "sensitive to initial conditions", meaning Lyapunov unstable. By "Lyapunov instability" we refer to the pervasive growth, exponential in time, of small perturbations of the initial conditions. More details are given in Sections 1.5 and 8.4.

In principle Lyapunov instability implies that choosing the "right" initial conditions requires an *infinite* amount of information. This requirement, a consequence of precisely defined trajectories, is *meaningless* from any operational point of view. In practice, the double-precision simulations (53 significant bits with the Gnu Fortran compiler that we use) are limited to an uncertainty of about $10^{-16} \times 10^{-16}$ per coordinate-momentum pair, similar to the quantum uncertainty which governs real experiments when the masses, lengths, and times are respectively of order kilograms, meters, and seconds. This latter observation follows from Heisenberg's "uncertainty principle", and guarantees that any precise effect on the atomic level, due to initial conditions, would disappear in just a few interatomic collision times. A perceptive discussion appears in Ruelle's delightful book *Chance and Chaos*. Double precision is coincidentally also the right choice for continuum simulations, in which microscopic fluctuations of order 10^{-12} are to be ignored.

1.2 Time Reversibility of Irreversible Processes

"Sensitivity to initial conditions" suggests, first of all, that no theory can be judged reasonable if its physical predictions depend sensitively upon the initial data. "Sensitivity" also suggests that a changed less-sensitive theory, perhaps with a more limited goal, might well do better. Such theories can be developed by focusing on another aspect of the solution process, the interaction of the system with its surroundings, as modeled in computer simulations by boundary conditions, constraints, or driving forces.

Nonequilibrium systems are typically *open* or *driven* systems, with external sources and sinks of mass, momentum, and energy. Though such systems are more complex than the idealized isolated systems of Newtonian mechanics, the cost of describing this additional complexity is justified by the wide scope of new problem areas such a generalized approach can explore. The simplest useful microscopic models for understanding the everyday world of human experience are time-reversible dynamical theories which include nonequilibrium boundary conditions, constraints, and driving forces. From both standpoints, computational and analytical, the best physical theories from which to start are (i) the simplest microscopic theory of particle motions, classical mechanics, and (ii) the macroscopic theory of continuum mechanics. Classical mechanics, with boundary conditions, constraints, and driving forces, presently provides a comprehensive description of a wide range of motions, generally thought of as including both the reversible and the irreversible types, and ranging in scale from the microscopic level of atomistic mechanics to the macroscopic levels of structural engineering and astronomy. Classical mechanics is a fairly faithful description of physical reality, simpler, and closer to human experience, in most cases, than is its quantum cousin.

A growing appreciation of the complexity inherent in simple theoretical structures has revealed that the task of constructing a comprehensive "unified theory" of "everything" is unattainable. For this reason it is logical to follow Occam's lead, using the philosophical principle of "Occam's Razor" to cut away all but the simplest parts of the candidate theories describing the phenomena of interest. Mechanics, when coupled with boundary conditions, constraints, and driving forces, is enough to explain the symmetry-breaking associated with irreversible processes, and to resolve the conceptual problems associated with conservative mechanics. We like to call the augmented mechanics "thermomechanics" to emphasize its link to thermodynamics and nonequilibrium flows through the explicit incorporation of thermal effects.

Thermomechanics is a direct outgrowth of computation and simulation. When fast computers became generally available, in the 1960s, new problem areas opened up and old analytic approaches could be gracefully abandoned.

By the early 1970s, thermomechanics had come into its own as a direct result of computation. Nonequilibrium molecular dynamics was developed in 1972. Nosé's discovery of time-reversible thermostats matching Gibbs' canonical ensemble came in 1984. Much of the subsequent work was devoted to checking that the new methods agreed with Gibbs' statistical equilibrium predictions, as augmented by Green and Kubo's exact formulation of linear transport processes.

Direct computer simulations replaced virial series for the pressure and integral equations for the distribution of particle pairs as the simplest path to equilibrium properties. Likewise, computer *algorithms* largely replaced the construction and analysis of "kinetic equations" for *non*equilibrium problems. The resulting extensions of mechanics to the definition and exploration of *nonequilibrium* systems with special boundary conditions, constraints, and driving forces, would have been incomplete and unrewarding without the computers necessary to solve (and to display the solutions of) the underlying differential equations. Let us begin to explore computer simulation by describing the application of fast computers to the task of solving the mechanical motion equations for both microscopic and macroscopic systems.

1.3   Classical Microscopic and Macroscopic Simulation

In classical continuum mechanics, the usual space-and-time-dependent variables are the mass density, velocity, and energy per unit mass,
$$\{\rho(r,t), u(r,t), e(r,t)\} \ .$$
The motion of a continuum is in principle more complex than that of a system of particles because the dependent variables $\{\rho, u, e\}$ must be known *everywhere*. The time evolution of this set reflects the interdependent flows of mass, momentum, and energy in response to the fields and gradients driving them.

Typical macroscopic computer simulations contain irreversible "constitutive relations". There are two different reasons for this. First, much of the irreversibility we see around us can be explicitly and accurately simulated by including Newtonian viscosity and Fourier's heat conductivity. Second, an enhanced *artificial* irreversibility must often be used (artificial viscosities and conductivities are examples) to stabilize numerical techniques. In either case, with "realistic" or "artificial" irreversibility, the simulations are complicated whenever nonlinear effects, leading to chaos, are included. The solutions of the irreversible macroscopic equations can closely resemble the results of laboratory experiments. But, due to their intrinsic irreversibility these macroscopic simulations are often viewed as "less fundamental" than time-reversible microscopic simulations based on particle mechanics. The main criticism leveled at the macroscopic approach is its lack of time reversibility. A subsidiary and related aspect of the macroscopic approach is its exclusion of certain fluctuations. The averaging which results in this exclusion has two effects: besides destroying time reversibility it eliminates the complexity associated with extraneous microscopic degrees of freedom. It is only in problems where this complexity is important, like turbulence, that pursuit of the macroscopic approach is bogged down by the complexity characteristic of microscopic representations. The probabilistic nature of quantum mechanics suggests a kind of "averaging" too, but, unlike macroscopic mechanics, Schrödinger's quantum mechanics is completely time-reversible.

Computer simulation solves problems in a way which was novel at the time of the Second World War, and which still meets occasional pockets of resistance. The analytical textbook style of problem solving gives a "solution" described by orthogonal polynomials or series expansions. The computational approach *simulates* the *evolution* of a physical system. The polynomials and expansions are replaced by computer *algorithms*. The computational solution is most likely a time-ordered sequence of coordinate data, supplemented with the evolving values of field variables (stress, heat flux, temperature, and the like). In classical particle mechanics, the numerical trajectories $\{r(+t)\}$ describing a solution of Newton's equations $\{F = m\ddot{r}\}$ provide also a reversed, second solution, of the *same* equations, obtained by tracing out exactly the same coordinate values, but in a time-reversed order. In such a time-reversed solution, $\{r(-t)\}$, the particle velocities $\{v \equiv \dot{r} \equiv (dr/dt)\}$ all change sign, but still obey Newton's equations linking the forces, masses, and accelerations.

How could such a symmetric time-reversible situation reliably describe the irreversible phenomena of the real world? There are several approaches to answering this paradoxical question. But, since the only missing ingredient is the set of initial conditions from which the solution is to be continued, it has been common to "explain" the irreversible behavior by pointing to the special nature of the initial conditions. There is a flaw to this misguided explanation. That flaw is chaos, introduced in the next two Sections and discussed at greater length in Chapter 8.

1.4 Continuity, Information, and Bit Reversibility

Newton's ordinary differential equations of motion describe the motion of mass points subject to forces. The motions which result, $\{r(t)\}$, are typically continuous flows with smooth time derivatives, $\{v(t) = \dot r\}$, which obey Newton's differential equations of motion: $\{a = (F/m) = \dot v = \ddot r(t)\}$. Because the equations are typically nonlinear, and beyond the reach of analytic techniques, a closed-form solution of these equations, giving the particle coordinates as explicit functions of the time, is generally not possible. As discussed in Section 1.2, the presence of chaos in the solution suggests an explanation for the lack of analytic solutions. As time goes by, the "information"—the number of binary bits required for an accurate analytic solution—grows linearly with time. Eventually the required information lies beyond the capabilities of analysis and computation. The gross features of the present and future come to depend upon finer and finer features of the far distant unknowable past. No conceivable improvement of the spatial and temporal resolutions can overcome this problem. The fundamental reason is sobering. It is characteristic of chaotic systems that the most precise experiments or most-carefully-designed simulations cannot probe the future reliably for more than a few collision times. Thus the "determinism" of mechanics is an *illusion*, as was well known to Maxwell and Poincaré. It is the *irrelevance* of the initial conditions (due to the randomizing effects of chaos) which makes the systematic study of physics possible. This randomizing—also referred to as "mixing"—corresponds to information loss. As time goes by the link between present and past becomes more tenuous.

A numerical solution of Newton's equations is typically approximate, with limited sixteen-digit 53-bit precision. The contrast between an ideal continuous solution $\{r(t)\}$, with both the coordinates and the time continuous and precisely known, and a doubly-discretized computer-generated numerical approximation is sharpest if one imagines a solution space in which points are restricted to a regular spatial grid and are evenly spaced in time $\{r(n\Delta t)\}$. In Chapter 2 we discuss and illustrate Levesque and Verlet's construction of "bit-reversible" doubly-discretized solutions which are *rigorously* time-reversible. Any such numerical trajectory is necessarily periodic, while a continuous trajectory would have to satisfy very special initial conditions in order to achieve periodicity. The *inevitable* periodicity associated with a discrete solution space suggests that such a space cannot be used to describe irreversible flows in isolated systems. Order of magnitude estimates correctly suggest that Poincaré recurrence times are of order $\sqrt{e^{S/k}} = \sqrt{\Omega}$ in a discretized space with Ω accessible states. This estimate is a generalized version of the "Birthday Problem": in a group of N people there are $N(N-1)/2$ *pairs* of birthdays. Evidently two of them are likely to match when $N(N-1)/2 \simeq 365 \rightarrow N \simeq 27$, actually 23. Similarly a system exploring its Ω states is likely to repeat a state when the number explored is of order $\sqrt{\Omega}$. The times required for such an exploration are effectively infinite (exceeding the Age of the Universe) once the number of particles is of order ten to a hundred. Despite the formal periodicity it is quite possible to describe Lyapunov instability, the sensitivity to small perturbations called "chaos", using the bit-reversible approach.

1.5 Instability and Chaos

Turbulence has long been singled out as a specially "difficult" subject. This characterization of turbulence has arisen from the continuing failure of attempts to predict, or at least to understand, the long-time behavior of complex flows, such as our weather, despite the well-recognized importance of the task. Turbulent instability occurs whenever the decay rate associated with fluid deformations (changes in shape) is sufficiently small. In 1963 Lorenz described his efforts to continue the numerical solution of his now-famous set of three differential equations,

$$\dot{x} = -\sigma(x - y);\ \dot{y} = Rx - y - xz;\ \dot{z} = xy - bz\ .$$

starting out from intermediate values. Lorenz' equations are a rough description of convective fluid flow driven by a temperature gradient in the presence of gravity. In his words,

> "In these equations x is proportional to the intensity of the convective motion, while y is proportional to the intensity of the temperature difference between the ascending and descending currents $\cdots$ z is proportional to the distortion of the vertical temperature profile from linearity".

Lorenz found that his second solution failed to agree with his original one after a fairly short time. Further investigation showed that the mechanism for the disagreement was the *exponentially* unstable loss of information, with the precision required to reproduce a solution of fixed accuracy increasing *exponentially* with the required time, corresponding to the required number of decimal *digits* or binary *bits* increasing *linearly* with time. With Lorenz' work it became "widely known" (to the experts) that *most* interesting flows contain this same sensitivity, "Lyapunov instability", to small changes in initial conditions.

Quite typically, both microscopic and macroscopic equations of motion are Lyapunov unstable, meaning that their solutions are very sensitive to small perturbations, so sensitive that such perturbations grow *exponentially* fast, in time. Though it is completely deterministic, and in principle reproducible, the chaos which characterizes Lyapunov instability means that particular *precise initial conditions are not a useful concept*. This is just as well, inasmuch as the concept of a completely isolated system has no sound basis in physics, where everyday gravitational forces have infinite range. On the other hand, *gross* characteristics of flows with reasonable initial conditions *do* describe the spatial variations of macroscopic features, such as the temperature or velocity field. The gross features lead to the reproducible averaged behavior described by the macroscopic physical theories. They also suggest using statistical ensembles to describe the time development of similar systems.

The details of microscopic initial conditions cannot be to blame for irreversible behavior, for the time-reversibility property of microscopic theories is completely independent of these conditions. The only likely possibility for breaking the apparent symmetry of future and past solutions lies in their relative *stabilities*. These stabilities can be analyzed in detail in terms of the Lyapunov spectrum, as is discussed in detail in Section 8.4. The Lyapunov exponents which make up the spectrum describe the *global* time-averaged rates of growth, or decay, of perturbations in the initial conditions. The complete spectrum gives a description of these rates for *all* perturbation directions, not just the most-rapidly-growing one. For *transient* irreversible flows (the inelastic collision of two blocks, for instance) the *time-dependent* exponents in the forward and backward directions can be quite different. For nominally stationary flows likewise following Newton's equations of motion, both the global and the local spectra proceeding forward in time are exactly equal to their time-reversed global or local analogs, regressing backward, in time.

Krylov emphasized that for this reason—the symmetry linking the past and the future—irreversibility cannot be understood from the standpoint of Newton's equations. More recently, Smale and Ruelle have echoed this view, suggesting that the description of irreversible processes requires new generalizations of the classical equations of motion. With the advent of computers making simulations possible, such generalizations were not long in coming. Analyses of the results have revealed an interesting ubiquitous breaking of time symmetry. *Typically*, the *forward* nonequilibrium computer trajectory is *less* sensitive to perturbations and is thus *more* stable than is the time-reversed backward one. This difference in sensitivity leads both to a symmetry breaking and to a simple geometric understanding of the irreversibility which lies all around us. It also gives rise to singular *fractal* distributions. These are distributions which have *no well-defined gradients*. Locally smooth in some directions while wildly singular in others, fractals display a pervasive power-law structure on *all* length scales. We will consider the first of many such examples in the next Chapter.

Loschmidt emphasized the paradoxical aspect of time-reversible Newtonian trajectories. And classical mechanics, as originated by Newton, but generalized to include boundaries, constraints, and driving forces, will be the main focus of our interest in reversibility. In relativity theory, electromagnetism, and quantum mechanics, time reversibility is less apparent in the fundamental equations, but is nevertheless present. Schrödinger was intrigued by this problem too. In a 1931 lecture described in his *Science, Theory, and Man* he publicized Exner's skeptical criticism of the view that conventional perfectly-deterministic, and time-reversible, classical mechanics is the only possible model describing "classical" phenomena. So long as energy and momentum are conserved in collisions, a small stochastic contribution to the dynamics could also be present, accounting for irreversibility. Exner's explanation, though technically possible, seems implausible, because it fails Occam's test of simplicity. Apart from integer algorithms, like Levesque and Verlet's, finite precision results in computational roundoff error. It seems to us very unlikely that stochastic low-level noise differs in any *significant* respect from this computational error.

1.6 Simple Explanations of Complex Phenomena

Time itself *can* be viewed as a puzzle, but we choose not to do so. And we also choose to ignore the couplings between mass, space, and time revealed by relativity. For us, Time is a primitive intuitive notion, like Space and Place. We think of time in purely-classical nonrelativistic terms. Time's passage can then be quantified through any periodic motion. It is the result of experience that the exact nature of that motion is immaterial. The difficulties involved in finding a precise and general definition of "time" are not important to an understanding of time reversibility. Simplicity dictates an understanding based on nonrelativistic classical concepts.

At the end of the nineteenth century, and again, toward the middle of the twentieth, some vocal physicists looked forward to finding a unified view of nature. In addition to linking our sensations to the physical world, through understanding consciousness, such a unified view would also require a consistent mathematical description of complex phenomena. The emergence of chaos and complexity renders such a goal obsolete. Gödel showed that most interesting purely-mathematical theories are intrinsically incomplete, unable to decide the truth or untruth of definite statements.

The premature announcements, around 1900 and again around 1950, that "classical mechanics is dead" evidently stemmed from this same obsolete viewpoint of a "complete" theory. If there were some complete and unified view of nature, then more-specialized and restricted special cases of it could perhaps be thought of as second-rate, even if their structures were simpler. Chaos limits the ability of the various theories to overlap.

Physics, chemistry, and biology are intrinsically *different* subjects, rather than special cases of a unified theory. Quantum mechanics is unable to select a particular evolving path. Simulations of *real-world* chaotic processes must *invariably* do just that. Organized biological activity is too complex for a description at the atomistic level. In view of the unattainability of a unified theory, classical mechanics furnishes the best possible basis for understanding problems on its borders with thermodynamics and irreversible fluid and solid mechanics. The exploration and penetration of this artificial perimeter is the main subject of this book.

By now, it is both necessary and commonplace to subdivide knowledge, separating biology from economics and engineering. Within physics classical, quantum, and relativistic mechanics all have their own idealizations, with none of them describing our experiences perfectly. We take the point of view that mechanics is an imperfect, but educational model. It is because mechanics' consequences have apt real-life analogs, that this subject is worth learning and knowing. The classical mechanics of isolated systems can be profitably generalized, to describe the interaction of systems with their surroundings, nearing the realm of thermodynamics. But the lack of fluctuations in thermodynamics prevents the agreement from being perfect. The only way to distinguish a better theory from its competitors describing the same phenomena, is to wield Occam's Razor, shaving away irrelevant assumptions, and leaving the *simplest* possible explanation as the best. The *simplest* theories can and do lead to the discovery of complexities as absorbing and interesting as those created by Bach, Brubeck, Mingus, and Monk.

Engineers deal with the application of physical theories to real problems. Their approach is typically totally different to the time-reversible approach of basic physics. Engineering problems include viscosity, heat conductivity, plasticity, and other patently irreversible phenomena based on observation. How does it happen that the engineers' alternative approach has lost the fundamental time symmetry of microscopic physics? For one thing, the macroscopic theories used by engineers incorporate averaging, both in space and in time. Their theories are continuum field theories which ignore short-ranged and high-frequency fluctuations. For another, the dependent variables are different too. They also reflect averaging—averaging over microscopic details and the process of measurement. Typical variables are temperature, strain rate, and stress, rather than position and momentum, or the wave function. Finally, the systems considered by engineers are seldom isolated. Ordinarily external sources and sinks for heat and work are included. The microscopic analogs of these sources and sinks require generalizing the purely-Newtonian mechanics familiar to physicists.

The equally familiar irreversible processes which are all around us are not at all similar to the near-equilibrium fluctuations exploited by linear-response theory. Linear-response theory deals with ensemble-averaged infinitesimals. Macroscopic irreversible processes are individual and strongly driven, far from equilibrium, and inherently complex. These far-from-equilibrium conditions require special computer simulation techniques.

1.7 The Paradox: Irreversibility from Reversible Dynamics

The conflict between basic time-reversible physics and applied irreversible engineering is the "reversibility paradox". For gases, Boltzmann clarified this paradox by showing that averaging, justified by collisional chaos, was an essential part of its resolution. He showed that a statistical averaging of collisions, which ignores any pre-existing correlations and fluctuations, converts the reversible equations governing low-density gas dynamics to the irreversible equations of continuum mechanics. His approximate Boltzmann equation is

$$\dot{f} = (\partial f/\partial t) + \dot{q}(\partial f/\partial q) + \dot{p}(\partial f/\partial p) = (\partial f/\partial t)_{\text{collisions}} ,$$

where the collisional part is *quadratic* in f, reflecting the need for two bodies for each two-body collision. As usual, we use the superior dot for a comoving time derivative, following the flow. The flow occurs in a one-particle phase space, with $q = (x, y)$ or (x, y, z) and likewise for p. Solutions for the

evolution of the single-particle probability density, $f(r, v, t)$, make detailed predictions for the approach to equilibrium, and for the velocity distributions characterizing systems undergoing diffusive, viscous, and conductive dissipation. For dilute gases, the time-development of Boltzmann's approximate single-particle entropy,
$$S_B(t) = -Nk\langle \ln f_1 \rangle = -k \int dr \int dv f_1(r, v, t) \ln f_1(r, v, t) ,$$
agreed with the predictions of irreversible thermodynamics, opening the way for Gibbs' formulation of statistical mechanics for general systems, but restricted to equilibrium.

Green and Kubo showed that Gibbs' averaging links the irreversible transport coefficients of phenomenological continuum theory to the decay of equilibrium fluctuations. For dilute gases, these results are also equivalent to Boltzmann's. After Green and Kubo discovered linear-response theory, theoretical progress was stalled, awaiting the development of fast computers. The need to understand complex chaotic behavior frustrated the attempts of analysts. Computers made progress possible again. Recent numerical work has shown that even *few*-body systems show irreversible behavior, on the average, even with rigorously time-reversible motion equations. The irreversibility emerges with great clarity and precision when the small-system results are time-averaged.

Computers made it possible to simulate both reversible mechanics and irreversible flows. In the latter case it was necessary to impose boundary conditions or constraints, driving the system away from equilibrium. Heat and work had to be incorporated explicitly into the programming. Handily, all this could be done without sacrificing the time reversibility of the underlying equations! Let us turn now to a particularly useful algorithm for solving the differential equations which make simulations possible.

1.8 Algorithm: Fourth-Order Runge-Kutta Integrator

Because most physical models involve solving differential equations we outline here the construction and application of a particularly useful integrator. This "fourth-order" Runge-Kutta integrator evolves differential equations for a timestep Δt correct through terms of order Δt^4. To illustrate it we apply the fourth-order Runge-Kutta integrator to the two harmonic-oscillator differential equations (with the mass and force constant both set equal to unity):
$$\{\dot{q} \equiv dq/dt = +p; \; \dot{p} \equiv dp/dt = -q\} .$$

The underlying idea is to combine together the time derivatives at four nearby (q, p) phase-space points in such a way as to generate the next trajectory point at time Δt, accurate through terms of order Δt^4.

The Runge-Kutta integration algorithm evolves a general vector $x(t)$ forward in time according to the differential equations $\dot{x} = y(t, x)$. An advantage of Runge-Kutta integration is its self-starting nature. For notational simplicity we consider the evolution from time 0 through the timestep Δt. The algorithm generates four related vectors, $\{x_1, x_2, x_3, x_4\}$ through the series of four steps:

$$x_1 = x_0 + (\Delta t/2) y(0, x_0) \; ;$$

$$x_2 = x_0 + (\Delta t/2) y((\Delta t/2), x_1) \; ;$$

$$x_3 = x_0 + (\Delta t) y((\Delta t/2), x_2) \; ;$$

$$x_4 = x_0 + (\Delta t/6)[y_0(0, x_0) + 2y_1((\Delta t/2), x_1) + 2y_2((\Delta t/2), x_2) + y_3(\Delta t, x_3)] \; ,$$

where $x(0) \equiv x_0$ and $x(\Delta t) = x_4$. Notice that for the purely time-dependent case, where y is independent of x, this integrator reduces to Simpson's Rule.

Let us now apply the integrator to the two-variable harmonic oscillator problem, arbitrarily choosing to begin at the phase point $(q = 1, p = 0)$. The series of four steps proceeds as follows [Note $x = (q, p)$ here.]:

$$x_0 = (1, 0) \rightarrow y_0 = (0, -1)$$

$$x_1 = (1, 0) + (\Delta t/2)(0, -1) = (1, -(\Delta t/2)) \rightarrow y_1 = (-(\Delta t/2), -1) \rightarrow$$

$$x_2 = (1, 0) + (\Delta t/2)(-(\Delta t/2), -1) = (1 - (\Delta t^2/4), -(\Delta t/2)) \rightarrow$$

$$y_2 = (-(\Delta t/2), -1 + (\Delta t^2/4)) \rightarrow$$

$$x_3 = (1, 0) + \Delta t(-(\Delta t/2), -1 + (\Delta t^2/4)) \rightarrow$$

$$y_3 = (-\Delta t + (\Delta t^3/4), -1 + (\Delta t^2/2)) \rightarrow$$

$$x_4 = x(\Delta t) = (1 - (\Delta t^2/2) + (\Delta t^4/24), -\Delta t + (\Delta t^3/6)) \; ,$$

agreeing through fourth order in Δt with the exact solution

$$q = +\cos(t); \; p = -\sin(t) \; .$$

A four-step subroutine implementing this integrator takes the form:

$$\{\text{find } y(x_0) \to x_1; \text{ find } y(x_1) \to x_2;$$

$$\text{find } y(x_2) \to x_3; \text{ find } y(x_3) \to x(\Delta t)\}$$

with each of the evaluations of the derivative vector y carried out by a righthand side subroutine containing the differential equations to be solved. It is a useful exercise to write out the integrator routine sketched above along with the righthand side routine for the oscillator special case.

A bit of work shows that the analytic form of the harmonic oscillator solution is:

$$q(n\Delta t) = [1 - (\Delta t^6/72) + (\Delta t^8/576)]^{n/2} \cos(n\Delta\theta),$$

where $\Delta\theta$ can be obtained from the equation

$$\tan(\Delta\theta) = [\Delta t - (\Delta t^3/6)]/[1 - (\Delta t^2/2) + (\Delta t^4/24)].$$

In the last of the four example problems which make up the next Section we use this same Runge-Kutta integrator to solve a *nonequilibrium* heat-conducting oscillator problem involving *four* ordinary differential equations. That example also details two alternative centered-difference algorithms for obtaining a solution. We believe the reader will agree that using the Runge-Kutta integrator requires much less time than devising special purpose integrators for different problems as they arise.

Here follows a complete Fortran program designed to solve a three-variable nonlinear nonequilibrium problem, the dynamics of a Nosé-Hoover oscillator with a coordinate-dependent temperature $T = 1 + \epsilon \tanh(q)$. Depending on the initial conditions and the parameter ϵ (equal to the maximum value of the temperature gradient), the solution can be a one-dimensional limit cycle, or a torus, or a chaotic strange attractor. Despite the simplicity of this problem the results can be complicated, even for the equilibrium case with $\epsilon = 0$. For that case see the 1986 paper by Posch, Hoover, and Vesely. The Nosé-Hoover oscillator is described by a coordinate q, momentum p, and friction coefficient ζ. The three motion equations solved by the program "rk.f" are as follows:

$$\dot{q} = p; \; \dot{p} = -q - \zeta p; \; \dot{\zeta} = p^2 - T.$$

Here is the program[1]:

```
call this program rk.f [Nonequilibrium Nosé-Hoover Dynamics]
implicit double precision(a-h,o-z)
```

---

[1] We have omitted the six blanks that precede a standard line of Fortran code.

```
dimension x(3),y(3)
q = 1.0d00
p = 0.0d00
z = 0.0d00
x(1) = q
x(2) = p
x(3) = z
time = 0.0d00
dt = 0.001d00
do it = 1,100000
call rk(x,y,time,dt)
time = time + dt
q = x(1)
p = x(2)
z = x(3)
write(6,6) q,p,z,time
enddo
stop
end
subroutine rk(x,y,time,dt)
parameter(neq = 3)
implicit double precision(a-h,o-z)
dimension x(neq),y(neq)
dimension y1(neq),y2(neq),y3(neq),y4(neq)
call rhs(x,y1)
do i = 1,neq
x1(i) = x(i) + (dt/2.0d00)*y1(i)
enddo
call rhs(x1,y2)
do i = 1,neq
x2(i) = x(i) + (dt/2.0d00)*y2(i)
enddo
call rhs(x2,y3)
do i = 1,neq
x3(i) = x(i) + dt*y3(i)
enddo
call rhs(x3,y4)
do i = 1,neq
x(i) = x(1) + (dt/6.0d00)*(y1(i) + 2*y2(i) + 2*y3(i) + y4(i))
```

```
enddo
return
end

subroutine rhs(x,y)
parameter(neq = 3)
implicit double precision(a-h,o-z)
dimension x(neq),y(neq)
epsilon = 1.0d00
q = x(1)
p = x(2)
z = x(3)
T = 1 + epsilon*dtanh(q)
y(1) = +p
y(2) = -q - z*p
y(3) = p*p - T
return
end
```

1.9 Example Problems

To illustrate the concepts of time reversibility and chaos we consider here four examples. The first of them is a two-dimensional area-preserving map. It is a caricature of equilibrium flows obeying Liouville's incompressible theorem $\dot{f} \equiv 0$. The second is a three-dimensional continuous flow, but with *discontinuous* forces and a simple phase-space structure. Third is a three-dimensional flow with continuous forces and a *complicated* phase-space structure. Last, we apply the fourth-order Runge-Kutta algorithm of the last Section to a harmonic oscillator interacting with a coordinate-dependent heat reservoir. All four problems can exhibit chaotic behavior, with small changes in initial conditions growing exponentially with time. All four are relatively easy to simulate and to visualize. Each is a building block in creating an understanding of the nonequilibrium systems which are emphasized in the following Chapters of the book. More of the underlying background in mechanics and numerical algorithms, required to generate numerical solutions for the last three problems, is given in Chapter 2.

1.9.1 Equilibrium Baker Map

If N coordinates suffice to represent a system's configuration then the configuration at any fixed time is represented by a single point in the corresponding N-dimensional space. The time development of that point defines a one-dimensional "trajectory"—a line—in that same space. Imagine the repeated intersections of such a trajectory with some fixed $(N-1)$-dimensional surface embedded in the N-dimensional configuration space. The successive intersections with such a surface provide a "Poincaré Section" defining a "mapping" on the surface coordinates. The mapping links each intersection to the next. Such "maps" can be viewed as simplified caricatures of flow problems. Their advantage is a reduction (but only by one!) in the number of coordinates which has to be considered.

Because a *one*-dimensional *reversible* map can only alternate between *two* values of the dependent variable, the simplest interesting reversible map is necessarily two-dimensional. Such a map is analogous to an ordered series of snapshots, each related to the previous by an integration of the motion equations over the time interval Δt by which the snapshots are separated. More complicated maps, with irregular time intervals, correspond to the crossings of particular surfaces in the phase space.

If the underlying motion equations are reversible, the map must likewise be reversible. The two-dimensional Baker Map which we consider here is often used to exhibit chaos, "sensitive dependence on initial conditions", because the separation of two nearby points increases exponentially, in a particular direction, the "unstable" direction. The map was popularized by Hopf in his 1934 paper. Provided that the initial point is chosen sufficiently randomly, with irrational coordinates for instance, the time-reversible Baker Map shown in Figure 1.1 eventually provides an "ergodic" coverage (coming arbitrarily close to all the points) of the 2×2 square, whether the map goes forward or backward in time. In both directions the map has the same form, but with x and y permuted in the backward map:

$$[-1 < x < +1;\ 0 < y < +1] \longrightarrow (x,y)' = [(x+1)/2, (2y-1)] \ ;$$
$$[-1 < x < +1;\ -1 < y < 0] \longrightarrow (x,y)' = [(x-1)/2, (2y+1)] \ .$$

This map *reduces* point-to-point separations in the x direction and increases them in the y direction, in both cases by a factor 2. The map is said to be "Lyapunov unstable" because small separation differences in the y direction double at each iteration, corresponding to a maximum Lyapunov exponent of $\ln 2$:

$$\delta y_{n+1} = 2\delta y_n \longrightarrow \lambda_1 = \langle \ln(|\delta y_{n+1}|/|\delta y_n|)\rangle = \ln 2 \ .$$

To attain a fixed level of precision in specifying the future iterates of an initial point, additional information would have to be added, in the y direction, and could be discarded, in the x direction, one binary bit in each direction, per iteration.

Overall, the "equilibrium" Baker Map conserves area, with the stretching (in the y direction) exactly compensating for the shrinking (in the x direction). Evidently the action of the map is "sensitive" to small differences in the stretching direction, so that *not even in principle* could the "true" motion be followed for long. This Baker Map is a rough caricature of motions governed by Lyapunov-unstable Hamiltonian mechanics.

When the exponential instability associated with stretching is represented by a limited-precision computer, it can give rise to a fixed y coordinate, at $y = \pm 1$, and a fixed x coordinate, $x = \pm 1$, so that the resulting "fixed point" repeats forever. Likewise, there are particular initial conditions which give rise to (unstable) periodic cycles of two or three or four, ... points. For example, the twice-iterated Baker Map has a cycle connecting the two (x, y) points $(-1/3, +1/3)$ and $(+1/3, -1/3)$. On the other hand, adding a small amount of random noise, in the eighth, tenth, or twelfth significant figure, is enough to destroy these artificial cycles.

This equilibrium Baker Map, though simple, already illustrates a disquieting feature, due to its singular nature along the line $y = 0$: any *rational* value of y, of the form $\pm n/(2^m)$, is eventually mapped to the singular line. For any fixed m these solutions, like those of the periodic cycles, have measure zero in the two-dimensional space. General *irrational* points are instead mapped in an "ergodic" manner, eventually coming arbitrarily close to any (x, y) point in the square. Despite this uniform coverage of the square, the Baker Map is a *very poor* random number generator. In both the (x, y) and the rotated (q, p) representations, as in Figure 1.1, it is easy to show that the products of successive iterates are strongly correlated:

$$\langle xx' \rangle = (1/6); \ \langle yy' \rangle = (2/3); \ \langle qq' \rangle = \langle pp' \rangle = (5/12) \ .$$

All these correlations would vanish if the (x, y) points or the rotated (q, p) points could truly be chosen *randomly*.

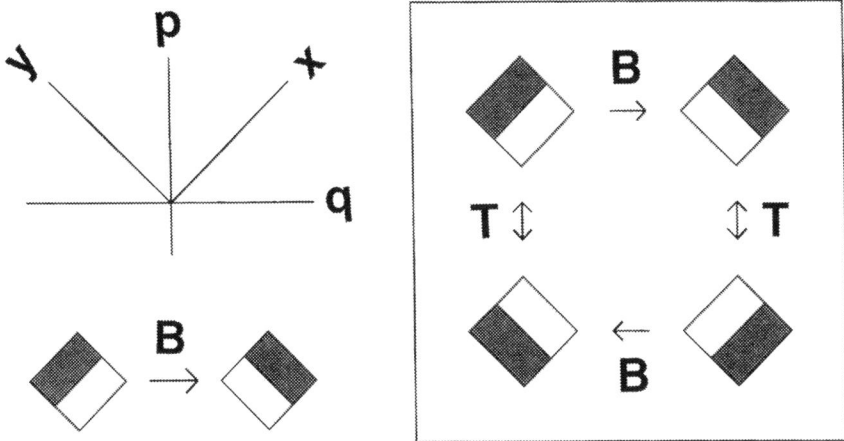

Figure 1.1. The Equilibrium Baker Map, B, in the usual (x,y) coordinates, is shown on the lower left. The mapping B relates the "new" coordinates to the old ones, $B_{xy}(x,y) = (x',y')$. The rotated map, $B_{qp}(q,p) = (q',p')$ in $(q,p) = \sqrt{1/2}(x-y, x+y)$ coordinates, is shown on the right. The time reversal operation T_{qp} shown in the figure corresponds to changing the sign of the "momentum" p.

The odd sensitivity to infinitesimal insignificant "information" which has no significance, being infinitesimal in scale, can plainly have nothing to do with physics. The sensitivity is a consequence of the vagaries of the real number system, in which any two rational numbers, no matter how close together, are separated by an infinite continuum of irrationals as well as by a set with the lesser cardinality, $\aleph_0$ in Cantor's theory of infinite sets, of rational numbers. Because the Baker Map is so simple, some computer programs, on some computers, will not generate sufficient internal noise to avoid settling onto one of the Map's fixed points at $(x,y) = \pm(1,1)$. In a more nearly accurate calculation this could not happen because these fixed points are unstable, with displacements in the y direction doubling at each iteration.

The disparity in evolution, between the rational periodic orbits and general irrational initial points, seems less serious if we consider exactly the same map, expressed now in terms of the new horizontal and vertical coordinates, $\{q,p\}$. The motivation for this rotation is the physicist's notion of time reversibility: coordinate values $\{q\}$ should be traced out backward

in time simply by (i) reversing the signs of the corresponding momenta $\{p\}$ and (ii) choosing the right initial conditions. The rotated (q,p) coordinates are related to the original (x,y) coordinates by a 45° clockwise rotation:
$$q = (x-y)/\sqrt{2}; \; p = (x+y)/\sqrt{2} \;.$$
Thus the periodic cycle of the twice-iterated map links the two (q,p) points $(\pm\sqrt{2/9},0)$. In the new (q,p) coordinate system the map satisfies the usual physicist's expectation for a time-reversible map:
$$B_{qp}(q,p) = (q',p') \longrightarrow B_{qp}(q',-p') = (q,-p) \;.$$
Explicitly, this rotated Baker's Map has the piecewise-linear, but irrational form:
$$q < p \longrightarrow \{q,p\}' = \{+(5q/4) - (3p/4) + \sqrt{9/8}, -(3q/4) + (5p/4) - \sqrt{1/8}\} \;;$$
$$p < q \longrightarrow \{q,p\}' = \{+(5q/4) - (3p/4) - \sqrt{9/8}, -(3q/4) + (5p/4) + \sqrt{1/8}\} \;.$$

See Figure 1.2 for a sequence of points generated using this map. We will come back to simple mappings again, in discussing dissipative analogs of the present area-preserving conservative Baker Map. Though simple, these maps have considerable pedagogical value.

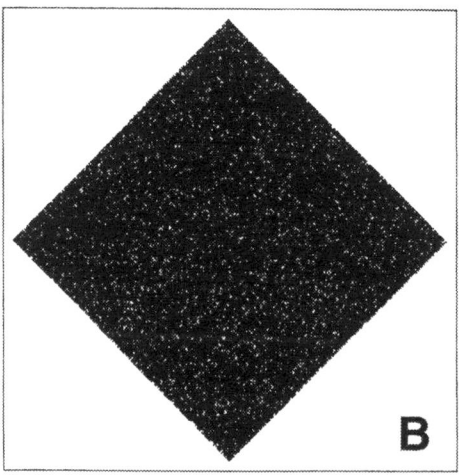

Figure 1.2. 100,000 points generated by the (q,p) Baker Map beginning with initial point at $(0.50, 0.00)$.

Evidently a probability density $f(x, y)$ which is constant is preserved exactly by the Baker Map. Each area element $dxdy$ is mapped to a new element with exactly the same area, $dx'dy' = (dx/2)2dy$. Thus the constant density solution $f(-1 < x < +1, -1 < y < +1) \equiv (1/4)$ is *stationary*.

Apart from families of lines which have special rational y coordinates $y = n/(2^m)$, along which trajectories approach the pair of fixed points $(-1, -1)$ and $(+1, +1)$, and the various unstable periodic cycles, a general irrational point in the 2×2 square eventually converges, in a "coarse-grained" sense, to the constant-density solution. This means that the integrated density inside *any* square box of sidelength ϵ (that is, the probability of occupying that box) approaches the limiting value $(\epsilon^2/4)$.

Because in principle the mapping process does not destroy any information, and can certainly be inverted—even precisely reversed, in the (q, p) representation—the solution becomes constant only in the limit of an infinite number of iterations. In practice, with a fixed number of significant figures, $\log(1/\epsilon)$, the mapping eventually produces a probability density which is essentially constant in cells of width and height exceeding ϵ. Increasing the precision of the representation and the mapping reduces the discrepancy between averages computed with the constant density $f = (1/4)$ and the coarse-grained approximation.

We will see, in the next Chapter, that a dissipative version of the Baker Map, a caricature of *non*equilibrium systems, provides an infinitely-detailed "fractal" solution, rather than the constant density f found here. Fractals display structure on all scales, no matter how fine. Because this behavior is not only unpredictable in principle, but is also too detailed for accurate description, averaging, in both space and time, is inevitably required. This averaging leads to the "coarse-grained" representations envisioned by Gibbs.

1.9.2 *Equilibrium Galton Board*

A more realistic model of an irreversible process has continuous "chaotic" trajectories, so that numerical solutions are "Lyapunov unstable". The "Galton Board" is such a model. It is the simplest analog for field-driven diffusion. The Galton Board is named for Sir Francis Galton, who used a similar laboratory device, beginning in 1873. He demonstrated the familiar binomial distribution, which results when falling particles are scattered to the right or left as they fall through a "Board" of scatterers. The probability distribution which results is the familiar binomial distribution:

$$\text{prob}(n_r, n_l) = [(n_r + n_l)!/(n_r! n_l!)](1/2)^{n_r+n_l} .$$

In both Galton's laboratory demonstrator and the corresponding computational model, a single particle moves through an array of hard-disk scatterers. In computation it is convenient to treat the moving particle as a mass point, scattering from a lattice of fixed particles of *radius* σ. Equivalently, it is possible to treat this problem as the spatially-periodic dynamics of *two* moving particles, each of *diameter* σ, with one particle accelerated to the right (or down) and the other to the left (or up). Figure 1.3 illustrates the first of these arrangements, a moving point, with the scatterers arranged in a regular "triangular" lattice.

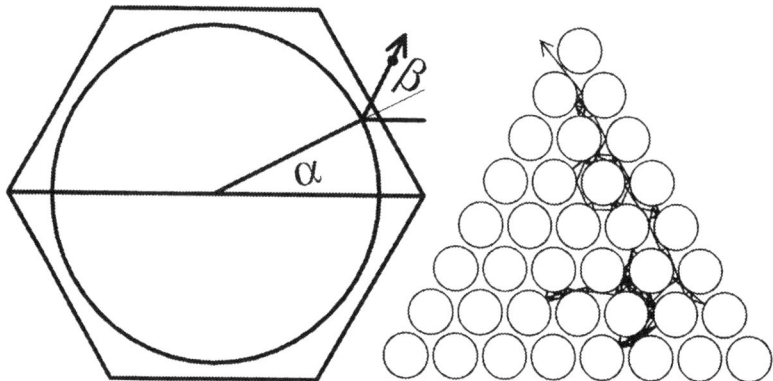

Figure 1.3. Portion of a Triangular Lattice with an equilibrium trajectory with $g \equiv 0$, where g is the field strength. Each of the 157 collisions shown defines two angles, (α, β), giving its location and the post-collision direction of motion, as indicated in the hexagonal "unit cell" on the left.

It is convenient to describe the collisions with two angles: α gives the location of a collision relative to the field direction; β gives the direction of the scattered trajectory, relative to a vector *from* the fixed scatterer *to* the moving point, just *after* each collision. The sequence of collisions is chaotic. The curved surfaces of the scatterers provide chaos. This chaotic instability means that small perturbations in the initial coordinates and momenta of the moving particle, $\{\delta q, \delta p\}$, will grow exponentially in the time, as $e^{\lambda t}$. Just as in the Baker Map, there is both a growing "unstable" direction and a shrinking "stable" one, so that a continuous flow of new information (in the growing direction) is required to operate the map. This quite typical "sensitive dependence on initial conditions" causes the particle to experience, as time goes on, all possible collision types:

$$\{0 < \alpha < \pi;\ -1 < \sin(\beta) < +1\}\ .$$

If all of space is represented by a single unit cell, the limiting spatial distribution within that cell becomes completely uniform in the absence of any accelerating field. These rather obvious features were proved by Sinai, and can be convincingly demonstrated by a short computer program simulating the motion. The simplest computer program simulating the time development of the Galton Board advances the moving particle step-by-step along the field-free straight line trajectory,

$$x_+ = x_0 + \dot{x}_0 \Delta t;\ y_+ = y_0 + \dot{y}_0 \Delta t;\ \dot{x}_+ = \dot{x}_0;\ \dot{y}_+ = \dot{y}_0\ ,$$

or the field-induced parabola,

$$x_+ = x_0 + \dot{x}_0 \Delta t + (g/2)\Delta t^2;\ y_+ = y_0 + \dot{y}_0 \Delta t;\ \dot{x}_+ = \dot{x}_0 + g\Delta t;\ \dot{y}_+ = \dot{y}_0\ ,$$

checking to see whether or not a collision, with either a periodic boundary or the scatterer boundary, has occurred.

Let us consider a cell which is the *top half* of the hexagon, less the top half of the scatterer disk, shown in Figure 1.3. This cell, in which the motion occurs, has five periodic line-segment boundaries and a reflective half-disk boundary. In the event of a "collision" with a periodic boundary the moving particle is simply replaced at the appropriate spot in the cell, with a new velocity, $(\pm\dot{x}, \pm\dot{y}) \to (\pm\dot{x}, \mp\dot{y})$, and the trajectory is continued. In the event of a scatterer collision, the particle is returned to its location at the previous timestep, and the *radial* momentum component (in the frame of the scatterer) is reversed. Figures 1.4 and 1.5 show the time development of two problems: (i) the uniform field-free distribution of collision types, which also corresponds to a uniform distribution in space; (ii) a *nonuniform* distribution of collision types induced by an accelerating field g. In both cases we use $\sin(\beta)$ rather than β to catalog the calculated collision sequences. This gives a *uniform* distribution in the field-free case because the *relative* collision rate is proportional to $\cos(\beta)d\beta = d\sin(\beta)$.

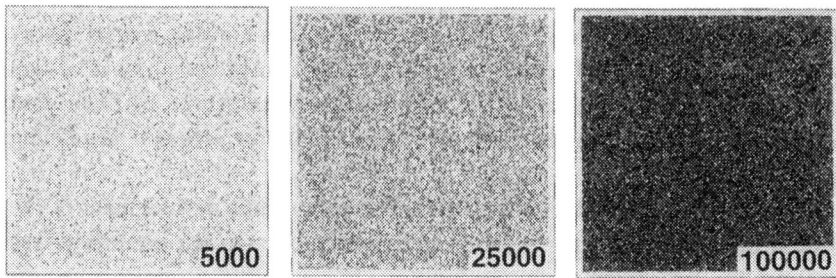

Figure 1.4. $[\alpha, \sin(\beta)]$ pairs for 5000, 25,000, and 100,000 successive collisions in the field-free case, $g = 0$. The scatterer density is (4/5) of close-packing. The moving particle has unit mass and kinetic energy 0.50.

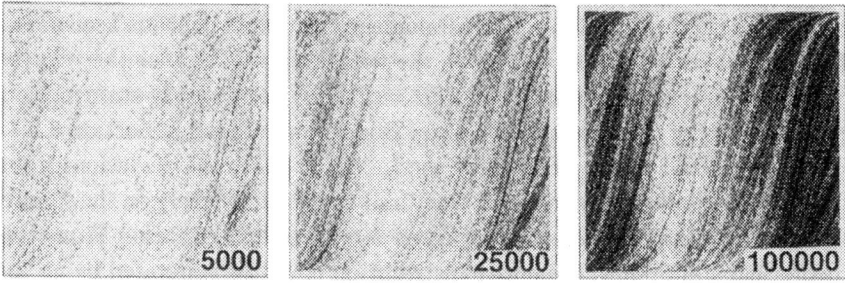

Figure 1.5. $[\alpha, \sin(\beta)]$ pairs for 5000, 25,000, and 100,000 collisions generated from randomly chosen zero-velocity initial conditions but with a nonzero field, $g = 0.5$. The density is 4/5 of the close-packed density. The scatterer diameter is unity and the moving particle has unit mass. Those collisions are shown which had kinetic energies between 0.495 and 0.505.

The steady-state motion characteristic of a *thermostated* Galton Board is simulated in Section 5.9.1. For any fixed coarse-graining of the $[\alpha, \sin(\beta)]$ space the deviations from the field-free uniform distribution eventually depend inversely on the square root of the sampling time, as expected for a random process. From the theoretical standpoint, the entire distribution could be generated by an alternative, much more cumbersome, approach. We could solve Liouville's partial differential equation,

$$\dot{f} \equiv (\partial f/\partial t) + \dot{q}(\partial f/\partial q) + \dot{p}(\partial f/\partial p) \equiv 0; \; f \equiv f_N(q, p, t)$$

for the time-development of the probability density. For the present model it is quite clear that trajectory dynamics provides a much simpler route to the steady-state distribution.

In the absence of "thermostat" or "control" forces, discussed more fully in Chapter 2, the field-driven distribution *never* converges. The field-driven Newtonian motion of the Galton Board has no stationary state. The collision pairs $[\alpha, \sin(\beta)]$ shown in Figure 1.5 were generated by advancing coordinates chosen randomly within the unit cell, with initial velocity zero, and plotting the resulting scattering collisions whenever the kinetic energy was within the limits $0.495 < K < 0.505$. The resulting distribution, though generated with classical Newtonian mechanics, shows a strong family resemblance to the "fractal" distributions characteristic of the nonequilibrium steady-state systems discussed in this book.

How could one make an *integer* model of the Galton Board, with a discrete coordinate space? The periodicity associated with a discrete coordinate space seems fundamentally opposed to chaos, but, as we shall see, Levesque and Verlet were able to construct a strictly and precisely time-reversible *algorithm* for chaotic problems by using an *integer* coordinate space and a continuous potential. How could one construct a discrete integer-space version of the Galton Board? Evidently this could be done, by using a steepened version of the potential—like that used in Section 2.12.2—rather than the discontinuous hard-disk scattering law considered here.

1.9.3 *Equilibrium Hookean Pendulum*

The simplest chaotic problems, suitable for straightforward computer simulation, and free of the explicit bifurcation singularities of the Baker Map and Galton Board, involve pendula. A double pendulum is relatively easy to build. Ted Hillyer, at UCDavis, built one for us. Its analytic counterpart is a familiar chaotic problem in a four-dimensional space free of any singularities. A conceptually simpler model, which is advantageous for studying the dependence of the local (instantaneous) Lyapunov exponents on the coordinate system, is a "Hookean pendulum" with its length governed by a Hooke's-Law spring. See Tél and Gruiz' 2006 book, *Chaotic Dynamics*, for a discussion of Poincaré sections for this model. The phase space for the Hookean pendulum is four-dimensional, with two coordinates and two momenta. The motion is restricted to a three-dimensional subspace by the constancy of the total energy, the value of the Hamiltonian. The Hamiltonian governing the motion can be written in either Cartesian coordinates

$\{x, y\}$ or plane polar coordinates $\{r, \theta\}$:
$$\mathcal{H}_C = [p_x^2 + p_y^2]/(2m) + mgy + (\kappa/2)(\sqrt{x^2 + y^2} - d)^2 \ ;$$
$$\mathcal{H}_P = [p_r^2 + (p_\theta/r)^2]/(2m) + mgr\sin(\theta) + (\kappa/2)(r - d)^2 \ .$$

As is explained more fully in the next Chapter, the equations of motion, in either coordinate system, follow from the Hamiltonian flow equations:
$$\dot{q} \equiv +(\partial \mathcal{H}/\partial p); \ \dot{p} \equiv -(\partial \mathcal{H}/\partial q) \ .$$

Here q is a coordinate in the Hamiltonian (x or y or θ or r), and p is the corresponding momentum (p_x or p_y or p_θ or p_r).

For convenience, choose the pendulum mass m, the rest length d, and the gravitational field strength g equal to unity with the force constant κ equal to four. This last choice promotes the Lyapunov-unstable coupling between the vertical and horizontal oscillations of the pendulum. The equations of motion follow by differentiation of the Hamiltonian $\mathcal{H}$:

$$\{\dot{x} = p_x; \ \dot{y} = p_y; \ \dot{p}_x = -(4x/r)(r - 1); \ \dot{p}_y = -1 - (4y/r)(r - 1)\} \ ;$$
$$\{\dot{r} = p_r; \ \dot{\theta} = (p_\theta/r^2); \ \dot{p}_r = (p_\theta^2/r^3) - \sin(\theta) - 4(r - 1); \ \dot{p}_\theta = -r\cos(\theta)\} \ .$$

The two sets of motion equations provide alternative descriptions of the *same* motion, so that a comparison furnishes a good check of the numerical integrator. The Runge-Kutta integrator algorithm described in the previous Section is the simplest useful choice. Both energy conservation:
$$\dot{\mathcal{H}} \equiv \sum [\dot{q}(\partial \mathcal{H}/\partial q) + \dot{p}(\partial \mathcal{H}/\partial p)] \equiv 0 \ ,$$

and time reversibility—(i) running forward for a time t, (ii) reversing the momenta, (iii) advancing the dynamics (and so going "backward") for time t, and (iv) reversing the momenta once again, finally arriving at the initial configuration with the initial momenta:
$$\{+q, +p\}_0 \stackrel{(i)}{\to} \{+q', +p'\}_t \stackrel{(ii)}{\to} \{+q', -p'\}_t \stackrel{(iii)}{\to} \{+q, -p\}_0 \stackrel{(iv)}{\to} \{+q, +p\}_0 \ ,$$
are additional useful checks of the numerical work.

Any solution of the pendulum problem can be visualized as a trajectory in the four-dimensional space, "phase space", with axes $\{q_1, q_2, p_1, p_2\}$, where a single "point" specifies the entire state of the system. For certain initial conditions the Hookean pendulum's motion is chaotic. If the evolution of a second pendulum problem is followed in the same phase space, with nearly the same chaotic initial condition, the magnitude of the "offset vector" connecting the two nearby trajectories,
$$\delta_C \equiv (\delta x^2 + \delta y^2 + \delta p_x^2 + \delta p_y^2)^{1/2}$$

in the Cartesian case, and

$$\delta_P \equiv (\delta r^2 + \delta\theta^2 + \delta p_r^2 + \delta p_\theta^2)^{1/2}$$

in the polar case, will then eventually grow as $e^{\lambda t}$. The long-time averages, indicated by $\langle\ldots\rangle$, of the two (different) local growth rates, λ_C and λ_P, define the "largest Lyapunov exponent", $\lambda_1 \equiv \langle\lambda_C\rangle = \langle\lambda_P\rangle$. It characterizes the chaotic motion. Although this growth rate must be independent of the choice of coordinates, when averaged over a sufficiently long time interval, the instantaneous values of the growth rate *do* depend upon the chosen coordinate system, so that the *local* Lyapunov exponents are properties of the description rather than the physical system. The mean values of λ_2 and λ_3 vanish. This is because there is no long-term expansion or contraction parallel to the trajectory direction. By symmetry, with expansion forward in time corresponding to contraction, when time is reversed: $\text{prob}(\pm\lambda_1) = \text{prob}(\mp\lambda_4)$; $\text{prob}(\pm\lambda_2) = \text{prob}(\mp\lambda_3)$. Figure 1.6 shows probability densities for "local" (time-dependent) values of the Lyapunov exponents $\{-5 < \lambda < +5\}$, using both the Cartesian and the polar-coordinate systems. Computation of complete Lyapunov spectra, one exponent for each phase-space dimension, is discussed in Section 8.4.

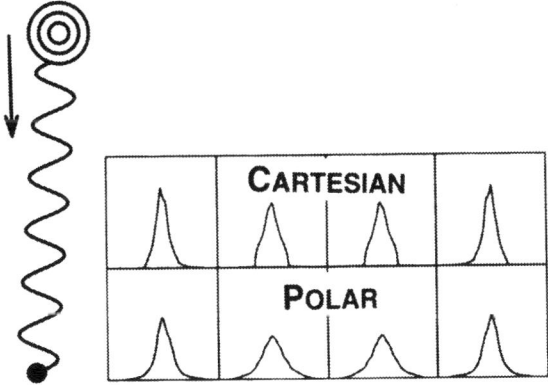

Figure 1.6. Probability distributions for the four Lyapunov exponents of the Hooke's-Law pendulum shown at the left. The corresponding Hamiltonian has the value, $\mathcal{H} = 1$, with the initial conditions $(x, y) = (0.00001, 1.0)$; $(p_x, p_y) = (0.0, 0.0)$.

1.9.4 Nosé-Hoover Oscillator with a Temperature Gradient

This nonequilibrium heat-transfer problem makes use of the kinetic temperature of a harmonic oscillator, $T \equiv (p^2/mk)$, where p and m are respectively the oscillator's (one-dimensional) momentum and mass, and k is Boltzmann's constant. The oscillator is governed by an external nonNewtonian feedback force, $-\zeta p$, controlling its temperature:

$$\{\dot{q} = p;\ \dot{p} = -q - \zeta p;\ \dot{\zeta} = p^2 - T(q)\} \ .$$

Assuming (correctly!) that the control variable ζ is bounded it is evident that the righthand side of the $\dot{\zeta}$ equation must have an average value 0:

$$\langle p^2 - T \rangle \equiv 0 \longleftarrow \zeta \text{ bounded} \ .$$

The motion is time-reversible in the sense that changing the signs of (t, p, ζ) lead to a reversed trajectory $q(+t) \to q(-t)$ satisfying *exactly the same* motion equations. Notice that on reversal, $\dot{q}$ changes sign, while $\dot{p}$ and $\dot{\zeta}$ do not.

The notion of time reversibility is tricky. Although the equations of motion are certainly time-reversible, so that the reversed trajectory satisfies them, the reversed trajectory can be made relatively unstable, simply by allowing the temperature T to depend upon the coordinate q. Consider the simplest case,

$$0 < T = 1 + \tanh(q) < 2 \ ,$$

where the temperature varies smoothly from 0 to 2 as q increases. This temperature variation opens up the possibility for a nonequilibrium flow of heat from higher to lower temperatures, in accord with Fourier's Law and the Second Law of Thermodynamics. The reversed motion, which would violate the Laws, is relatively *unstable* and *unobservable*. To see this, carry out a numerical trajectory integration for long enough for the system to reach its limit cycle. A time of 100 is enough. Then reverse the motion, by suddenly changing the signs of p and ζ. What happens is shown in Figures 1.7-1.9 for three *different* approximate solutions of the oscillator's equations of motion, all for the *same* initial conditions. In all three cases, after a time of order 30 the reversed motion returns to the Second-Law-abiding limit cycle. The details of the transition depend sensitively upon the particular integrator, and so are not physical, but the overwhelming power of the Second Law of Thermodynamics refuses to allow such a reversed nonequilibrium trajectory to go the wrong way for very long. In Chapter 8 we will analyze the mechanism for this behavior.

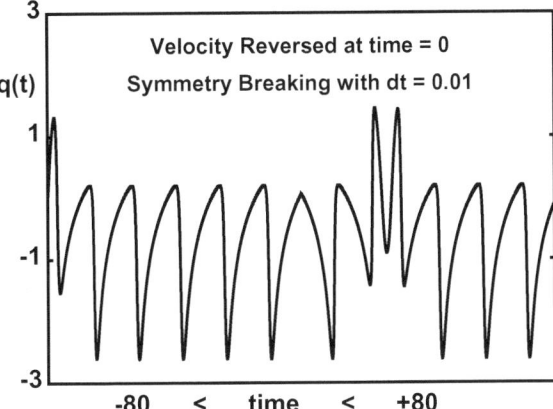

Figure 1.7. Time dependence of the Nosé-Hoover oscillator with velocity and friction coefficient reversal at time 0. In a time of order 30 the unstable expanding flow comes back to the attractive contracting limit cycle. Here $T = 1 + \epsilon \tanh(q)$ with $\epsilon = 1$ and the Runge-Kutta timestep $dt = 0.01$.

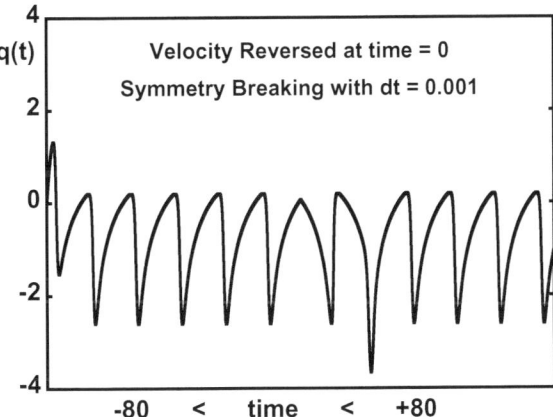

Figure 1.8. Time dependence of the Nosé-Hoover oscillator with velocity and friction coefficient reversal at time zero. Here the Runge-Kutta timestep dt is 0.001 and the details of the transition are quite different. Nevertheless the time required to regain the contracting attractor is about the same as in Figure 1.7.

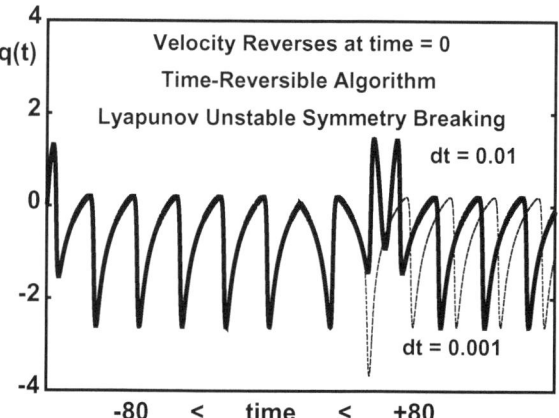

Figure 1.9. Time dependence of the Nosé-Hoover oscillator with velocity and friction coefficient reversal at time zero. Here two timesteps, $dt = 0.01$ and $dt = 0.001$, are compared. The time-reversible integration algorithm is used here rather than the Runge-Kutta one used in Figures 1.7 and 1.8. Nevertheless, the times required to regain the attractor are about the same as in those figures.

The three first-order differential equations for $\{\dot{q}, \dot{p}, \dot{\zeta}\}$ can be solved by an infinite variety of algorithms, not just the Runge-Kutta integration of Section 1.8 chosen for Figures 1.7 and 1.8. Consider, for instance, a centered-difference approach emphasizing the time-reversible character of the equations. By rearranging the coordinate equations of motion,

$$\left[\frac{(q_+ - 2q_0 + q_-)}{dt^2}\right] = -q_0 - \zeta_0\left[\frac{(q_+ - q_-)}{2dt}\right],$$

we can find q_+ in terms of $\{q_0, q_-, \zeta_0\}$.

With the new coordinate q_+ the corresponding new friction coefficient can be obtained from $\{q_+, q_0, q_-, \zeta_0, \zeta_-\}$ in either of two equally-good ways:

$$\zeta_+ = \zeta_- + 2dt\left[\left(\frac{q_+ - q_-}{2dt}\right)^2 - T_0\right] \text{ or}$$

$$\zeta_+ = \zeta_0 + dt\left[\left(\frac{q_+ - q_0}{dt}\right)^2 - T_{1/2}\right]; \; T_{1/2} \equiv (T_+ + T_0)/2.$$

Relative to the Runge-Kutta algorithm, these two finite-difference algorithms have the apparent advantage that they are explicitly *time-reversible* (apart from the unavoidable effects of computer roundoff errors). This advantage is more apparent than real, with the Runge-Kutta integration error only slightly greater than computer roundoff error for a timestep $dt = 0.001$. Additionally, the finite-difference expressions do require less storage than does the Runge-Kutta algorithm, with this advantage countered by the increased complexity of the finite-difference programming. Figure 1.10 shows another view of this same symmetry breaking. When the friction coefficient and velocity are simultaneously reversed so that the overall entropy production becomes negative, the motion quickly seeks out the original limit cycle so as to satisfy the Second Law of Thermodynamics.

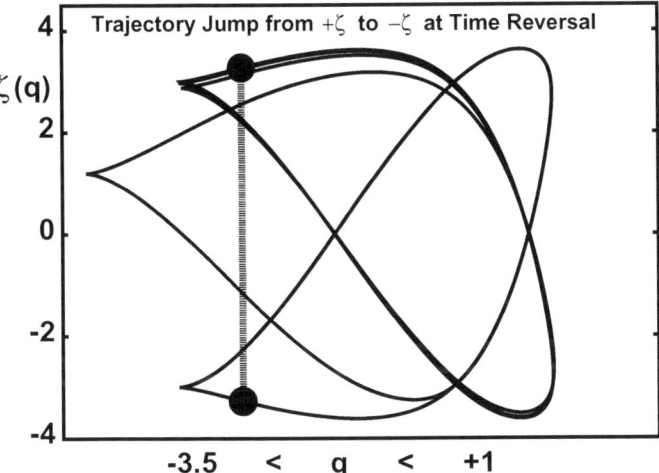

Figure 1.10. The stable dissipative limit cycle with $\epsilon = 1$ includes the upper filled circle. The motion around that cycle is *clockwise*. Near the top, with the oscillator moving toward the right the friction coefficient is positive, with heat flowing *from the oscillator* to the Nosé-Hoover reservoir. As the oscillator moves to the left (from hot to cold temperatures) the negative friction coefficient, changing sign near $q = -1.4$, indicates heat flow *from* the Nosé-Hoover reservoir to the oscillator. The mirror image unstable cycle includes the lower dot. Changing the signs of the velocity and friction coefficient (the vertical dot-to-dot downward jump) leads to a short transient before the motion returns to the dissipative stable cycle.

Whenever the temperature gradient is sufficiently large ($\epsilon > 0.44$) the time-reversible motion equations exhibit symmetry breaking. The solution

forward in time corresponds to an overall (stable) decrease in phase volume, with $\langle \zeta \rangle > 0$. Figure 1.7 illustrates the symmetry breaking with $\epsilon = 1$ and $dt = 0.01$. Figure 1.8 is the corresponding simulation with $dt = 0.001$, for which the Runge-Kutta integration errors are of the same order as the double-precision roundoff error. Though the details do depend upon the timestep, the time required for the unstable expanding motion to convert to stable contraction can be estimated from the largest Lyapunov exponent ($\lambda_1 = 1.22$ in the reversed flow, as calculated with 53-bit precision):
$$e^{1.22t} = 2^{53} \longrightarrow t = 30 .$$
This nonequilibrium oscillator example shows very nicely the consistency of time-reversible simulations with the inexorable dissipation described by thermodynamics' Second Law.

1.10 Summary and Notes

The time reversibility of fundamental physical theories might seem to preclude the irreversibility which is such an obvious and pervasive characteristic of physical reality. But the presence of chaos leads to two important effects, which together negate that idea. First, the chaotic Lyapunov instability inherent in single-trajectory solutions rules out any useful link between a long-time trajectory and the initial conditions. Boltzmann, Gibbs, and Maxwell all emphasized the importance of a probabilistic interpretation of dynamics reflecting this underlying instability. The *positive* Lyapunov exponents describe this instability, which corresponds to the continual creation of new information. Second, existing information is typically destroyed in chaotic evolutions. The *negative* Lyapunov exponents describe this stabilizing feature. Though both positive and negative Lyapunov exponents coexist in chaotic problems the negative exponents have the upper hand. They imply a stable convergence as a result of the overall destruction of system information.

Over a long time, the equilibrium situation corresponds to an exact balance of these tendencies, while the nonequilibrium situation invariably corresponds to an enhanced stability and to a net destruction of information, equivalent to a net increase of the external surroundings' entropy. These ideas will be illustrated in more detail in Chapter 5.

The irrelevance of the initial conditions, due to chaos, certainly suggests that an alternative to the study of detailed trajectories could be profitably taken up. Following Gibbs' successful formulation of equilibrium properties, as detailed in Chapter 3, perhaps we could follow *nonequilibrium* probability densities directly in time, making it unnecessary to follow detailed trajectories? If such an ensemble approach were successful, it would provide a description of an averaged approach to equilibrium, followed by Gibbsian fluctuations. Numerical work indicates instead that simple trajectory analyses provide the simplest route by far to nonequilibrium properties, despite the evident approximate nature of their resemblance to "real trajectories".

Real trajectories, as Joseph Ford was fond of pointing out, are figments of the imagination, once chaos is considered. Despite the promise of chaos' help in understanding the irrelevance of initial conditions, the impossibility of constructing a chaotic trajectory is a real fly in the ointment, imposed upon us by the real number system. Cantor's set theory established that the rational numbers which we use in computation are a negligibly-small, but uniquely important, subset. In a *mathematical* sense there is a qualitative distinction between systems with finite and infinite numbers of states. In a *practical* sense this distinction appears to be insignificant. Although it appears that this distinction has no practical importance, it is precisely as unsettling as the question of free will when we wish to relate computation to the real world around us. To date Gödel has come the closest to explaining difficulties of this kind (by pointing out that they are insurmountable).

1.10.1 *Notes and References*

There is no shortage of books describing classical time-reversible physics. *The Feynman Lectures on Physics*, Pars' *Treatise on Analytical Mechanics*, and Landau and Lifshitz' *Course in Theoretical Physics* are our standbys.

The fundamental basis of our microscopic understanding is Gibbs' *Elementary Principles in Statistical Mechanics*. For background in microscopic models of macroscopic behavior we have the Mayers' *Statistical Mechanics*, Hirschfelder, Curtiss, and Bird's *Molecular Theory of Gases and Liquids*, and Kubo's *Statistical Mechanics*, all by now somewhat dated. Frenkel's *Understanding Molecular Simulation: from Algorithms to Applications* and Evans and Morriss' *Statistical Mechanics of Nonequilibrium Liquids*, and Dorfman's *An Introduction to Chaos in Nonequilibrium Statistical Mechanics* are all useful. Bill's *Molecular Dynamics* and *Computational Statistical Mechanics* (both available at http://williamhoover.info) contain many of the problems (such as the Hookean Pendulum and the Nosé-Hoover oscillator) also treated in the present book.

There are lots of philosophically oriented books touching on the conflict between microscopic Time Reversibility and macroscopic Irreversibility from the standpoint of statistical physics. Sklar's *Physics and Chance: Philosophical Issues in the Foundations of Statistical Mechanics*, Price's *Time's Arrow and Archimedes' Point: New Directions for the Physics of Time*, Krylov's posthumous *Works on the Foundations of Statistical Physics*, and Zubarev's *Nonequilibrium Statistical Thermodynamics* fit that category.

To find out more about Boltzmann and Gibbs we recommend Cercignani's *Ludwig Boltzmann: the Man who Trusted Atoms*, Brush's *Lectures on Gas Theory*, his translation of *Vorlesungen über Gastheorie*, as well as Wheeler's *Josiah Willard Gibbs - the History of a Great Mind*.

Today the Press, Teukolsky, Vetterling, and Flannery compendium *Numerical Recipes* is still useful, but not so necessary as it was, because there is abundant free software on the internet.

For recreational reading try Pearson's *The Life, Letters, and Labours of Francis Galton*, Gleick's *Genius* (1992) [Feynman], and the many other books about Feynman and his adventures. Gleick's other contributions to the human side of our subject, *Chaos* (1987) and *The Information* (2011) make him our true friend and colleague. Ruelle's *Chaos and Chance* is another wonderful book. Joseph Ford's papers, such as "What Is Chaos, that We Should Be Mindful of It?" in *The New Physics* and Bridgman's "A Physicist's Second Reaction to Mengenlehre" should be included too.

On the technical side Levesque and Verlet's 1993 paper "Molecular Dynamics and Time Reversibility" introduces the simplest "bit-reversible algorithm", which nicely connects the reversiblity paradox to computer simulation. A classic sample of Lorenz' work, "Deterministic Nonperiodic Flow", is just one of the many thought-provoking papers available at his MIT website. Liouville's theorem is almost unrecognizable in the 1838 original, "Sur la Théorie de la Variation des Constantes". For a more modern approach see Bill's "Liouville's Theorems, Gibbs' Entropy, and Multifractal Distributions for Nonequilibrium Steady States". Nosé's highly original papers are worth a look. One of them is his "Unified Formulation of the Constant Temperature Molecular Dynamics Methods". A fruitful example, the Nosé-Hoover oscillator, is explored in Posch, Vesely, and Bill's 1986 Physical Review A paper, "Canonical Dynamics of the Nosé Oscillator: Stability, Order, and Chaos".

Chapter 2

Time-Reversibility in Physics and Computation

*To see a world in a grain of sand,
and eternity in an hour.*

Wm. Blake

2.1 Introduction

All of the fundamental differential equations of mathematical physics—Einstein's, Hamilton's, Lagrange's, Maxwell's, Newton's, Schrödinger's—are "time-reversible". The dependent variables can be the particle or continuum coordinates of classical mechanics, or the electromagnetic field, or the wave function underlying the probabilities of quantum mechanics. The reversibility of all these theories differs qualitatively from the *irreversibility* of thermodynamics. In this book we consider "time reversibility" in detail, for systems based on classical mechanics and thermomechanics. What does time reversibility mean for these systems? Simply that all possible "solutions" of the fundamental equations—time histories of particle or field variables—can be followed either forward or backward in time *without any change in the equations themselves*. Reversing time requires only a change in the initial conditions.

By contrast, precisely reversible *computation* is quite rare. Computational roundoff errors accumulate. Typically there is no simple relation linking the errors in a reversed trajectory to those of the forward trajectory. The exponential growth of these differences frustrates attempts to reverse trajectories for more than a few collision times. In parallel computation, where a problem is divided up among several processors, the results can lack reproducibility for a different reason. The dependence of sums on the *order* of their evaluation (with $a + b + c$ differing from $c + b + a$, for instance) can thwart reproducibility. Similar disparities occur between different compilers and different hardware, even when only a single processor is used.

We know of only one way to avoid the irreversibility inherent in this general sensitivity to roundoff error. This is to use a scheme like Verlet and Levesque's "bit-reversible" "leapfrog" dynamics, in which *integer arithmetic* coupled with a time-symmetric algorithm gives trajectories rigorously reversible, "to the very last bit". It appears that this approach is both stable and accurate for the equilibrium flows described by Liouville's incompressible theorem. The bit-reversible algorithm is fully described in Section 2.3 with a corresponding Fortran program in Section 2.11. Because the underlying Newtonian leapfrog dynamics is completely isoenergetic, as at equilibrium, there is none of the dissipative contraction which characterizes stationary nonequilibrium systems.

Nevertheless, the bit-reversible algorithm can also be used to generate patently-nonequilibrium, or at least *apparently-nonequilibrium* situations, including the formation of strong shockwaves. Consider the N-body bit-reversible Fortran algorithm we detail in Section 2.11. The problem used there to demonstrate the algorithm describes the symmetric collision of two cold many-body blocks with $(N/2)$ particles in each. First, as the blocks collide, they merge and are compressed and heated by shockwaves. Next, the resulting compressed heated fluid expands at constant energy, with the drop in potential energy triggering further heating, as the fluid fills the (periodic) container homogeneously. In the end the initial kinetic energy of the two colliding cold blocks is converted into the equilibrium thermal energy of a hot fluid. This interesting problem delivers another reversibility bonus due to the use of periodic boundaries. Reversing the initial velocities in this shock-compression free-expansion problem traces out an exact mirror image of the forward-in-time history. Here the "past" is identical to the "future"! This unusual behavior is crucially dependent on the isolation of the system from the external world, from the destruction of information, and from the Second Law of Thermodynamics. This problem illustrates both

Loschmidt's reversibility paradox and the Poincaré recurrence of Zermélo's paradox.

We will see that in only slightly more general situations, both at and away from equilibrium, phase volume is *not* conserved, even in principle. Stationary *non*equilibrium states are quite typically characterized by *many-to-one* attractive mappings in phase space. Evidently no such contracting flow could be described by a bit-reversible algorithm. Thus the motion equations of thermostated *non*equilibrium systems which specially interest us in this book *cannot* be solved by bit-reversible methods.

Computer simulations can extract detailed "experimental data" from physical theories in a way which physical experiments cannot. But in order to relate the results of these simulations to the world around us it is necessary that the simulation variables have real-world physical analogs. So far we have discussed mechanics. Forces and stresses in mechanical simulations correspond quite naturally to their experimental analogs. In broadening mechanics to include thermodynamics *temperature* is the essential new concept. In Section 2.8 we will discuss temperature, and its introduction into mechanics, through thermostats, from the perspective of time reversibility. Though there is nothing specially peculiar about heat transfer or thermostated trajectories, the nonequilibrium probability distributions to which the trajectories correspond *are* profoundly different to their equilibrium counterparts. As a consequence of an overall contractive character, these nonequilibrium distributions become *fractal* attractors. Their fractal nature means that their phase-space probability densities have a singular structure on all length scales.

2.2   Time Reversibility

Though time reversibility would seem to be a familiar intuitive notion, the details can be confusing, and the definitions can seem all too vague, when it comes to applications. In part, this stems from the fact that dynamics is described by time-reversible *differential* equations and is imagined as the continuous time development of a set of *precisely-defined* coordinates. Because analytic approaches fail for chaotic problems (because the chaos produces so much new information), practical "solutions" are necessarily numerical and discontinuous, with a typical 53-bit precision of sixteen decimal digits. Real continuity, in both space and time, and with infinite precision, is never present in such a "solution" of the equations because numerical solutions are necessarily finitely-expressible, or "coarse-grained". Such finite approximations nearly always lack the time reversibility of the underlying differential equations.

The usual numerical solutions, representing attempts to describe functions of a continuous time variable, inevitably involve some approximation or truncation. Series of coordinate sets, given at discrete times, truncated Taylor or Fourier series in the time, and movies or tapes of the motion, are example representations. It appears that the time reversibility of a particular approximate numerical solution depends not only on the form of the underlying differential equations, but also on the way in which the solution is approximated and presented. For a clear example, see again the three time-reversed Nosé-Hoover oscillator trajectories in Section 9.4 of Chapter 1. The picture becomes cloudier still when the evolution of not just a trajectory, but rather a fractal-density *ensemble* of systems, is described.

Let us focus on the simpler case, not an ensemble, but rather the evolution of a single dynamical system. Consider the following representative forms which dynamical "solutions" of the system trajectory can take:
(i) A set of movie frames;
(ii) An ordered set of integer coordinates;
(iii) An ordered set of floating point coordinates;
(iv) A truncated Fourier time series;
(v) A truncated Taylor time series.

Evidently any movie is "time-reversible" in a simple nontechnical sense, whether or not the underlying equations or map (if any) are reversible, because the movie frames could equally well be projected in reverse order. For differential equations this type of reversal is equivalent to changing the sign of the timestep $(+\Delta t \to -\Delta t)$ in using a computer algorithm to approximate the solution of a differential equation. This type of reversal has been called "invertibility" and "reversibility with time inversion" by Illner and Neunzert. Such reversals are not at all what physicists mean by "time reversibility". For a physicist, the fundamental differential equations describing the forward evolution in time are necessarily *identical* to those going backward in time. Only the initial conditions, *not* the equations, can differ in the "reversed" flow. To clarify this observation consider the approximate harmonic-oscillator trajectory shown in Figure 2.1 (Segment #1, at the *top* right of the figure).

Time-Reversibility in Physics and Computation 43

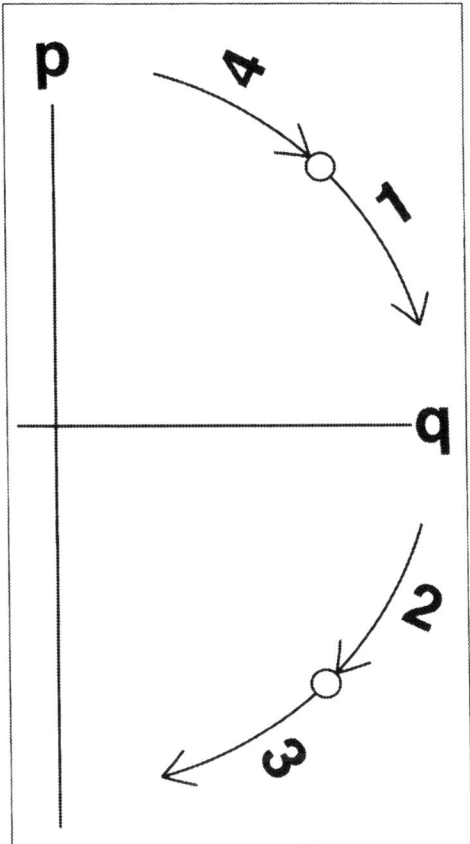

Figure 2.1. Harmonic oscillator trajectory segments. The time-reversal of any trajectory segment is a two-step process: (i) Reflect the segment vertically: $\{+q, +p\} \to \{+q, -p\}$; (ii) Reverse the time-ordering.

The oscillator coordinate q is advanced in time from its initial value q_0 and the momentum p, initially p_0, equal to $m\dot{q}$, changes accordingly. Newton's formulation of the oscillator motion can be written as a single second-order ordinary differential equation for q, or as an equivalent set of two coupled first-order equations for (q, p):

$$m\ddot{q} = -\kappa q \leftrightarrow \{\dot{q} = v = (p/m); \; m\dot{v} = \dot{p} = -\kappa q\} \; .$$

The time-reversed trajectory (Segment #2 at the *bottom right* of the figure) goes through *exactly the same* coordinate values, but in reversed time order, and with momenta which are changed in sign at each coordinate value,

$$+p_{\text{forward}}(q) = -p_{\text{reversed}}(q) \longleftrightarrow +v_{\text{forward}}(q) = -v_{\text{reversed}}(q) \ .$$

To recapture the "past" history of the oscillator, Segment #4, requires a two-step process. First, integrate forward from the time-reversed initial point $(+q_0, -p_0)$, to get Segment #3. Last, reverse both the momenta and the time-ordering of these points to get the desired "past", Segment #4.

2.3 Levesque and Verlet's Bit-Reversible Algorithm

Let us choose a simple example problem governed by Newtonian mechanics, with $\{a = (F/m)\}$. Discretize the time evolution of the chosen Newtonian system, and represent its spatial coordinates as *integers*, given at equally spaced times. Levesque and Verlet pointed out that such sets of integer coordinates, describing the "time" development of a dynamical system, can be generated with a special "bit-reversible" algorithm, such as the simple Störmer-Verlet centered-difference scheme, but with *integer*-valued coordinates:

$$\{a_0(\Delta t)^2 \equiv q_+ - 2q_0 + q_- = [(F/m)_0(\Delta t)^2]_{\text{integer}}\} \ .$$

Here the subscripts indicate three successive times, separated by timestep intervals Δt. *The righthand side is to be truncated to an integer.* This algorithm is exactly "bit-reversible". This means that the full—but finite—precision of the coordinate data is regained in the reversed trajectory. The only price to be paid for this exact coordinate reversibility is *approximate* forces and energies. All the combinations $\{(F/m)_0(\Delta t)^2\}$ are rounded off to integer values. With this approximation for the forces, the reversal of the numerical coordinate sets is *exact*, to the very last bit. Levesque and Verlet's symmetric algorithm can evidently be extended, equally well, either forward or backward in time, starting with any *two* successive sets of coordinates—equivalent to specifying the velocities as well as the coordinates. The algorithm is patently time-reversible. The numerical values which it generates are replayed *exactly*, rather than approximately, in any time-reversed trajectory.

As an example, consider again the one-dimensional harmonic oscillator, with the combination $(\kappa/m)_0(\Delta t)^2$ arbitrarily set equal to unity. A sample bit-reversible solution of the centered-difference scheme is the repeating

sequence of six single-digit integer coordinates:
$$\{q\} = \{\ldots, -2, -1, +1, +2, +1, -1, -2, -1, +1, +2, +1, -1, \ldots\}.$$
Very few sets of dynamical equations are precisely reversible, "bit-reversible" as algorithms. But there are a few other examples:
$$\{q_{++} - q_+ - q_- + q_{--} \equiv [(\Delta t/2)^2(5a_+ + 2a_0 + 5a_-)]_{\text{integer}}\}.$$
It is much more common to find "time-reversible" algorithms which are *not* bit-reversible. Consider the set of approximate first-order implicit harmonic oscillator equations:
$$\{q_+ = q_0 + (\Delta t/2)(v_+ + v_0); \ v_+ = v_0 - (\Delta t/2)(q_+ + q_0)\}.$$
This approximation to the differential system $\{\dot{q} = +v; \ \dot{v} = -q\}$, though patently *time*-reversible, is not *bit*-reversible. This is apparent from the analytic solution of the implicit difference equations:
$$q(n\Delta t) \propto e^{in\alpha}; \ \cos(\alpha) = [1 - (\Delta t/2)^2]/[1 + (\Delta t/2)^2].$$
A sample solution of the implicit oscillator equations, for the special choice $(\Delta t)^2 = (4/3)$, is the repeating sequence of six (q, v) pairs:
$$\{\ldots, (-2, 0), (-1, +\sqrt{3}), (+1, +\sqrt{3}), (+2, 0), (+1, -\sqrt{3}), (-1, -\sqrt{3}), \ldots\}.$$
Oddly enough, the energy corresponding to this solution, $(q^2 + v^2)/2$, is constant.

With stationary boundary conditions any bounded bit-reversible solution must also be, in principle, periodic. Its time-reversed twin is likewise bit-reversible and can, in simple cases, be an *identical twin*, not just a mirror image. In either case, because the number of integer-valued state points is finite, the initial state *must* repeat exactly. Thus the bit-reversible "dynamics" *always* consists of a *single* periodic orbit. Figure 2.2 shows snapshots from the time evolution of a 36-particle system generated in this way. A reversal of all the momenta after another 100,000 time steps—interchanging $\{q_-\}$ and $\{q_+\}$ while leaving $\{q_0\}$ unchanged—leads, precisely and exactly, to the time-reversed initial condition 100,000 time steps later. See the Figure.

Suppose that the floating-point version of the leapfrog algorithm, rather than the bit-reversible integer algorithm, had been applied to this same many-body problem. Then Lyapunov instability would immediately have destroyed the exact step-by-step reversibility of the numerical trajectory.

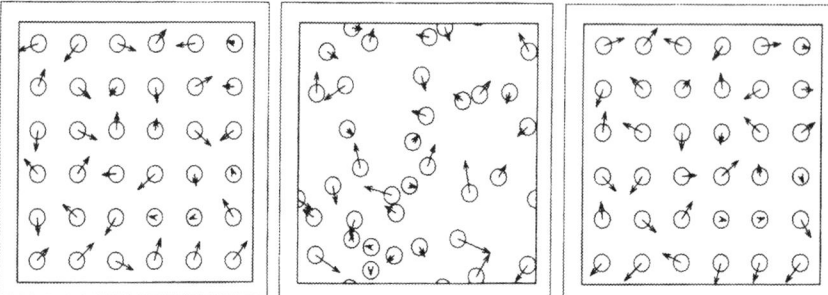

Figure 2.2. Initial, final, and time-reversed 36-particle bit-reversible configurations. The initial configuration is the regular structure shown on the left. The central "final" configuration occurs 100,000 time steps later. That same configuration, but with the velocities reversed, gives the rightmost configuration after another 100,000 iterations. Note that the initial *configuration* is recovered exactly, but with all the velocities changed in sign from the initial values. See Kum and Hoover (1994).

This destruction of information would have been caused by the *exponential* amplification of the inevitable roundoff errors. Whether or not the distinction between *exactly periodic* bit-reversible orbits and their irreversible relatives has any practical consequences is not known. The overall usefulness of computer simulations, *exactly* reversible or not, argues strongly that it does not.

Although in practice simple stepwise algorithms are usual, in principle there are many alternative representations of dynamical trajectories. Truncated Fourier series suggest the absence of Lyapunov instability (which is actually present in real problems) because such series imply periodicity. Truncated Taylor's series are also limited in usefulness, by their finite convergence radii. High-order series solutions are not very useful when discontinuities are present. Hard-sphere collisions are discontinuous, for instance. *Any* useful forcelaw *necessarily* has singular high-order derivatives in order to keep the range of the interaction finite.

2.4 Lagrangian and Hamiltonian Mechanics

Lagrange's and Hamilton's generalizations of Newtonian mechanics are specially useful for describing *constrained* systems. Coordinate-dependent constraints can be incorporated in a governing Lagrangian $\mathcal{L}(q,\dot{q})$ automatically. Thus, *constrained* systems (such as those composed of molecules whose fixed shapes define a rigid structure) are best described by *Lagrangian* mechanics. Systems for which *energy* is more fundamental than *force* (quantum mechanics is the best example) are best treated by *Hamiltonian* mechanics. For the *nonequilibrium* thermomechanical systems either of these classical forms of conservative mechanics can furnish a useful beginning.

Let us begin with Lagrangian mechanics. Lagrangian mechanics *extends* the set of Cartesian coordinates $\{x, y, z\}$ and velocities $\{\dot{x}, \dot{y}, \dot{z}\}$ of Newtonian mechanics. It includes "generalized" coordinates and velocities, $\{q, \dot{q}\}$. Generalized coordinates can, for example, include bond lengths, angles, or normal-mode vibrational amplitudes. Lengths and angles are particularly useful in formulating geometric constraints. Usually the Lagrangian $\mathcal{L}$ governing the motion is the difference between the kinetic and potential energies, $\mathcal{L}(q, \dot{q}) = K - \Phi$. The Lagrangian can be used to solve for the time development of the generalized coordinates and the corresponding conjugate momenta, $\{p \equiv (\partial \mathcal{L}/\partial \dot{q})\}$:

$$\{p = +(\partial \mathcal{L}/\partial \dot{q}); \dot{p} = +(\partial \mathcal{L}/\partial q)\} \ .$$

These same Lagrangian equations of motion can then be equally well expressed in terms of the equivalent Hamiltonian $\mathcal{H}(q, p)$,

$$\mathcal{L}(q, \dot{q}) \longrightarrow \mathcal{H}(q, p) = K + \Phi \ .$$

As indicated here, the Hamiltonian usually corresponds to the *total* system energy, kinetic plus potential.

Flows generated by Hamiltonian dynamics are naturally described in $\{q, p\}$ "phase space", rather than coordinate space. Hamilton's equations of motion give time derivatives for both the coordinates and the momenta. The "flow" $\{\dot{q}, \dot{p}\}$ of the variables $\{q, p\}$, is then given in terms of explicit derivatives of the Hamiltonian $\mathcal{H}(q, p)$:

$$\{\dot{q} = +(\partial \mathcal{H}/\partial p); \dot{p} = \ (\partial \mathcal{H}/\partial q)\} \ .$$

The simplest system for which the advantages of generalized coordinates are clear is a particle constrained to circle the origin at a radius of unity. The motion can be so constrained by using a Lagrangian with a "Lagrange Multiplier" $\Lambda(x, y, \dot{x}, \dot{y})$:

$$\mathcal{L}(x, y, \dot{x}, \dot{y}, \Lambda) = [(\dot{x}^2 + \dot{y}^2)/2] + [\Lambda(x^2 + y^2 - 1)/2] \ .$$

The resulting equations of motion are $\{\ddot{x} = \Lambda x; \ddot{y} = \Lambda y\}$. There are no contributions involving derivatives of Λ because all such contributions are multiplied by the *vanishing* constraint: $(x^2 + y^2 - 1)/2 = 0$. The *second* time derivative of the constraint,

$$x\ddot{x} + \dot{x}^2 + y\ddot{y} + \dot{y}^2 = 0 = \Lambda(x^2 + y^2) + v^2 ,$$

can then be solved for the Lagrange multiplier. Λ then takes on the value, $-(v^2/r^2)$, necessary to maintain the constraint. Exactly the same motion would result under the influence of a proper attractive central force with the magnitude $-(v^2/r)$. If we choose *harmonic* forces, with a quadratic potential $(r^2/2)$, and unit mass, then the corresponding Lagrangian can be written in either Cartesian or polar coordinates:

$$\mathcal{L}(x, y, \dot{x}, \dot{y}) = (\dot{x}^2 + \dot{y}^2 - x^2 - y^2)/2 \longrightarrow$$

$$\{\ddot{x} = \dot{p}_x = -x; \ddot{y} = \dot{p}_y = -y\} .$$

$$\mathcal{L}(r, \theta, \dot{r}, \dot{\theta}) = (\dot{r}^2 + r^2\dot{\theta}^2 - r^2)/2 \longrightarrow$$

$$\{\ddot{r} = \dot{p}_r = r\dot{\theta}^2 - r; r^2\ddot{\theta} = 2r\dot{r}\dot{\theta} = d(r^2\dot{\theta})/dt = \dot{p}_\theta = 0\} ,$$

where $p_r \equiv (\partial \mathcal{L}/\partial \dot{r}) = \dot{r}$ and $p_\theta \equiv (\partial \mathcal{L}/\partial \dot{\theta}) = r^2\dot{\theta}$. In the polar-coordinate form the conserved nature of the angular momentum, $r^2\dot{\theta}$ is apparent. The two corresponding forms—Cartesian and polar—of the Hamiltonian for a two-dimensional oscillator, follow from the definitions $\{p \equiv (\partial \mathcal{L}/\partial \dot{q})\}$:

$$\mathcal{H}(x, y, p_x, p_y) = (p_x^2 + p_y^2 + x^2 + y^2)/2 \longrightarrow$$

$$\{\dot{x} = p_x; \dot{y} = p_y; \dot{p}_x = -x; \dot{p}_y = -y\} .$$

$$\mathcal{H}(r, \theta, p_r, p_\theta) = [p_r^2 + (p_\theta/r)^2 + r^2]/2 \longrightarrow$$

$$\{\dot{r} = p_r; \dot{\theta} = (p_\theta/r^2); \dot{p}_r = (p_\theta^2/r^3) - r; \dot{p}_\theta = 0\} .$$

Evidently the mechanics of a Lagrangian or Hamiltonian system can be described either as a coordinate-space trajectory $\{q(t) \to \dot{q}(t)\}$ or as a phase-space trajectory $\{q(t), p(t)\}$. The phase-space trajectory view can be generalized to follow the flow of a probability density $f(\{q, p\}, t)$ through the space. This latter point of view is particularly useful in developing Gibbs' statistical mechanics from Hamiltonian mechanics. As we show in the following Section, Hamilton's flow equations do have an important consequence, *Liouville's Incompressible Theorem*, the best basis for Gibbs' statistical mechanics.

2.5 Liouville's Incompressible Theorem

Imagine a smoothly-varying probability density $f(q, p, t)$ in the $\{q, p\}$ phase space. Here we have in mind that q and p represent all the coordinates and momenta required to describe a system's state. The phase-space motion of f represents collectively all the motions of members of an "ensemble" of systems, all with the same Hamiltonian, but with different initial conditions. It is evident that this smoothly-varying phase-space density obeys the "continuity equation",

$$(\partial f / \partial t) = -\nabla \cdot (fv) ,$$

where ∇ is a generalized gradient, $\sum[(\partial/\partial q), (\partial/\partial p)]$ and v is the corresponding generalized velocity, with components $\{\dot{q}, \dot{p}\}$. Then, from this "Eulerian" fixed-frame form of Liouville's Theorems, the "comoving", or "Lagrangian", time derivative, following the motion, becomes

$$\dot{f}(q, p, t) = (\partial f / \partial t) + v \cdot \nabla f = -f \nabla \cdot v .$$

This more general "compressible" form of Liouville's Theorem is particularly useful *away* from equilibrium, where the Nosé-Hoover thermostat forces $\{-\zeta p\}$, which we will discuss in Section 2.8, can give rise to a nonvanishing divergence of the flow velocity, $\nabla \cdot v \equiv -\sum \zeta \neq 0$. With Hamilton's equations of motion the divergence vanishes, so that no compressibility can occur:

$$\nabla \cdot v = \sum[(\partial \dot{q}/\partial q) + (\partial \dot{p}/\partial p)] = \sum[(\partial^2 \mathcal{H}/\partial q \partial p) - (\partial^2 \mathcal{H}/\partial p \partial q)] \equiv 0 .$$

In the Hamiltonian case the probability density $f(q, p, t)$ flows through the phase space unchanged, just as does the mass density $\rho(r, t)$ of an incompressible fluid flowing in ordinary three-dimensional space. The corresponding incompressible Liouville Theorem describes the equilibrium situation. As a corollary, the comoving differential phase-space volume element, which we denote as $\otimes$, is conserved by Hamilton's equations of motion.

The approximate explicit finite-difference algorithm:

$$\{q_+ = q_0 + (p/m)_0 \Delta t; \; (p/m)_+ = (p/m)_0 + (F/m)_+ \Delta t\} ,$$

produces coordinate sequences *identical* to those from the leapfrog algorithm:

$$\{(q_+ - q_0) - (q_0 - q_-) = [(p/m)_0 - (p/m)_-]\Delta t = (F/m)_0 (\Delta t)^2\} .$$

It is striking that this finite-difference algorithm satisfies a finite-difference form of Liouville's incompressible Theorem *exactly*:

$$\otimes_+ = \otimes_0 = \otimes_- .$$

⊗, the *comoving finite element* of phase-space hypervolume, represents what Gibbs called "extension in phase [space $\{dq, dp\}$]". It is unknown if the underlying "symplectic property" of the leapfrog algorithm has any significant effect on the "quality" or the "utility" of the resulting trajectory. For many problems the fourth-order Runge-Kutta algorithm yields more nearly accurate results for the same amount of computer time.

Phase-space flow incompressibility—conserving the "extension in phase" ⊗—is an important defining characteristic of isoenergetic Hamiltonian systems. The incompressible phase-space flow establishes that the corresponding equilibrium phase-space probability density $f(q, p)$ must certainly be time-independent. Consequently an equilibrium f must have the same *constant* value along any phase-space trajectory. If there is sufficient *mixing* to cover the entire energy surface, then necessarily the probability density is constant along such a surface. The simplest such constant density f defines Gibbs' "microcanonical distribution", or "microcanonical ensemble", which contains all phase-space states close to a specified energy E,

$$E - (dE/2) < \mathcal{H}(q, p) < E + (dE/2),$$

all with equal weights.

2.6 What *Is* Macroscopic Thermodynamics?

From the detailed standpoint of microscopic mechanics an isolated system is continually undergoing fluctuations and changing state, as is described by the equations of motion, or, equivalently, by Liouville's incompressible flow theorem. Individual particles can move about. The potential and kinetic parts of the total energy can vary though their sum remains fixed. Rather than considering or even imagining all of these microscopic details for specific phase-space flows it is often worthwhile to ignore them. One can then adopt an alternative *macroscopic* view, analogous to assessing only "average" behavior and focusing on total energy as a *global* "state variable". Thermodynamics takes this macroscopic point of view.

Macroscopic thermodynamics evolved from efforts to design and understand "heat engines"—machines for converting heat into useful work. As a result, heat, work, and energy are among the primitive concepts of thermodynamics. "Thermal variables" needed to describe temperature and heat transfer are also included. In thermodynamics the *efficiency* of macroscopic heat engines converting heat to work is important. Thermodynamic work and heat are analogs of the microscopic potential and kinetic energies. "Work" denotes an energy change in response to *coordinate* variations. Pushing on a piston or lifting a weight are ways of doing work. "Heat" denotes energy changes taking place in the absence of any related coordinate changes. The fraction of the heat taken in which is converted to work is the "thermodynamic efficiency" of a cyclic process. "Reversible" work and heat are abstract thermodynamic concepts having no direct connection to mechanical time reversibility. The thermodynamic reversibility concepts instead apply to energy changes or other state changes taking place *through a sequence of equilibrium states*, often called a "quasistatic process".

Overall, thermodynamics differs from mechanics in three ways. First, the description of a system's state is *broadened*, to include thermal variables, but also narrowed through the neglect of fluctuations in the mechanical and thermal variables. Second, *mechanisms* for change are omitted too. Thermodynamics describes which states are possible, but does not predict transformation *rates* or *mechanisms*. Finally, the thermodynamic description of processes, though including heat flow, is necessarily *macroscopic* and is therefore intrinsically less detailed than the microscopic many-body picture. Despite the limited nature of the thermodynamic view, that view does provide some extremely *useful* information. It emphasizes what is *possible*. The Laws of thermodynamics detailed in the following Section establish that it is illegal to have temperature drop in response to added heat, just as it is illegal to have density decrease in response to increasing pressure.

As is discussed more fully in the following Section, lacking any knowledge of detailed mechanisms, the *best* one can do, from the standpoint of increasing the thermodynamic efficiency, is to follow an idealized "reversible" or "quasistatic" process. Because the detailed *dynamic* mechanisms are omitted, thermodynamics is sometimes termed "thermostatics". It *is* perfectly possible to compute the rate-dependence of irreversible processes by carrying out more detailed simulations incorporating kinetic constitutive information—such as viscosity and plasticity—which are not present in thermodynamics. Simulation of transport processes in macroscopic continua is the subject of Chapter 7.

The wholly new thermal state variables provided by thermodynamics are *temperature* and *entropy*. *Differences* in temperature drive heat transfer. And the *integrated* effect of reversible heat transfer is entropy change. Entropy, like energy, is a thermodynamic state function, directly related to the microscopic dynamics. Boltzmann and Gibbs were able to show that

the macroscopic entropy of all those microscopic equilibrium states linked by a phase-space flow is simply related to the corresponding many-body phase-space hypervolume described in Section 3.5:
$$S_{\text{Gibbs}}(N, E, V) \equiv k \ln \Omega(N, E, V) .$$
Energy is a well-defined property for any $\{q, p\}$ phase-space state of a microscopic system. Unlike energy, *entropy* is instead a collective, or "ensemble" property, which can be thought of as a system property only through a time-averaging process. Entropy is somewhat subtle. Its value corresponds to the total number of microscopic states consistent with our knowledge. We discuss the connection of the thermodynamic entropy, from integrated heat transfer, to Gibbs' statistical state-counting entropy in Chapter 3.

2.7 First and Second Laws of Thermodynamics

The First Law of Thermodynamics has both global and local versions. These are often stated "The energy of the universe is fixed" or "The energy is a function of state". A system obeying the latter law suffers no net energy change in any cyclic process: $\oint dE = 0$. Thus the First Law prevents any system's acting as a perpetual energy source. The Second Law of Thermodynamics has been formulated in an even greater variety of equivalent ways. Here are four of them: (i) "Heat cannot flow from a colder body to a hotter one"; (ii) "Entropy must increase"; (iii) "No *cyclic* process can convert heat entirely to work"; (iv) "In any *cyclic* process the heat Q transferred *to* the system *from* its surroundings at the temperature T must obey an inequality: $\oint dQ/T < 0$."

These four statements, though equivalent from a macroscopic point of view, are not equally useful on the more detailed microscopic level. The first of these statements is inconsistent with the existence of fluctuations— despite the unquestioned fact that heat conductivity is positive, it is always possible, for sufficiently small temperature gradients in sufficiently small volumes, to find heat traveling in the wrong direction, for a while.

The second of the Second Law statements is likewise inconsistent with a time-reversible dynamics. Leaving aside the problem of *defining* nonequilibrium entropies, any situation in which the entropy "correctly" increases could be made "incorrect" by reversing the velocities. The last two statements, dealing with *cyclic* processes, are more promising. Long time averages, over *many* cycles, can be thermodynamically-correct, both microscopically and macroscopically.

The Second Law of Thermodynamics, stated as a long-time average, either over a large number of cycles or for a stationary state, need not conflict with microscopic mechanics, once that mechanics is generalized to include the concepts of heat and temperature. The cyclic-process form of the Second Law does *not* apply to isolated systems, which are both artificial and typically uninteresting. It is evident that, with rather mild assumptions (that the accessible phase space is bounded), an isolated system will eventually return arbitrarily closely to its initial state ("Poincaré recurrence"). It is also evident that any evolution thought to correspond to an entropy increase would necessarily correspond also, when reversed, to a compensating entropy decrease ("Loschmidt's paradox"). Thus it is unreasonable, on both counts, to expect an isolated system to show a systematic entropy increase.

But, *when it is properly time-averaged, the Second Law is true*, $\langle \oint (dQ/T) \rangle < 0$. This inequality states that cyclic processes generate entropy, *provided* that an average over many repetitions of the cycle is implied. In this time-averaged form the Second Law of Thermodynamics can be proven as a theorem in thermomechanics. It applies to simple prototypical stationary processes, like shear flow and heat flow, as well as to more complex flows. The most interesting, puzzling, and enduring aspect of time reversibility is the connection between atomistic microscopic dynamics and the macroscopic Second Law of Thermodynamics. Most explanations for this connection, though without doubt correct, are somewhat limited in scope.

Boltzmann's dilute-gas approach, taken together with Loschmidt's and Zermélo's objections to it, mentioned in Chapter 1, is the most familiar. Green and Kubo's much more recent linear-response theory of transport is a first-order perturbation theory. It applies to liquids and solids as well as to gases. Objections to Green-Kubo theory are less convincing and not so well organized as those Boltzmann faced. For more mathematical discussions, see the reviews of Zwanzig and Ichiyanagi. Perhaps these objections are also less important, because the Green-Kubo-Onsager approach is limited to states which are close to equilibrium and because the results of the theory can be obtained in *so many* alternative ways.

It might well be thought that time-reversible equations of motion are inconsistent with the symmetry-breaking inherent in the dissipative shrinking flows called "strange attractors" that characterize nonequilibrium problems. The ways in which time-reversible dissipation can, and does, occur have become clear with the development and analysis of computer algorithms, particularly during the past 25 years. These algorithms, and related simulations, represent and elucidate new paths to understanding the reversibility paradox faced earlier by Boltzmann, Green, and Kubo. In order to describe the new paths we first must introduce temperature, and thermostats, into the microscopic equations of motion. The distribution functions which then result are *fractal* distributions, which lead directly to an understanding of nonequilibrium steady states and to the time-averaged micromechanical version of the Second Law.

2.8 Temperature, Zeroth Law, Reservoirs, Thermostats

In order to simulate processes involving thermodynamic work and heat transfer it is useful to adopt a microscopic *definition* of temperature, corresponding to the temperature measured by an ideal-gas thermometer. The usefulness of temperature as a state variable can be stated as the *Zeroth Law of Thermodynamics*, "Two bodies in thermal equilibrium with a third are also in thermal equilibrium with each other", meaning that there is no net transfer of heat between any two bodies with the same temperature T. This macroscopic thermodynamic concept ignores fluctuations. It is certainly false, on a local and instantaneous level. But an equivalent microscopic concept, for stationary equilibrium boundary conditions, can be expressed as a time average: "Two bodies, each with no long-time tendency to transfer heat with a third, have the same time-averaged temperature, $\langle T \rangle$".

Such a microscopic Zeroth law provides no operational definition for the equilibrium temperature common to the three bodies. Among many possibilities, by far the most *natural*, and most *convenient* microscopic definition of T, or $\langle T \rangle$, is the classical ideal-gas temperature, given in terms of the (time-averaged) mean-squared velocity: $kT_{xx} \equiv \langle mv_x^2 \rangle$. In a stationary equilibrium situation it is not necessary to distinguish between T and $\langle T \rangle$. Likewise, at equilibrium T_{xx} and T_{yy} are equal. This temperature definition follows automatically if one considers the "third body" of the zeroth law to *be* an ideal gas thermometer, with enough degrees of freedom to make the fluctuating difference, $T(t) - \langle T \rangle$, negligible.

Such a thermometer is most simply conceived of as containing many infinitesimal particles. Their interaction is very weak, $\Phi \simeq 0$, but strong enough to provide the equilibrium Maxwell-Boltzmann distribution proportional to $e^{-(mv^2/2kT)}$ and to establish the isotropic ideal gas law within the

thermometer:
$$P_{xx}V = P_{yy}V = P_{zz}V = NkT.$$

With the ideal-gas temperature scale defined, it is possible to model both equilibrium and nonequilibrium thermostats in either of two ways: by constraining the instantaneous kinetic energy or by constraining its time average. The *first* choice gives an "isokinetic" thermostat. The thermostat has to be implemented by imposing an additional thermostat force, which constrains the kinetic energy to a fixed value. The potential energy is left free to fluctuate. For a sufficiently mixing system this approach leads to Gibbs' canonical distribution for the potential energy, $\propto e^{-\Phi/kT}$. Finding the required constraint force is a straightforward application of Gauss' mechanical "Principle of Least Constraint" or of the closely-related variational principle considered by Gibbs.

Gauss' Principle states that any required constraint forces $\{F_C\}$ are best chosen so as to *minimize* the corresponding sums, over all of the constrained degrees of freedom: $\sum\{F_C^2/(2m)\}$. See, for further details, Pars' *Treatise on Analytical Dynamics* or Bill's *Molecular Dynamics*. By applying Gauss' Principle to the constraint of fixed kinetic energy, $\dot{K} = 0$, an additional "frictional force" results, $-\zeta p$ for each thermostated degree of freedom. The friction coefficient ζ varies with time and needs to be chosen so as to keep the kinetic energy constant. For an otherwise isolated system, with equations of motion $\{\dot{p} = F - \zeta p\}$, it is evident that the choice $\zeta \equiv \sum F \cdot (p/m)/(2K)$ implies a fixed kinetic energy, $K(t) = K_0$.

The resulting isokinetic dynamical system is not among the usual types represented by Gibbs' ensembles. More usual is the canonical case corresponding to thermal equilibrium at a temperature T, with the phase-space density $f(q,p) \propto e^{(A-\mathcal{H})/kT}$. Helmholtz' free energy $A \equiv E - TS$ provides a normalization for the "Boltzmann factor" $e^{-\mathcal{H}/kT}$ in Gibbs' formulation of the canonical ensemble and its "partition function" $Z(N,V,T)$:

$$(1/N!)\prod[\int\int dqdp/h]e^{-\mathcal{H}/kT} \equiv e^{-A/kT} = e^{+S/k}e^{-E/kT} = Z(N,V,T).$$

In 1984 Shuichi Nosé extended Gauss' and Gibbs' work. Nosé developed a more general approach to the canonical ensemble through his thermostated mechanics. Fluctuations were included, corresponding to the *second* choice of thermostat, one in which the *time average* of the ideal-gas thermometer temperature, $\langle T \rangle = \langle mv_x^2/k \rangle$, is constrained. In his mechanics the instantaneous temperature is allowed to fluctuate in such a way as to fill out the entire canonical distribution *if* the dynamics is sufficiently mixing. Nosé's original formulation included an extraneous "time-scaling" variable s. After meeting with Nosé to discuss his work, Bill suggested that s be omitted. The resulting simpler formulation, free of time scaling or effective masses, is usually called "Nosé-Hoover mechanics". Nosé-Hoover mechanics includes additional constraint forces, $\{-\zeta p\}$, ensuring that the long-time-averaged kinetic energy $\langle K \rangle$ approaches the canonical equipartition value from Gibbs' statistical mechanics:

$$\langle K \rangle_{t \to \infty} \equiv K_0 = DNkT/2 ,$$

where T is the temperature and D is the dimensionality of the system. The Nosé-Hoover feedback forces $\{-\zeta p\}$ include a characteristic thermostat relaxation time τ. For an otherwise isolated system, the equations of motion take the form of "integral control" equations, with the friction coefficient proportional to the time-integrated deviation of $K(t)$ from K_0:

$$\{\dot{q} = (p/m); \; \dot{p} = F - \zeta p\}; \; \dot{\zeta} = [(K/K_0) - 1]/\tau^2 .$$

Taking the long-time average of the last equation, with the assumption that the mean value of the friction coefficient $\langle \zeta \rangle$ is bounded, establishes that the mean value of the kinetic energy must approach the constant K_0:

$$\langle \zeta \rangle \text{ constant} \to \langle \dot{\zeta} \rangle = 0 \propto \langle \, [(K/K_0) - 1] \, \rangle \to \langle K \rangle = K_0 .$$

Thus the Nosé-Hoover approach (i) fixes the *time-averaged* kinetic energy, rather than the instantaneous value and (ii) introduces a useful phenomenological relaxation time τ with which to simulate *rates* of heat transfer.

At equilibrium these motion equations are exactly consistent with Gibbs' canonical distribution. That is, *assume* that $f(q, p, \zeta, t)$ *has* the canonical form, augmented by a Gaussian distribution for the $\{\zeta\}$. Then the Nosé-Hoover motion equations imply that $f(q, p, \zeta)$ is stationary:

$$f(q, p, \zeta, t) \propto e^{-\mathcal{H}/kT} e^{-\sum \zeta^2 \tau^2 / 2} \longrightarrow$$

$$(\partial f / \partial t)_{\text{Nosé-Hoover}} \equiv 0 \longrightarrow f = f(q, p, \zeta) .$$

Nosé-Hoover mechanics is the simplest *dynamical* analog of Gibbs' statistical mechanics. It is an example of "thermomechanics", an augmentation of Newtonian mechanics which includes temperature control for selected degrees of freedom. Though his original derivation was unnecessarily complicated, with "time scaling" included, Nosé's idea led to something new and useful—a time-reversible dynamics in which the phase-space probability density $f(q,p)$ changes in response to heat transfer, $d \ln f / dt \equiv -\nabla \cdot v = \sum \zeta$, as was mentioned in Section 2.5.

There is another elegant way to avoid Nosé's "time scaling", while retaining the flexibility of the relaxation time τ, the physical interpretation of phase-space compressibility, and an exact link to Hamiltonian mechanics. This approach was discovered by Carl Dettmann, in 1996. For a system with # thermostated degrees of freedom, Dettmann showed that the Nosé-Hoover equations of motion also follow naturally when a special Hamiltonian, resembling that which Nosé had used, is set equal to zero:

$$\mathcal{H}_{\text{Dettmann}} = s\mathcal{H}_{\text{Nosé}} = \sum(p^2/2ms) + s[\Phi + \#kT \ln s + (p_s^2/2M)] \equiv 0 \ .$$

In the absence of the special choice, $\mathcal{H} \equiv 0$, Dettmann's Hamiltonian bears a superficial resemblance to Nosé's, in which s is a time-scaling variable or a reduced mass. But Dettmann's s is simply a new "thermostat variable" (its logarithm is proportional to the time integral of ζ). Dettmann's approach completely avoids the need for any scaling of time. His motion equations are:

$$\{\dot{q} = (p/ms); \ \dot{p} = sF\}, \ \dot{s} = s(p_s/M) \ ;$$

$$\dot{p}_s = -(\partial \mathcal{H}/\partial s) = \sum(p^2/2ms^2) - [\Phi + \#kT \ln s + (p_s^2/2M)] - \#kT \ .$$

Evidently the essential constraint $\mathcal{H} = 0$ allows the uninteresting variable s to be omitted and permits the (q, p, ζ) phase-volume to change with time, *despite* the Hamiltonian basis. Further, the arbitrary restriction $\mathcal{H} = 0$ makes it possible to simplify the evolution equation for the friction coefficient, $p_s \propto \zeta$:

$$\dot{p}_s = \sum(p^2/ms^2) - \#kT; \ \zeta = (p_s/M) = \int_0^t [(K/K_0) - 1]dt'/\tau^2 \ .$$

If we abbreviate the combination $(p/ms) \equiv v$, for velocity, the complete scheme reduces exactly to the Nosé-Hoover equations:

$$\{\dot{q} = v; \ \dot{v} = (F/m) - \zeta v\}; \ \dot{\zeta} = \sum[(mv^2/kT) - 1]/\tau^2 \ .$$

Although this approach sometimes fails to generate the entire canonical density distribution, relatively simple modifications, illustrated on page 89, can accomplish this. Frictional forces, of the form $-\zeta p$, are common to *all* computationally-useful deterministic approaches to thermomechanics. Gauss' isokinetic form also corresponds to the simple idea of velocity rescaling used in early computer simulations. Gauss' isokinetic thermostat is the instantaneous $\tau \to 0$ limit of Nosé-Hoover mechanics.

2.9 Irreversibility from Stochastic Irreversible Equations

A longstanding route to simulating irreversible behavior is the Langevin equation of motion for a "heavy" or "Brownian" particle of mass M,

$$M\ddot{r} = \dot{p} = F - (p/\tau) + F_{\text{stochastic}} .$$

The characteristic relaxation time τ is a constant and the random numbers underlying the "stochastic" force are to be chosen so as to recover the desired temperature. This approach contains two irreversible ingredients: the drag force $(-p/\tau)$, which if reversed would lead to exponentially-rapid divergence of the kinetic energy, and the stochastic forces, which introduce a stream of "information" into the dynamics through the underlying random numbers. Although both sources of irreversibility seem to be far from fundamental physics, it is not difficult to "derive" such an irreversible equation—with a drag force—by carrying out an average over collisions. For simplicity, we demonstrate this for a one-dimensional problem.

Consider a large heavy particle with mass M and velocity V undergoing a collision with a small light particle representative of a "heat reservoir": an ideal-gas thermometer made up of particles with mass m. The velocity v is chosen from an equilibrium Maxwell-Boltzmann distribution at the temperature T. *Heat reservoir* particles are always at *equilibrium*. In the center-of-mass frame, the velocity of the heavy particle necessarily changes sign in such a way as to conserve momentum and energy:

$$V - (MV + mv)/(M + m) \longrightarrow (MV + mv)/(M + m) - V .$$

We can analyze collisional effects systematically, as a series in the square root of the mass ratio (m/M). It is necessary to keep in mind that the equilibrium ratio of (v/V) is of order $\sqrt{(M/m)}$. Thus the two velocities in the "first term" of an explicit expansion in powers of (m/M),

$$\Delta V = -2(V - v)(m/M) ,$$

actually differ, on the average, near equilibrium, by the factor $\sqrt{(m/M)}$.

To begin, we calculate the average effect of light-particle collisions with the heavy particle. Choose an ideal-gas number density n with a thermal distribution of velocities $\{v\}$ corresponding to the temperature T. Because the heavy-particle collision rate is proportional to the relative speed, the average requires a collision probability proportional to the relative velocity, $V - v$. The mean collision rate $\langle \Gamma \rangle$ follows directly from the Maxwell-Boltzmann distribution. Assuming that the massive particle velocity V is negligibly small, the result is

$$\langle \Gamma \rangle = n(m/2\pi kT)^{1/2} \int_{-\infty}^{+\infty} |v - V| e^{-mv^2/2kT} dv \longrightarrow$$

$$n(m/2\pi kT)^{1/2} \int_{-\infty}^{+\infty} |v| e^{-mv^2/2kT} dv = n(2kT/\pi m)^{1/2},$$

where the integral ranges over all values of v so as to include collisions from both left and right.

The calculation of the *averaged* heavy-particle momentum change,

$$\langle V \rangle \to \langle V + 2(v - V)[m/(M + m)] \rangle,$$

proceeds in a similar way. To first order in the mass ratio (m/M) the mean rate at which the velocity changes is

$$\langle (dV/dt) \rangle \longrightarrow \langle 2\Gamma(v - V)(m/M) \rangle =$$

$$n(m/2\pi kT)^{1/2} \int_{-\infty}^{+\infty} 2(v - V)(m/M)|v - V| e^{-mv^2/2kT} dv.$$

Introducing the relative velocity, $\alpha \equiv v - V$, the integral can be rewritten as an integral over α, and the exponent can be expanded:

$$e^{-mv^2/2kT} = e^{-m\alpha^2/2kT}[1 - (m\alpha V/kT) + \ldots].$$

The first term in the square brackets does not contribute to the integral, where its contribution is odd in α. The second term gives the final result:

$$\langle \dot{V} \rangle \to n(m/2\pi kT)^{1/2} \int_0^\infty \frac{-4\alpha^3 m^2 V}{MkT} e^{-m\alpha^2/2kT} d\alpha = -4\langle \Gamma \rangle (m/M) V.$$

The averaged effect is a friction coefficient $4\langle \Gamma \rangle (m/M)$. These straightforward calculations can be carried out in two or three dimensions with similar results. They also show that collisions with the bath particles provide a frictional damping force, proportional to the heavy particle velocity V and the mean collision rate $\langle \Gamma \rangle$.

The mean *energy* change for the heavy particle, due to the bath collisions, can be calculated in a similar way. Expressing the *energy* change in terms of the relative velocity, $\alpha \equiv v - V$, gives terms both linear and quadratic in α:

$$\langle \dot{E} \rangle = (M/2) \langle [4\alpha V m/(M + m)] + [4\alpha^2 m^2/(M + m)^2] \rangle,$$

where the average on the righthand side is over collisions. Using again an expansion of $e^{-mv^2/2kT}$, and keeping only terms of order (m/M) the mean energy change rate becomes

$$\langle \dot{E} \rangle = 4\langle \Gamma \rangle (m/M)[kT - MV^2].$$

This extremely interesting result shows that the kinetic energy of the massive particle relaxes smoothly toward $(kT/2)$. This relaxation is quite general. It is easy to show, for a two-dimensional hard disk or a three-dimensional hard sphere, interacting with a two- or three-dimensional ideal gas thermometer, that the kinetic energy of the hard particle, $(MV^2/2)$, relaxes toward kT in two dimensions or $(3kT/2)$ in three.

The argument just given closely resembles the statistical arguments used by Boltzmann in deriving the Boltzmann equation. In both cases irreversible behavior is the result of statistical averaging over the completed collisions in an underlying time-reversible mechanics. This same irreversible relaxation can also be represented by a much less cumbersome model, the phenomenological Langevin equation. Here, the viscous drag force, plus a "stochastic" noise force sufficient to maintain the kinetic energy are added to the single-particle equation of motion:

$$\dot{p} = F - (p/\tau) + F_{\text{stochastic}} .$$

The stochastic noise force is usually chosen from an appropriate Gaussian distribution. Klages, Rateitschak, and Nicolis have considered the use of deterministic time-reversible maps as thermostat forces. More realistic forms could be chosen provided they incorporate many oscillations during the decay time of the drag force, $\tau = (1/\zeta)$. The resulting heavy particle "temperature" will then reproduce that of the bath. In some cases it is desirable to consider a very low-temperature bath, giving the simpler *noise-free* motion equation $\dot{p} = F - \zeta p$. Provided that the force F represents a *driven* system such a low-temperature bath leads to a simple nonequilibrium steady state with $T_{\text{system}} >> T_{\text{bath}}$. It has been shown that this low-temperature limit of the Langevin equation leads to the fractal attractors usually associated with dissipative irreversible flows. See Bill's 1999 paper on "Steady States" in Physics Letters A.

2.10 Irreversibility from Time-Reversible Equations?

In the remainder of this book, we discuss links between the microscopic time reversibility of atomistic dynamics and the macroscopic thermodynamic irreversibility described by the Second Law of Thermodynamics. From the computational standpoint, the simplest way to forge such a link makes use of the mechanical ideal-gas temperature concept, with heat reservoirs based on Gauss' or Nosé-Hoover mechanics. This approach leads to irreversible dissipative solutions from time-reversible equations of motion.

Temperature can easily be introduced into computer simulations by using special forces, of the kind developed by Nosé. In principle, the special forces could have been avoided, at least temporarily, by instead studying "open systems" attached to large reservoirs of mass, momentum, and energy. However, the sources and sinks which drive open systems must ultimately be given operational significance too, and their own thermostats. It is plainly better to honor Occam, taking reservoirs with a few degrees of freedom rather than many. For *linear* transport processes it can be confidently expected that the values for closed systems and open systems will agree. This is a requirement for any useful computer algorithm.

We will see that solutions following the *time-reversible* friction-coefficient approach share many features with solutions of *irreversible* dissipative equations of motion. Both approaches produce nonequilibrium multifractal attractor states with a family resemblance to the Lorenz attractor of Section 4.11.1 and to the Sinai-Ruelle-Bowen states mentioned in Section 9.12.

Some of the formal difficulties encountered in describing nonequilibrium flows with multifractal distributions are fundamental. It is hard to imagine a useful mathematical description of a fractal flow through phase space, particularly in the case that the initial distribution is fractal too. Formally, this apparent difficulty can only be avoided by considering coarse-grained distributions, reflecting the limited resolution, or information content, of any conceivable measurement or simulation.

2.11   An Algorithm Implementing Bit-Reversible Dynamics

Levesque and Verlet's observed that a *perfectly reversible* dynamics can be developed by combining *integer* arithmetic with the Leapfrog Algorithm:
$$\{[q(+\Delta t) - 2q(0) + q(-\Delta t)] = [F(q_0)\Delta t^2/m]_{\text{integer}}\} \ .$$
The *integer* subscript indicates that the integer part of the righthand side is chosen. The time symmetry of the algorithm guarantees reversibility. The only question is whether or not the algorithm is stable. We investigate this by considering the dynamics of 1600 smoothly-repulsive particles in a periodic box. The pair potential, for particles close enough to interact, is
$$\phi(r < 1) = (1 - r^2)^4 \ ,$$
and the particle mass is unity. We choose a timestep of $\Delta t = 0.01$ and map the interval $[-20 \cdots + 20]$, for the coordinates and the $(F\Delta t^2/m)$ combination, onto the nine-digit interval $[1 \ldots 10^9]$. We choose a square lattice

initially, with small random "thermal" velocities added onto a systematic *highly nonequilibrium* velocity. The leftmost 800 particles move right with $\dot{x} \simeq +0.5$ and the rightmost 800 move left with $\dot{x} \simeq -0.5$. See Figure 2.3.

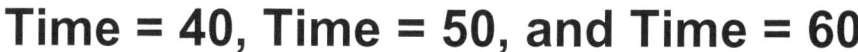

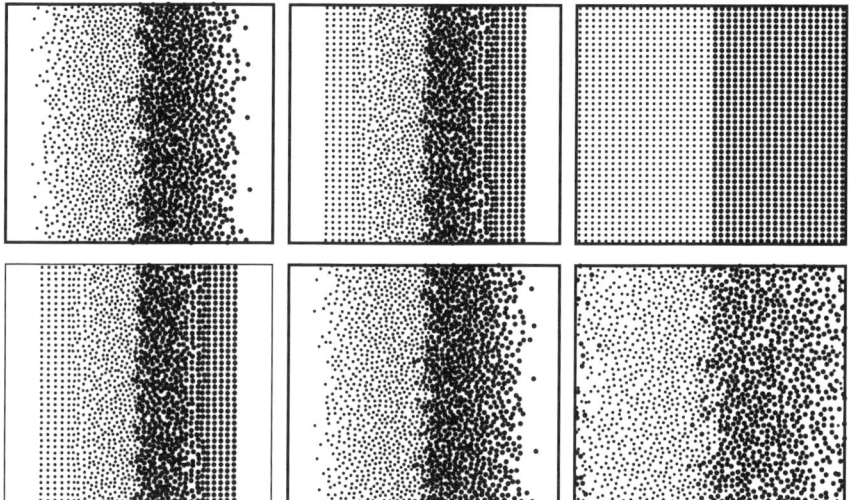

Figure 2.3. Six snapshots, equally spaced in time, from the shockwave compression of 1600 particles interacting with the short-ranged repulsion $\phi(r < 1) = (1-r^2)^4$. The final picture, in the upper righthand corner, shows that the initial configuration is recovered perfectly, following a time reversal at $t = 30$. The initial velocities were $\simeq -0.5$ for the 800 rightmost particles and $\simeq +0.5$ for the leftmost. Although all particles are identical, those initially at the right and left of center are shown here as larger and smaller disks for clarity.

The additional "random" parts of the velocities can be selected using the random number generator rand(intx,inty). This subroutine requires two integer "seed variables" {intx,inty}. For instance, {0,0} can be used. Here is the simple random-number subroutine:

```
function rand(intx,inty)
implicit double precision (a-h,o-z)
i = 1029*intx + 1731
j = i + 1029*inty + 507*intx - 1731
intx = mod(i,2048)
j = j + (i - intx)/2048
inty = mod(j,2048)
rand = (intx + 2048*inty)/4194304.0d00
return
end
```

This routine goes through *all* of the rational fractions with denominator 4,194,304 before repeating.

Because two colliding blocks of material generate a pair of shockwaves we call the following Fortran program implementing Levesque and Verlet's idea Shock.f given here for 400 particles. The variable iround is set so as to implement either conventional leapfrog dynamics or Levesque and Verlet's bit-reversible dynamics. The function itgr(gug) maps numbers on the 20×20 domain to integers from 1 to 10^9 (nmax = 10**9); [-10<gug<+10].

```
call this program Shock.f [bit-reversible shock dynamics]
implicit double precision(a-h,o-z)
parameter(N = 400)
dimension xm(N),ym(N),xo(N),yo(N),xp(N),yp(N)
dimension fx(N),fy(N)
iround = 1
dt = 0.01d00
time = 0.0d00
elx = 20.0d00
ely = 20.0d00
intx = 0
inty = 0
index = 0
do ix = 1,20
do iy = 1,20
index = index + 1
xo(index) = (ix - 10.5d00)
yo(index) = (iy - 10.5d00)
xm(index) = (ix - 10.5d00)
```

```
ym(index) = (iy - 10.5d00)
create initial speed of 1/2
speed = 0.5d00
if(ix.lt.11) xm(index) = xm(index) - speed*dt
if(ix.gt.10) xm(index) = xm(index) + speed*dt
create initial speed of 1/2
create initial temperature
dx = 1.0d00*(rand(intx,inty) - 0.5d00)*dt/100.0d00
xm(index) = xm(index) - dx
dy = 1.0d00*(rand(intx,inty) - 0.5d00)*dt/100.0d00
ym(index) = ym(index) - dy
create initial temperature
enddo
enddo
itmax = 5000
do it = 1,itmax
time = it*dt
do i = 1,N
fx(i) = 0.0d00
fy(i) = 0.0d00
enddo
epot = 0.0d00
ekin = 0.0d00
etot = 0.0d00
do i = 1,N-1
do j = i+1,N
dx = xo(i) - xo(j)
dy = yo(i) - yo(j)
if(dx.lt.-elx/2.0d00) dx = dx + elx
if(dx.gt.+elx/2.0d00) dx = dx - elx
if(dy.lt.-ely/2.0d00) dy = dy + ely
if(dy.gt.+ely/2.0d00) dy = dy - ely
rr = dx*dx + dy*dy
if(rr.lt.1.0d00) then
epot = epot + (1.0 - rr)**4
fx(i) = fx(i) + 8.0d00*dx*(1.0 - rr)**3
fy(i) = fy(i) + 8.0d00*dy*(1.0 - rr)**3
fx(j) = fx(j) - 8.0d00*dx*(1.0 - rr)**3
fy(j) = fy(j) - 8.0d00*dy*(1.0 - rr)**3
```

```
      endif
      enddo
      enddo
      if(iround.eq.0) then
C     Conventional Calculation
      do i = 1,N
      xn(i) = 2.0d00*xo(i) - xm(i) + fx(i)*dt*dt
      yn(i) = 2.0d00*yo(i) - ym(i) + fy(i)*dt*dt
      enddo
C     Conventional Calculation
      endif
      if(iround.eq.1) then
C     bit-reversible Calculation
      do i = 1,N
      i1 = itgr(xm(i))
      i2 = itgr(xo(i))
      i3 = nmax*0.05d00*fx(i)*dt*dt
      xn(i) = flt(i2 + i2 - i1 + i3)
      i1 = itgr(ym(i))
      i2 = itgr(yo(i))
      i3 = nmax*0.05d00*fy(i)*dt*dt
      yn(i) = flt(i2 + i2 - i1 + i3)
      enddo
C     bit-reversible Calculation
      endif
C     Calculation of Kinetic Energy
      ekin = 0.0d00
      ekin = ekin + 0.125d00*dt*dt*(xm(i) - xn(i))**2
      ekin = ekin + 0.125d00*dt*dt*(ym(i) - yn(i))**2
C     Calculation of Kinetic Energy
      etot = ekin + epot
C     CONSIDER PRINTING ENERGIES, COORDINATES, GRAPHICS FILES HERE
      if(it.eq.itmax/2) then
C     Change +dt to -dt at midpoint of the run
      a = xm(i)
      b = xn(i)
      c = ym(i)
      d = yn(i)
      xm(i) = b
```

```
xn(i) = a
ym(i) = d
yn(i) = c
Change +dt to -dt at midpoint of the run
endif
Check Boundary Conditions and Update the Time
do i = 1,N
xm(i) = xo(i)
xo(i) = xn(i)
ym(i) = yo(i)
yo(i) = yn(i)
if(xo(i).gt.+elx/2) then
xo(i) = xo(i) - elx
xm(i) = xm(i) - elx
endif
if(yo(i).gt.+ely/2) then
yo(i) = yo(i) - ely
ym(i) = ym(i) - ely
endif
if(xo(i).lt.-elx/2) then
xo(i) = xo(i) + elx
xm(i) = xm(i) + elx
endif
if(yo(i).lt.-ely/2) then
yo(i) = yo(i) + ely
ym(i) = ym(i) + ely
endif
enddo
Check Boundary Conditions and Update the Time
enddo
stop
end
```

Note that the lines beginning with c or C are "comments", not to be used by the program's compiler, but included here to help the reader. Notice also that the "integer" part of the program requires two functions: one, itgr(gug) which takes a floating point number between -10 and $+10$ and converts this to an integer, and two, a function flt(item) which converts an integer back to the floating point form. For a maximum integer of 10^9 the first of these functions is

```
function itgr(gug)
implicit double precision(a-h,o-z)
nmax = 1 000 000 000
itgr = ((gug + 10.d00)*nmax/20.0d00) + 1
end
```

The function flt(item), reverses the process, taking an integer no larger than 10^9 and converting it to a floating point number in the range from -10 to +10. The function has the following form:

```
function flt(item)
implicit double precision (a-h,o-z)
nmax = 1 000 000 000
flt = 20.0d00*((item - 0.5d00)/nmax) - 10.0d00
return
end
```

This algorithm can be scaled up, as in Figure 2.3, by increasing {N, elx, ely}. Even this small 400-particle program shows very convincingly the mixing nature of the dynamics, the generation of two shockwaves emanating from the midplane and the confined expansion of the hot fluid, evolving to fill the periodic square more or less uniformly.

2.12 Example Problems

The first example we consider here is the two-dimensional dissipative Baker Map. This is an instructive caricature of flows which incorporate a changing phase volume, $\langle \dot\otimes \rangle \neq 0$. Such flows obey the local *compressible* form of Liouville's Theorem for the time-development of f and an infinitesimal extension in phase, the comoving hypervolume $\otimes$:

$$(\dot f / f) \equiv -(\dot\otimes/\otimes) = -\nabla \cdot v .$$

Compare the singular structure shown in Figure 2.5 and generated by the dissipative map, to the smooth and continuous structure generated by the equilibrium incompressible Baker Map considered in Chapter 1.

The other example is a continuous flow, in two space dimensions to facilitate visualization, but complex enough to reproduce Gibbs' microcanonical distribution. It is the motion of a mass-point particle through a periodic two-dimensional array of soft scatterers.

2.12.1 *Time-Reversible Dissipative Map*

In Chapter 1 we considered a caricature of equilibrium systems, with their incompressible phase-space flows, $\dot{\otimes} = 0$. This was an area-conserving Baker Map B with a stationary ergodic solution $f = (1/4)$ within a 2×2 square. Here we consider a "dissipative" modification D of that map. The dissipative map, $D_{xy}(x,y) = (x',y')$ or the rotated version $D_{qp}(q,p) = (q',p')$ is *compressible*, with $\dot{\otimes} \neq 0$. As is shown in Figure 2.4, the upper leftmost *third* of the same 2×2 square is mapped into the upper rightmost *two thirds* of that same square:

$$+(1/3) < y < +1 \rightarrow \{x,y\}' = \{(2x+1)/3, (3y-2)\} \ ;$$

$$-1 < y < +(1/3) \rightarrow \{x,y\}' = \{(x-2)/3, (3y+1)/2\} \ .$$

Written in terms of a (horizontal) "coordinate" $q \equiv \sqrt{1/2}(x-y)$ and a (vertical) "momentum" $p \equiv \sqrt{1/2}(x+y)$, within the square $|\pm q \pm p| < \sqrt{2}$, the rotated map D_{qp} becomes:

$$q < (p - \sqrt{2/9}) \rightarrow q' = (11q/6) - (7p/6) + \sqrt{49/18} \ ,$$

$$p' = (11p/6) - (7q/6) - \sqrt{25/18} \ ;$$

$$(p - \sqrt{2/9}) < q \rightarrow q' = (11q/12) - (7p/12) - \sqrt{49/72} \ ,$$

$$p' = (11p/12) - (7q/12) - \sqrt{1/72} \ .$$

This time-reversible but dissipative map produces ergodic steady-state solutions which are distinctly different to those of the Equilibrium Baker Map of Chapter 1. This *dissipative* map includes area *changes*, leading to a *fractal* distribution. The distribution is actually "multifractal", meaning that the singular power-law dependence of density on distance varies with position. More details of such chaotic systems are given in Chapter 8. This problem is a useful introduction to dissipative chaos. The map is termed "dissipative" meaning that information is *lost* in the contraction of the mapping in the stable x direction. The simultaneous spreading, in the unstable y direction, smoothes and destroys any irregularities in that direction, and creates information, with formerly insignificant differences growing and governing present and future behavior.

This dissipative map includes periodic cycles analogous to those found in solving the equilibrium Baker Map. The two-iteration cycle links the points $(x, y) = (-5/7, +5/7)$ and $(x, y) = (-1/7, +1/7)$. These correspond to the *rotated* cycle linking the points $(q, p) = (-\sqrt{50/49}, 0)$ and $(q, p) = (-\sqrt{2/49}, 0)$.

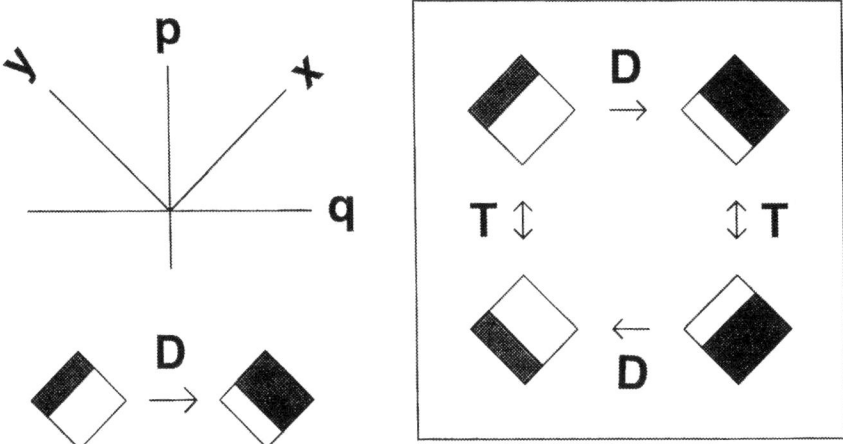

Figure 2.4. The Dissipative Baker Map, in (x, y) coordinates, is shown on the left. The rotated map, in (q, p) coordinates, is shown on the right. The mapping operation is denoted by D; the *time-reversal* operation T corresponds to changing the sign of the "momentum" p with the coordinate q fixed.

The inhomogeneous probability density can be characterized through a "coarse-grained" evaluation of an "entropy", $-k \langle \ln \text{prob} \rangle$. The probability "prob" is the product of the box area and the probability density f. This coarse-grained entropy *diverges* as the box width is decreased. Because the stretching in the y direction is Lyapunov unstable, eventually leading to a smooth, constant probability density in that direction, it is evident that the two strips $\{-1 < x < -1/3, -1/3 < x < +1\}$ are eventually occupied with probabilities of $(2/3)$ and $(1/3)$. A numerical calculation, with fifty million points, gave probabilities for occupying three strips of width $(2/3)$ which were consistent with this expectation:

$$\{\text{prob}\} = \{0.66664, 0.29512, 0.03817\} \rightarrow \sum -\text{prob} \times \ln(\text{prob}) = 0.6874 \;$$

$$\{f\} = \{0.99996, 0.44268, 0.05725\} \,.$$

For nine strips of width $(2/9)$ the probabilities were:

$$\{0.4443, 0.1968, 0.0255, 0.1968, 0.0255, 0.0729, 0.0254, 0.0110, 0.0017\}$$

$$\longrightarrow \sum -\text{prob} \times (\ln \text{prob}) = 0.6973 \,.$$

With nine strips, the coarse-grained probability density f ranges from 2, in the first strip, to 0.008 in the last. As the strip width is reduced further, slowly but surely $\langle \ln \text{prob} \rangle$ diverges, the signature of a fractal distribution.

The two "Lyapunov exponents" which characterize this dissipative map follow from the uniform distribution in the y direction. The map is unstable in that direction, expanding a small perturbation Δy by a factor 3 with probability $(1/3)$, and by a factor $(3/2)$ with probability $(2/3)$. Thus the larger exponent is

$$\lambda_1 = (1/3)\ln(3/1) + (2/3)\ln(3/2) = +0.63651 \,.$$

Similarly, in the stable x direction Δx shrinks by a factor 3 with probability $(2/3)$ and by a factor $(3/2)$ with probability $(1/3)$, leading to

$$\lambda_2 = (2/3)\ln(1/3) + (1/3)\ln(2/3) = -0.86756 \,.$$

Unlike the equilibrium Baker Map the stationary solution of this dissipative (but time-reversible) Baker Map is necessarily *singular everywhere*. This is because repeated iteration of the map continually magnifies the separation of nearby points in the y direction, eventually leading to a bifurcation between y coordinates which are greater than $(1/3)$ and those which are less. The result is the fractal distribution shown in Figure 2.5. This distribution, despite its ergodic behavior, has an "information dimension"—described below—of only 1.734, significantly less than that of the space in which the distribution is embedded.

The entire 2×2 square constitutes an *attractor* with two thirds of the square [y values less than $(1/3)$] contracting and the remaining third expanding. Any sequence of points $\{(\pm q, +p)\}$ produced by the mapping can *formally* be converted to another satisfactory sequence $\{(\pm q, -p)\}$ by reversing the *order* of the points as well as the sign of p. Reversing the sequence of the points changes the signs of both Lyapunov exponents (by replacing the expanding and contracting processes with their inverses) so that the sum,

$$(\lambda_1 + \lambda_2)_{\text{reversed}} = 0.86756 - 0.63651 = 0.23105 \,,$$

is positive, corresponding to an unstable—and therefore unobservable—repeller. We will see that this same structure, attractor-repeller pairs, occurs in time-reversible many-body systems, away from equilibrium. The *inevitable* contracting attractor states obey the Second Law of Thermodynamics. The *unobservable* repeller states would violate that Law.

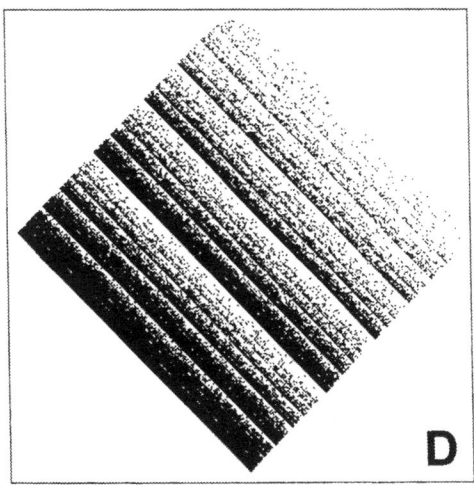

Figure 2.5. 100,000 points from the time-reversible dissipative Baker Map shown in Figure 2.4. Evaluation of the "information dimension" for this multifractal distribution gives $D_I = 1.734$.

The dimensionality of fractal distributions can be characterized in a variety of ways. The "box-counting dimension" and the "information dimension" are the most useful. The *box-counting* dimension gives the limiting power-law dependence of the number of occupied boxes # on the (sufficiently small) box size ϵ:

$$D_{\rm BC} \equiv \ln(\#)/\ln(1/\epsilon) \ .$$

Direct computation indicates that this time-reversible dissipative Baker Map is *ergodic*. That is, with enough data, the points generated by the map eventually occupy *every* square box, so that the *box-counting* dimension of the fractal is exactly 2. The *information* dimension describes the limiting power-law dependence of the small-ϵ box probabilities,

$$D_I \equiv \langle \ln {\rm prob} \rangle / \ln(\epsilon) \ .$$

Using 3^n square boxes of size $\{\epsilon = 2(1/3)^n\}$ provides a series of estimates for the information dimension, as n increases and ϵ decreases:

$$n = 1 \to 1.6874;\ n = 2 \to 1.6973;\ \ldots;\ n = \infty \to 1.734\ .$$

The Gibbs' entropy associated with the Baker Map *diverges* as the box size ϵ is reduced, as is shown in Figure 2.6:

$$S_G/k \equiv -\langle \ln f \rangle \simeq 0.266 \ln \epsilon\ .$$

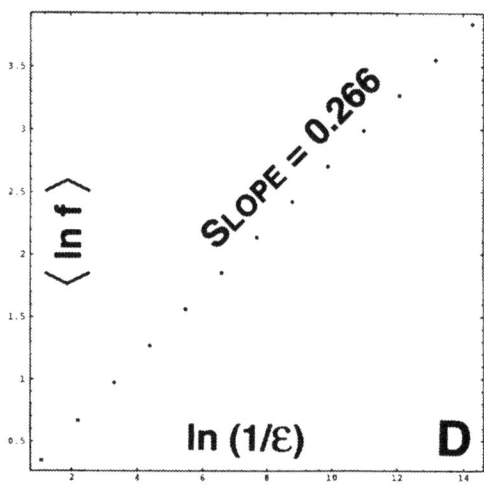

Figure 2.6. Coarse-grained Gibbs entropy, relative to that of the equilibrium Baker Map. The data are 40,000,000 points, which show the divergence of $\langle \ln f \rangle$, varying as $0.266 \ln(1/\epsilon)$ for $3^{-13} < \epsilon < 3^{-1}$.

The power-law divergence of the Gibbs entropy is closely related to the information dimension of the fractal:

$$D_I = \langle \ln \text{prob} \rangle / \ln \epsilon = \langle (\ln f\epsilon^2) \rangle / \ln \epsilon = 1.734 - 0.06(1/n)^{0.20}\ .$$

Thus the estimate for the information dimension of the two-dimensional dissipative Baker Map is 1.734. This corresponds to the small-ϵ divergence of the probability density as follows:

$$f_{\epsilon \to 0} \propto \epsilon^{1.734-2.000}\ .$$

This divergence of f indicates that "almost all" of the probability is found in a "negligibly small" fraction of the boxes, the attractor "core". For many interesting illustrations of fractal dimensionality, and a clear overview of the geometrical significance of attractor dimensions, see the 1983 paper by Farmer, Ott, and Yorke. Schröder's and Sprott's books are both particularly helpful in this area.

Kaplan and Yorke pointed out that the information dimension should correspond to that linearly-interpolated dimension at which the partial Lyapunov-exponent sum *changes sign*. With only two Lyapunov exponents the one-term "sum" is $\lambda_1 > 0$ while the two-term sum is $\lambda_1 + \lambda_2 < 0$. The "Kaplan-Yorke dimension" is $1 + |\lambda_1|/|\lambda_2|$. For the Baker Map the information and Kaplan-Yorke dimensions are *identical*:

$$D_I = D_{KY} = 1 - (\lambda_1/\lambda_2) = 1 - (+0.63651/-0.86756) = 1.73368 ,$$

as is known to hold exactly in this case.

Similar singular fractal characteristics are shared by the more complicated nonequilibrium many-body phase-space distributions. The probability density typically diverges, indicating the concentration of the probability onto a fractal set, but with a box-counting dimension characteristic of equilibrium. The simple Baker Map example is a useful caricature of real flows. The analysis of many-body flows relies on the Kaplan-Yorke conjecture because binning and box-counting operations are impractical in spaces of more than a few dimensions.

2.12.2 A Smooth-Potential Galton Board

The Galton Board problem—the dynamics of a single hard disk moving through a periodic array of scatterers—is an excellent vehicle for understanding the contributions of nonlinearity and chaos to equilibrium and nonequilibrium problems. In the absence of external forces the ergodic and chaotic Galton Board problem of Section 1.9.2 is probably the simplest system for which Gibbs' equilibrium statistical mechanics applies.

This dynamical system can be viewed as a single particle moving through a periodic array of fixed hard-disk scatterers, as is illustrated in Section 1.9.2 or, alternatively, as a *two-disk* problem with a fixed center of mass and square or parallelogram periodic boundaries. Analysis of the free volume available to the moving disk provides an interesting analytic caricature of van der Waals' loop, which links a high-density solid phase to a lower-density fluid.

With an external field added, along with a compensating thermostat or ergostat, the Galton Board provides a field-dependent Lyapunov spectrum as well as a nonlinear diffusivity. Adding a systematic shear velocity to the periodic lattice makes it possible to study viscosity for a two-disk system. Such nonequilibrium steady state problems generate interesting fractal distributions in the phase space, resembling the equilibrium distributions shown in Figure 1.4 on page 28 but having an irregular fractal structure even at infinitesimal resolution scales.

The impulsive hard-disk forces which punctuate the dynamics can be viewed as a limiting case of *smooth* repulsive forces. The smooth models, with short-ranged repulsive potentials such as

$$\phi(r < 1) \propto (1-r)^n \text{ or } \phi(r < 1) \propto (1-r^2)^n ,$$

are a specially good choice because the numerical integration errors are minimized when the potential cutoff has several vanishing derivatives at the cutoff distance.

Trajectory for t = 200

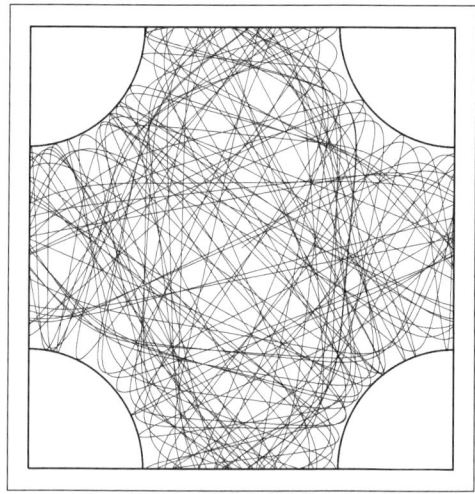

Figure 2.7. Runge-Kutta trajectory, with $\Delta t = 0.002$, for a soft-disk particle with four square-lattice scatterers and $\phi = 4(1-r^2)^4$. With initial velocity $(\sqrt{4/3}, \sqrt{2/3})$ the moving particle is excluded from the four quarter circles shown at the corners of the periodic box.

It is relatively easy to generate trajectories covering millions of scattering collisions. See Figure 2.7 for a trajectory segment of length 200, corresponding to about 200 soft "collisions". To ensure the *time-reversibility*

of the trajectories, the Störmer-Verlet "leapfrog" algorithm is the simplest choice:
$$\{r_+ - 2r_0 + r_- = (\Delta t)^2 (F/m)_0\} \ .$$
The Runge-Kutta method used here provides higher *accuracy*. To ensure *energy conservation*, for energies of order unity, an error analysis of the algorithm suggests, as simulations bear out, that timesteps in the range $0.001 \leq \Delta t \leq 0.01$ give reasonably accurate (from the standpoint of energy) accelerations over run lengths on the order of a million collisions.

Energy isn't a very useful criterion for *accuracy* if by accuracy we mean nearly correct values of the coordinates and momenta. Notice that the "local" single-timestep coordinate error is of order $(\Delta t)^5$ so that the *global* error, at a fixed time, is *formally* of order $(\Delta t)^4$, but *actually* of order $(\Delta t)^5 e^{\lambda t}$, where λ is the largest Lyapunov exponent. The exponential error growth of the fifth-order error is due to Lyapunov instability. Lyapunov instability ensures that the *accuracy* of the trajectory degrades rapidly.

Figure 2.8 on page 76 compares the exponential growths separating pairs of Galton-Board trajectories with
$$\phi(r<1) = 4(1-r^2)^{3 \text{ or } 4 \text{ or } 5} \ .$$
The initial separations, $\{\Delta x, \Delta y, \Delta p_x, \Delta p_y\}$, for the three pairs of simulations, were all chosen equal to 10^{-15}. The separation growth rate, from the initial value $\simeq e^{-34}$ is shown in Figure 2.8. The separation grows by a factor of e^{34} in a time of 25, verifying our expectation that the largest Lyapunov exponent for this system is approximately unity. Were we to make the Herculean effort of using hundred-digit or thousand-digit accuracy in solving this problem our accuracy would be completely degraded in a time of order $230 = 100 \ln(10)$ — a trajectory length corresponding to Figure 2.7 — or $2300 = 1000 \ln(10)$, just ten times longer.

As Levesque and Verlet pointed out, the integer version of the leapfrog algorithm provides strict time reversibility. Such a problem has, in principle, a periodic solution (provided that the algorithm does not, eventually, diverge) but present day computers are still not fast enough for a straightforward investigation of this periodicity. Numerical work shows that at least eight-digit accuracy is required, for which the determination of Poincaré recurrence periods of length $\simeq 10^{16} \Delta t$ is an unappealing prospect. Probability densities, and the comparison of dynamical averages with Gibbs' ensemble theory, suggests that this problem is *ergodic* (covering all the available energy states) over a wide range of energies.

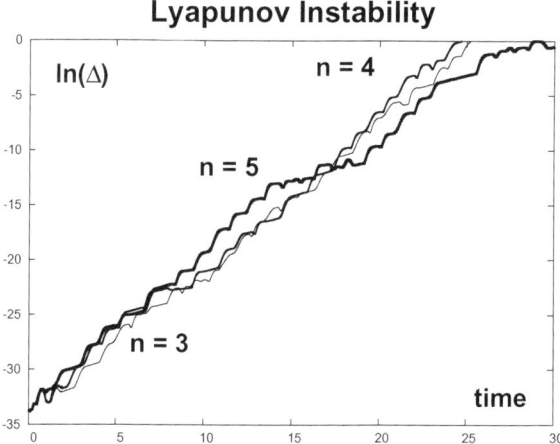

Figure 2.8. Exponential Lyapunov-unstable growth of the phase-space separation between *pairs* of nearby trajectories of the type shown in Figure 2.7. The results correspond to three different soft-disk force laws. The thinnest line represents $n = 3$ and the thickest line $n = 5$. The "offset" is $\Delta \equiv [\Delta x^2 + \Delta y^2 + \Delta p_x^2 + \Delta p_y^2]^{1/2}$.

For a more nearly accurate (but irreversible) dynamics the workhorse Runge-Kutta integrator, used to prepare Figures 2.7 and 2.8, is a good choice. Using a periodic 2×2 square unit cell, and a moving particle of unit mass, total energies less than 3 appear to provide an ergodic coverage of the available phase space. See the discussion of numerical ergodicity in the next Chapter. The values of the time averages $\langle \Phi = \sum \phi \rangle$ and $\langle \Phi^2 \rangle$ support the apparent ergodicity, agreeing well with the predictions of Gibbs' isoenergetic microcanonical-ensemble theory for this problem. Choosing four fixed neighbors at the square-lattice positions $\{\pm 1, \pm 1\}$, the forces and energy are computed in loops over the four neighboring particles:

```
energy = 0.5d00*(px*px + py*py)
do i = 2,5
dx = x(1) - x(i)
dy = y(1) - y(i)
rr = dx*dx + dy*dy
energy = energy + 4.0d00*(1.0d00 - rr)**n
fx = fx + 8.0d00*n*dx*(1.0d00 - rr)**(n-1)
fy = fy + 8.0d00*n*dy*(1.0d00 - rr)**(n-1)
enddo
```

At the end of each timestep periodic boundary conditions are applied:

```
if(x(1).lt.-1.0d00) x(1) = x(1) + 2.0d00
if(x(1).gt.+1.0d00) x(1) = x(1) - 2.0d00
if(y(1).lt.-1.0d00) y(1) = y(1) + 2.0d00
if(y(1).gt.+1.0d00) y(1) = y(1) - 2.0d00
```

We used a convenient initial condition, with an energy of unity, resulting from the choices

```
x = 0.0d00
y = 0.0d00
px = dsqrt(4.0d00/3.0d00)
py = dsqrt(2.0d00/3.0d00)
```

By doubling the number of differential equations from four (Figure 2.7) to eight (Figure 2.8), the Lyapunov instability governing the separation of two nearby trajectories can be studied as is illustrated by Figure 2.8.

2.13 Summary

The exact time-reversibility found in the classical equations of motion for isolated systems can be retained in thermomechanical computer simulations of systems which interact with their environments. Thermal environments can be characterized by time-reversible "thermostats" constraining the kinetic energy of selected degrees of freedom. A variety of time-reversible approaches to thermostats all lead to time-reversible forces linear in the momenta, $\{-\zeta p\}$, as does also the low-temperature limit of the *irreversible* stochastic approach summarized by the Langevin equation. In all of the time-reversible approaches to thermomechanics, the friction coefficient ζ changes sign along any time-reversed trajectory.

Computer simulation of chaotic trajectories necessarily requires a discrete truncated caricature of hypothetical continuous trajectories. In principle, the discrete and finite nature of the numerical state space would seem to make the dynamics vulnerable to the objections raised by Loschmidt and Zermélo. Although these objections are valid for any isolated system, adding an interaction with the surroundings provides a simple geometric understanding of a (time-averaged) Second Law of Thermodynamics.

The presence of chaos, or "sensitivity to initial conditions" does not itself affect trajectory reversibility, but, with the added influence of a system's surroundings, provides a new mechanism for irreversible behavior. This mechanism is the formation of fractal attractors with an information dimension which decreases as the departure from equilibrium increases. The combination of dissipation with time-reversibility yields strange attractors with a family resemblance to the fractal distribution generated by the dissipative Baker Map. The association of dissipation with fractals will be described in greater detail in Chapters 8 and 9.

2.13.1   *Notes and References*

The case made here for kinetic temperature is based on the Hoover, Holian, and Posch "Comment I on 'Possible Experiment to Check the Reality of a Nonequilibrium Temperature' " (1993). For a discussion of Braga, Jepps, and Travis' work on configurational temperatures see our 2008 Physical Review E work "Nonequilibrium Temperature and Thermometry in Heat-Conducting ϕ^4 Models". Nosé's pioneering approach to inserting canonical-ensemble temperature into mechanics is discussed in his "Constant Temperature Molecular Dynamics Methods". Carl Dettmann improved upon Nosé's derivation of Nosé-Hoover mechanics by eliminating the need for "time-scaling", a flaw embedded in the original treatment. For the history of Dettmann's discovery see Dettmann and Morriss' "Hamiltonian Formulation of the Gaussian Isokinetic Thermostat" along with Bill's "Mécanique de Nonéquilibre à la Californienne". For more on Langevin thermostating see Ian Snook's *The Langevin and Generalised Langevin Approach to the Dynamics of Atomic, Polymeric and Colloidal Systems.*

Figure 2.2 is taken from Oyeon Kum and Bill's paper on "Time-Reversible Continuum Mechanics", which was stimulated by the Levesque-Verlet work, "Molecular Dynamics and Time Reversibility". Illner and Neunzert's "The Concept of Irreversibility in the Kinetic Theory of Gases" contains their discussion of time inversion in connection with reversibility.

Farmer, Ott, and Yorke's "Dimension of Chaotic Attractors" and Sprott's *Chaos and Time-Series Analysis* are two useful references on multifractal dimensionality. For work on the Sinai-Ruelle-Bowen distributions, which are special strange attractors, have a look at the publications, many in electronic form, on David Ruelle's homepage. Our joint work with Posch and Codelli, "The Second Law of Thermodynamics and Multifractal Distribution Functions; Bin Counting, Pair Correlations, and the Kaplan-Yorke Conjecture" illustrates a failure [2.56 *versus* 2.80] of the conjecture made in Kaplan and Yorke's "Chaotic Behavior of Multidimensional Difference Equations" in Lecture Notes in Mathematics, volume 730. Their information dimension estimate seems likely to be within unity of the true value.

Ichiyanagi's "Conceptual Developments of Nonequilibrium Statistical Mechanics in the Early Days of Japan", and Zwanzig's "Time-Correlation Functions and Transport Coefficients in Statistical Mechanics" are interesting historical references detailing the Green-Kubo Linear-Response Theory.

Chapter 3

Gibbs' Statistical Mechanics

Our life is frittered away by detail ...
Simplify, simplify.

Henry David Thoreau

3.1 Scope and History

Energy is *the* basic state function in thermodynamics. For a given amount of material in a particular thermodynamic "state"—(V,T) or (P,T) or (P,V) for instance—the equilibrium energy is always the same. *Changes* of energy in mechanics involve work, the product of force and displacement. In thermodynamics the work done by a fluid expanding reversibly against a pressure P is the integral $\Delta W = \int P dV$. Thermodynamics describes not only such energy changes due to mechanical work, but also those due to heat transfer. A thermodynamic description of heat transfer requires the *two* additional state functions discussed in Chapter 2, temperature—defined in terms of the ideal-gas thermometer—and the entropy S. The *reversible* transfer of heat *to* a fluid *from* a heat reservoir at temperature T is the integral $\Delta Q \equiv \int T dS$. Thus thermodynamic entropy is defined by an integration linking an initial standard state to any other current state of interest. The integration must follow a thermodynamically *reversible* process:

$$\Delta S(N, E, V) \equiv \int_{\text{rev}} (1/T) dQ \ .$$

Though this thermodynamic approach to mechanical and thermal energy changes gives the appearance of generality, sufficiently complex systems frustrate the need for an underlying reversible process. Thermodynamically reversible processes must take place through a sequence of *equilibrium* states. *Slow* processes are not necessarily reversible. Bridgman emphasized the intrinsic *irreversibility* of dislocation motion, the mechanism for irreversible "plastic flow" in metals. Once the barrier to dislocation motion is surmounted, by applying stress, the main part of the stored energy is dissipated into heat during the sudden *irreversible* relaxation which results.

For fluids, and for solids with the defects frozen in, thermodynamics *is* a useful framework for describing changes of entropy and energy. Further, the usefulness of energy, as a potential for describing the equilibrium of isolated systems, can be extended to the enthalpy, for the equilibrium of adiabatically-isolated systems at fixed *pressure*, and to the two "free energies", Helmholtz' and Gibbs', for describing the equilibria of constant-volume or constant-pressure systems at fixed *temperature*.

More than a century ago, Gibbs followed Boltzmann's lead for gases, relating microscopic mechanics and (q, p) phase-space energy states to macroscopic thermodynamics through statistical mechanics. His linking the two approaches proceeds through a series of three steps: (i) Liouville's incompressible Theorem ($\dot{f}_N \equiv 0$), is used to motivate the "microcanonical ensemble" description of an isolated Hamiltonian system—a "closed" system in static equilibrium; (ii) this allows dynamical time averages to be replaced by phase-space averages using equal weights for all accessible (q, p) states of the same energy; (iii) *weakly* coupling a general system to an ideal gas, for which the number of phase-space states, Ω, can be calculated, allows both mechanical and thermal equilibrium conditions to be formulated, leading to a new relation for the entropy,

$$S_{\text{Gibbs}} = k \ln \Omega \longrightarrow S = -k \langle \ln f_N \rangle \ ,$$

valid for *all* equilibrium ensembles, not just the microcanonical one. We will consistently refer to this collective N-particle entropy as "Gibbs'", though the scanty evidence available today suggests that both Boltzmann and Gibbs discovered the new relation independently and at about the same time. The evidence—in particular Gibbs' 1884 Philadelphia abstract—is discussed in the biographies of Gibbs written by Klein and Wheeler. Cercignani discusses Boltzmann's 1884 paper. The N-body "Gibbs' entropy" resembles Boltzmann's one-body dilute-gas entropy, but applies to liquids and solids, as well as to gases, both dense and dilute. It would seem to conflict with Liouville's incompressible phase-space flow Theorem, which rules out a changing Gibbs' entropy for Hamiltonian flows. This difficulty led Gibbs to discuss irreversibility in terms of a "coarse-grained" approximation to the phase-space density f_N. It is this part of his approach which has generated the most criticism. Of course by now our digital implementations of mechanics are coarse-grained at the double-precision level. Gibbs' approach made it possible to compute equilibrium properties from phase-space averages, without the need for solving any dynamical equations.

50 years later, Green and Kubo followed Einstein and Onsager, in extending Gibbs' ideas to nonequilibrium systems. They treated linear transport processes in a convincing and complete manner. More recently chaos has provided an answer to those who had questioned the applicability of Gibbs' ensembles to transport theory. Our goal in this Chapter is to summarize statistical mechanics so as to connect thermomechanical microscopic simulations (including nonequilibrium molecular dynamics) with macroscopic descriptions (thermodynamics and continuum mechanics) of material properties.

3.2 Formal Structure of Gibbs' Statistical Mechanics

The fundamental mechanical result which Gibbs used in his equilibrium theory, in order to convert time averages to phase averages, was Liouville's incompressible Theorem. This phase-space flow Theorem is an exact rigorous consequence of Hamilton's equations of motion. Liouville's incompressible theorem makes it possible to introduce probability and to use it in the computation of simple equilibrium averages. Liouville's Theorem is *the* foundation of Gibbs' equilibrium statistical mechanics. The additional tool Gibbs needed was the concept of weak coupling—the idea that two systems could, when so coupled together, access *all* of their states, and in a way which could *violate* Liouville's Theorem, so as to *agree* with the phenomenological predictions of thermodynamics. No doubt Gibbs had in mind a small perturbation, such as a corrugated container, which would introduce what the mathematicians call "mixing", the breaking of correlations between the initial conditions and the current state. In his boxed notes, at the Yale University library, Gibbs suggests putting systems in thermal communication through "slight" gravitational interactions. His concept of "mixing" or "coarse-graining", with the analog of ink in milk, is familiar from his textbook description. We know today that Lyapunov instability provides a natural mechanism for this mixing without the need for any special corrugations, gravitational interactions, or other special mechanical constructions. Gibbs' cautious nature precluded a more complete description of this coupling, but without it, as Krylov was so fond of stating, Gibbs' work was at best incomplete, or at worst wrong. To avoid such criticisms of Gibbs let us grant him the favor of a small piece of conceptual copper wire, able to transfer energy between Gibbs' systems without perturbing their accessible states.

Consider then, with Gibbs, a two-part system. The larger part is an ideal *gas*, providing a heat reservoir for the smaller part, and coupled to it in such a way as to access all states of both, consistent with a fixed total energy E and total volume V. The ideal-gas reservoir is imagined to be so large that its temperature, T, has negligible fluctuations. The smaller of the two parts making up our combined system could be any small system whatever. For simplicity, we will imagine it to be a *fluid*, with an energy $E_{\text{fluid}} \ll E$ and a volume $V_{\text{fluid}} \ll V$. The ideal-gas heat reservoir, with which the fluid interacts, takes up the rest of the energy and volume:

$$E = \mathcal{H}_{\text{gas}} + \mathcal{H}_{\text{fluid}} \ ; \ V = V_{\text{gas}} + V_{\text{fluid}} \ .$$

Gibbs formulated the distribution of this two-part system among its energy states and was able to derive the properties of a "canonical" (constant temperature T) ensemble directly from this simple two-part "microcanonical" (constant energy) picture. The final result is the equilibrium phase-space probability density:

$$f_{\text{fluid}}(\{q,p\}) \propto e^{[A(N,V,T) - \mathcal{H}(q,p)]/kT} \ .$$

This "canonical" distribution applies to the smaller of the two linked systems—the fluid in the example just described. It follows from this probability density that the Helmholtz free energy $A \equiv E - TS$ can be determined by working out Gibbs' canonical partition function $Z(N,V,T)$:

$$Z(N,V,T) = e^{-A/kT} = (1/N!) \prod [\int \int dqdp/h] e^{-\mathcal{H}/kT} \ .$$

The development of this approach was purely theoretical. Neither Gibbs, nor most of his followers, considered the explicit construction of boundary conditions required to contain N particles in a volume V while rendering all states of energy E accessible. They likewise ignored mechanisms to implement the sources and sinks of energy needed to equilibrate systems at a temperature T. Operational computational analogs for these relatively unimportant "formal details" become all-important to any computational application of Gibbs' formulation. Furthermore, the lack of any specific mixing or thermal boundaries provided mathematically-inclined physicists with a quandary—how to be sure that a system described by the Hamiltonian $\mathcal{H}$ would be able to reach all those states included in Gibbs' partition function integral $Z(N, V, T)$.

This quandary launched a variety of investigations into "ergodic theory", a study of necessary and sufficient conditions for phase-space averages and time averages to agree. Sinai obtained a tantalizing result: hard disks and spheres *do* behave in an ergodic mixing manner, even with periodic boundaries. Kolmogorov, Arnold, and Moser obtained a more disquieting (and somewhat more obvious) result (their "Theorem"): systems described by potentials with a smooth stable bound state cannot generally access all the Gibbs' states at a fixed energy. The lack of any physical boundaries in both these demonstrations is quite unrealistic. Though there is no doubt that complicated boundaries enhance the mixing effects Sinai demonstrated, such boundaries would greatly restrict the applicability of the Kolmogorov-Arnold-Moser Theorem. For a physicist, the clear success of Gibbs' theory suggests that mixing and ergodicity be accepted as basic principles.

Apart from an additive constant E_0, Helmholtz' free energy $A = E - TS$ corresponds to the total energy E of a system—such as our fluid—together with that of a larger heat reservoir, with energy $E_0 - TS$. The temperature T characterizes *both* the system and the reservoir. S is the system entropy. The volume and temperature derivatives of $A(N, V, T)$ provide the system's pressure and entropy: $dA = -PdV - SdT$. Gibbs had to show that the Helmholtz free energy following from his canonical partition function integral agreed with experimental pressures and energies. In this he was completely successful. With this achievement, Gibbs' theory could be accepted as exact for the model of Hamiltonian systems. There remained the nagging need to resolve the "ergodicity" quandary, though there was never any real doubt that interesting physical systems are (i) sufficiently perturbed that the initial conditions are unimportant and (ii) sufficiently complicated (with recurrence times exceeding the age of the universe) that the concept of ergodicity is both irrelevant and misleading.

Gibbs' statistical mechanics differed from Boltzmann's kinetic theory by taking place in the full many-body phase space, "gas-space" or "γ-space", while Boltzmann's considerations were mostly restricted to densities in the phase space of a single molecule, "molecule-space" or "μ-space". But, unlike Boltzmann, Gibbs had no results *away* from equilibrium. Let us explore Gibbs' statistical mechanics from the perspective of computer simulations in γ-space. This will allow us to explore the interrelations among thermodynamics, statistical mechanics, and computer simulation. We begin with the initial and boundary conditions.

3.3 Initial Conditions, Boundary Conditions, Ergodicity

Computer simulation is a demanding discipline, in which *all* the "details" included in the computer algorithm must be precisely described. Typical details are (i) the initial conditions, (ii) the boundary conditions, and (iii) the equations of motion. In the atomistic equilibrium systems to which Gibbs' statistical mechanics can be applied, the number of particles, their type, positions, velocities, and the nature of their surroundings, must all be specified. The "surroundings" include (i) a container for the system, with the simplest choice being *periodic* boundaries, as used in an example problem in Section 3.11.3, as well as (ii) any necessary sources and sinks of energy, including forces to perform thermodynamic work and friction coefficients necessary to transfer heat, as described in Section 2.8.

An entire branch of mathematics, ergodic theory, arose in an effort to investigate whether or not a single dynamical trajectory could faithfully represent the states included in Gibbs' microcanonical ensemble. The Ehrenfests believed that it is meaningful to distinguish "quasiergodic" behavior (coming arbitrarily *close* to all states) from "ergodic" behavior (actually *reaching* all states). This distinction is based on the conceptual difficulty inherent in adequately covering a many-dimensional phase-space object with a one-dimensional trajectory. It is one of many examples in which the order of taking limits—small boxes and long times in this case—takes on a significance more apparent than real.

It is evident that ergodic theory has little to do with physics since a small but well-chosen perturbation, a container with a properly-corrugated surface or the gravitational interaction of the system with a small mass outside, would satisfy most physicists' need for an equilibration mechanism. Only in the very simplest of systems, with a few degrees of freedom, is there sufficient time ($\propto e^N$) to access *all* such states. For example, the time required for an eight-atom system of liquid argon to access all of its quantum states is already comparable to the age of the universe, 10^{17} seconds. Evidently, in view of the impossibility of observing ergodic behavior for most systems, it is actually only important that the fluctuations be sufficiently small. Were this not to be the case, then statistical mechanics would not be a correct description of equilibrium thermodynamics.

Computer simulation of many-body systems began at Los Alamos with the investigation of one-dimensional anharmonic chains of particles. The investigators, Fermi, Pasta, and Ulam, were surprised to find that typical trajectories scrupulously avoided most of the available phase-space energy shell corresponding to Gibbs' microcanonical ensemble. With the benefit of hindsight, we now know that in two- and three-dimensional systems such difficulties are less common, and usually result from an unfortunate combination of force laws and initial conditions. Of course, sufficiently simple systems—a single free particle is the best example—have no hope of passing through (or even near) *all* the states characterizing a complete Gibbs' distribution unless additional thermostat forces are used. An early remedy for the lack of ergodicity was to use the stochastic Langevin equation of Section 2.9, with its irreversible frictional force, $-(p/\tau)$, and its compensating high-frequency "noise"—usually Gaussian random forces refreshed at every time step—together providing the desired temperature. Much later, inspired by Nosé's work, Kusnezov, Bulgac, and Bauer invented relatively simple, robust, and purely-deterministic forces quite capable of imposing Gibbs' canonical distribution on an otherwise free particle.

Excepting such unusual cases, dynamical equilibrium simulations presented no special difficulties. Periodic boundaries could be used to eliminate the unacceptably large boundary effects of a physical container. At equilibrium, the statistical properties could be computed by using the "Monte Carlo" configurational sampling technique developed at Los Alamos by Metropolis, the Rosenbluths, and the Tellers. In this way, Gibbs' canonical distribution could be generated by temperature-dependent "moves" of particles. Provided that the initial conditions were wisely chosen, simulations following a few hundred particles for a few tens of thousands of moves accurately reproduced the thermodynamic properties from Gibbs' statistical mechanics.

The good agreement between molecular dynamics and Gibbs' statistics was gratifying. The outcome was not at all obvious until parallel investigations, at Livermore and Los Alamos, applied the two methods to exactly the same systems. Though computers were rapidly gaining speed, many-body simulations have voracious appetites for computer time. In the early 1950s hard elastic disks and spheres were the usual choice for simulation, in order to save computer time and to simplify theoretical interpretations. In the middle 1950s Bill Wood, at Los Alamos, characterized the fluid-solid phase transitions for both hard disks and hard spheres. His Monte Carlo work agreed nicely with simultaneous dynamical studies, at Livermore, carried out by Tom Wainwright and Berni Alder. Wood's mature reminiscences about this work, both scientific and sociological, as given in lectures at two Lake Como summer schools (1985 and 1996), make interesting reading. By 1959 George Vineyard and his colleagues at Brookhaven were analyzing fully continuous dynamical models, simulating the nonequilibrium relaxation of copper crystals exposed to high-energy radiation.

A generation after Fermi's work on anharmonic chains, dynamic simulations began to be carried out at constant *temperature*, rather than constant energy. The *kinetic* energy was controlled by feedback forces while the *total* energy was allowed to fluctuate. This development of isothermal methods was a response to two influences. First, isothermal canonical-ensemble Monte Carlo simulations were commonplace, so that comparison with a dynamical analog was desirable. Second, real laboratory experiments are seldom carried out on isolated systems. Instead they typically use *thermal* boundaries.

In all such *statistical* canonical-ensemble simulations, temperature was defined as usual, by the mean value of the fluctuating kinetic energy: $T \equiv \langle p_x^2/mk \rangle = \langle mv_x^2/k \rangle$. Isokinetic molecular dynamics calculations, with fixed kinetic energy, were the usual dynamical alternative. That approach had been used to model heat reservoirs, first at equilibrium and then in nonequilibrium steady states. The isokinetic dynamics produces an ensemble—different to any of Gibbs'—in which the kinetic energy is fixed while the potential energy is distributed canonically.

Nosé's more-elegant feedback method corresponds *exactly* to Gibbs' canonical distribution. As described in Section 2.8, Nosé's equations of motion, written in the simpler "Nosé-Hoover" form, are:

$$\{\ddot{r} = (F/m) - \zeta \dot{r}\} \; ; \; \dot{\zeta} = [(K/K_0) - 1]/\tau^2 \; .$$

The many-body thermal relaxation time is of order $\sqrt{N}\tau$. This Nosé-Hoover approach is "robust", in that the time-averaged kinetic temperature is maintained *exactly*, even in the presence of additional external driving forces. Such external nonequilibrium driving can convert the *smooth canonical distribution* to a geometrically complicated *multifractal attractor*.

Systems, such as the harmonic oscillator, which are even farther from ergodic than Fermi's nonlinear chains, require additional control variables to reach the complete canonical distribution. A simple workable approach is to use *two control* variables $\{\zeta, \xi\}$, rather than just one. Consider a simple one-dimensional harmonic oscillator. For convenience choose the mass, force constant, temperature, and characteristic thermostat times all equal to unity. Two alternative examples of this generalized approach to the oscillator's thermomechanics are the following:

$$\{\dot{q} = p \; ; \; \dot{p} = -q - \zeta p - \xi p^3 \; ; \; \dot{\zeta} = p^2 - 1 \; ; \; \dot{\xi} = p^4 - 3p^2\} \; ;$$

$$\{\dot{q} = p \; ; \; \dot{p} = -q - \zeta p \; ; \; \dot{\zeta} = p^2 - 1 - \zeta \xi \; ; \; \dot{\xi} = \zeta^2 - 1\} \; .$$

Numerical investigations of these models, dating back to the 1990s, have established that *both* these approaches provide the complete canonical distribution for the harmonic oscillator energy, $(q^2 + p^2)/2$, with independent Gaussian distributions for the two thermostat variables $\{\zeta, \xi\}$:

$$f \propto e^{-(q^2+p^2)/2} e^{-(\zeta^2+\xi^2)/2} \; .$$

Generalizing Nosé's approach to *quantum* systems remains a pressing need. A variety of approaches has been tried. Some approaches use feedback, adjusting the wavefunction or its gradient in such a way as to promote thermal equality between system and thermometer. Some use "Gaussian random matrices" to represent thermostats. See the articles by Jürgen Schnack and by Dimitri Kusnezov for accounts.

3.4 From Hamiltonian Dynamics to Gibbs' Probability

In phase space, the coordinates of a single multidimensional point give the complete microscopic description $(\{q\}, \{p\}, \{\zeta\}, \dots)$. For *Hamiltonian* systems without heat sources and sinks there is no expansion or contraction of the phase-space flow. The comoving probability density is incompressible. And, according to Gibbs' equilibrium statistical mechanics, the logarithm of this density, when properly averaged, gives the entropy. Without any further assumptions or considerations, it would appear therefore that Hamilton's mechanics, though fine at equilibrium, provides no *approach* to it. Gibbs' entropy cannot change with time.

It is Liouville's *incompressible* theorem which shows that both the comoving phase volume ⊗—the "extension in phase"—and the comoving phase-space probability density f flow through phase space unchanged. In the usual *Cartesian* case, all the components of the phase-space velocity divergence vanish:

$$\{\dot{x} = (p_x/m) \to (\partial \dot{x}/\partial x) = 0 \; ; \; \dot{p}_x = F(\{x\}) \to (\partial \dot{p}_x/\partial p_x) = 0\} \;.$$

Even with arbitrary *generalized* coordinates the *summed-up* terms vanish:

$$(\partial \dot{q}/\partial q) + (\partial \dot{p}/\partial p) = (\partial/\partial q)(+\partial \mathcal{H}/\partial p) + (\partial/\partial p)(-\partial \mathcal{H}/\partial q) \equiv 0 \;.$$

For either type of coordinates the generalized velocity of the flow through phase space, $v \equiv (\dot{q}, \dot{p})$ has no divergence. Thus *any* comoving element of phase-space volume $\otimes = \prod(dqdp)$, small or large, is conserved by the flow, and so must be the corresponding probability density $f(\{q, p, t\})$. Either of these two results implies the other because the product $(f \otimes)$ corresponds precisely to probability, and so must be conserved by *any* flow equations. Both components, f and $\otimes$ separately, are unchanged in Hamiltonian flows.

Complete thermodynamic equilibrium has three components, mechanical, thermal, and chemical. Equilibrium is characterized by constancy of pressure, temperature, and all the chemical potentials. No net accelerations, heat currents, or chemical reactions occur at equilibrium. This static equilibrium situation corresponds, in Gibbs' view, to a dynamic ensemble in an unchanging stationary state of flow, with a fixed probability density $f_{eq}(q, p, t) \to f(q, p)$ everywhere in phase space: $(\partial f/\partial t) \equiv 0$. In an ergodic isoenergetic mixing system, the conclusion that the probability density f does not change with time suggests that it can only be constant, in the entire accessible region. Otherwise, the density at a fixed phase-space point would have to change with time, and could not characterize equilibrium. A thin energy shell of the constant probability density which results defines Gibbs' "microcanonical ensemble". This "ensemble" can be viewed as a collection of systems distributed according to the constant invariant probability density $f_{eq}(q, p, t) = f(E)$. In order for Gibbs' ensemble to represent the long-time-averaged properties of an arbitrarily selected dynamical system—what Gibbs called the "time ensemble"—it is necessary that the dynamics be *mixing*, so that the correlation of long-time-averaged properties with initial conditions is eventually lost.

3.5 From Gibbs' Probability to Thermodynamics

Before Alder, Wainwright, and Wood's computer simulations of the 1950s it was entirely unknown whether or not Gibbs' static statistical ensemble averaging would agree with long-time dynamical averages based on Newton's equations of motion. Until then intuition was crippled by a lack of examples. Arguments raged over the existence and significance of "holes" and "cages" in liquids. Even the existence of a solid phase for hard spheres was controversial. The computer simulations dissipated this foggy atmosphere. Alder, Wainwright, and Wood found that time averages, for fluid and solid systems of a few hundred hard disks or spheres, were in excellent agreement with Gibbs' statistical phase-space theory, even including the "number-dependent" effects which cause small systems to deviate (relatively slightly) from large-system "thermodynamic behavior".

In Gibbs' theory entropy is of primary significance. Gibbs had established the identity of thermodynamic entropy with the microcanonical phase volume,

$$S = k \ln \Omega(N, E, V) = -k \langle \ln f_N \rangle .$$

He generalized this microcanonical relation to other equilibrium ensembles with different phase-space probability densities: $S_{\text{Gibbs}} = -k \langle \ln f \rangle$. His demonstration that thermodynamic entropy corresponds *generally* to phase-space probability density is a two-step process. First, the relation is established in the microcanonical constant-energy case. Second, the same relation is shown to hold in other ensembles.

Let us follow Gibbs' argument in the microcanonical case. We begin by considering the interaction of an ideal gas with another system. The ideal gas is chosen because its phase volume can be calculated analytically:

$$\Omega_{\text{gas}} \equiv (1/N!) \prod [\int \int dq dp/h] \propto (V/N)^N (E/N)^{ND/2} .$$

The First Law, $\Delta E = \Delta Q - \Delta W$, combined with the ideal-gas equation of state gives the ideal-gas pressure and temperature in terms of the volume and energy derivatives of Ω_{gas}:

$$PV = (2E/D) = NkT \rightarrow (N/V) = (P/kT) = (\partial \ln \Omega / \partial V)_E ;$$

$$\Delta E = T\Delta S - P\Delta V \rightarrow (ND/2E) = (1/kT) = (\partial \ln \Omega / \partial E)_V .$$

We omit subscripts on Ω here because the last relations, which link the pressure and temperature to the phase volume Ω_{gas}, actually hold for *any* fluid. To see this, imagine coupling the ideal gas, weakly, to an arbitrary equilibrium fluid, such that the total volume and energy are fixed:

$$V_{\text{total}} \equiv V_{\text{gas}} + V_{\text{fluid}} ; \quad E_{\text{total}} \equiv E_{\text{gas}} + E_{\text{fluid}} .$$

Assuming ergodic mixing, the states available to such a combined, but weakly-coupled, system are given by the product, $\Omega_{\text{gas}} \times \Omega_{\text{fluid}}$. Thus
$$\ln \Omega_{\text{Total}} = \ln \Omega_{\text{gas}}(N, E, V)_{\text{gas}} + \ln \Omega_{\text{fluid}}(N, E, V)_{\text{fluid}} \ .$$
It is clear that *maximizing* $\ln \Omega_{\text{total}}$, so as to find the equilibrium conditions relating $\ln \Omega_{\text{gas}}$ to $\ln \Omega_{\text{fluid}}$, gives *equal* derivatives at the maximum:
$$(\partial \ln \Omega_{\text{gas}}/\partial V)_E = (\partial \ln \Omega_{\text{fluid}}/\partial V)_E \equiv (P_{\text{eq}}/kT_{\text{eq}}) \ .$$
$$(\partial \ln \Omega_{\text{gas}}/\partial E)_V = (\partial \ln \Omega_{\text{fluid}}/\partial E)_V \equiv (1/kT_{\text{eq}}) \ .$$
These two conditions on the partitioning of the volume and energy between gas and fluid correspond to the thermodynamic conditions of mechanical and thermal equilibrium. Their identification from the maximum condition $\delta \ln \Omega = 0$ requires the neglect of fluctuations in the neighborhood of the maximum. This neglect is abundantly justifiable through the Central Limit Theorem, which is related to the fluctuations in Section 3.8.

Because $k \ln \Omega$ is a state function, depending upon E and V and with the *same derivatives* with respect to these variables as the thermodynamic entropy, it is clear that, apart from a conventional additive constant the entropy is *exactly* $k \ln \Omega$. $k \ln \Omega$ has *exactly the same* properties as the thermodynamic entropy S: (i) it is a state function; (ii) it is additive, for weak, but mixing, coupling; (iii) it is a maximum at equilibrium; (iv) its isoenergetic volume derivative is (P/T); (v) its isochoric energy derivative is $(1/T)$.

The correspondence between the microscopic and macroscopic formulations of entropy is complete. The only lingering question, to which we will return repeatedly, is the compatibility of Liouville's Theorem with the description of *irreversible* processes. Remember the bit-reversible shockwave-expansion problem of Section 2.11.

The *canonical* ensemble, with an isothermal phase-space probability density $f \propto e^{[A-\mathcal{H}]/kT}$, gives an alternative, and computationally more useful, relation between dynamics and thermodynamics. It comes from the imposition of "weak coupling" linking a large thermostat, or heat reservoir, to a relatively smaller fluid subsystem with Hamiltonian $\mathcal{H}_{\text{fluid}}$. The thermostat and fluid together make up a two-part system described by Gibbs' microcanonical ensemble.

3.6 Pressure and Energy from Gibbs' Canonical Ensemble

From the microscopic point of view, the energy is not at all mysterious. It is natural to write it, as we have done continually, as the sum of kinetic and potential parts, $\mathcal{H} = K + \Phi$. Here we consider the simplest case, in which the N-body potential energy is a sum of pair terms,

$$\Phi(\{r_i\}) = \sum_{i<j} \phi_{ij}(|r_i - r_j|) \ .$$

For simplicity in what follows, we leave off the explicit dependence of the summed pair potentials $\sum \phi = \Phi$ on the separations of all pairs of interacting particles $\{i < j\}$. Gibbs' canonical partition function,

$$Z(N, V, T) \propto \int \int e^{-\mathcal{H}/kT} dq^{3N} dp^{3N} \ ,$$

can be explicitly differentiated with respect to volume and temperature. The resulting expressions match the derivatives of the macroscopic thermodynamic *Helmholtz'* free energy,

$$d(A/kT) = -(P/kT)dV - (E/kT^2)dT \rightarrow dA(N, V, T) = -d(kT \ln Z) \ .$$

Apart from the conventional additive constant the microscopic analog of Helmholtz' free energy is the logarithm of the microscopic canonical partition function Z, multiplied by $-kT$. The temperature derivative of $\ln Z$ gives the expected expression for the thermodynamic energy:

$$E = kT(\partial \ln Z/\partial \ln T)_V = \langle \Phi \rangle + \langle K \rangle \ ,$$

where the averages are carried out with a probability density proportional to $e^{-\mathcal{H}/kT} = e^{-\Phi/kT} e^{-K/kT}$. The volume derivative gives the "virial theorem" expression for the hydrostatic pressure in a fluid with D space dimensions:

$$PV = (1/D) \sum_N [\langle -r \cdot \nabla \Phi \rangle + \langle p \cdot p/m \rangle] \ .$$

The tensor form of PV, with the centered dots, angular brackets, and D removed, follows much more directly from a purely *mechanical* evaluation of the instantaneous momentum flux. The latter pressure expression, and its analog for the heat flux, are invaluable in microscopic simulations.

3.7 Gibbs' Entropy *versus* Boltzmann's Entropy

Until Section 3 of his 1884 paper, "Über die Eigenschaften monozyklischer und anderer damit verwandter Systeme", Boltzmann's gas-phase discussions of entropy all took place in single-molecule phase space, "μ space" $\{x, y, z, p_x, p_y, p_z\}$. In μ space Boltzmann considered cells large enough to contain several molecules—so that Stirling's approximation could be used for the corresponding factorials, $\{N_{\text{cell}}!\}$. The μ-space "Boltzmann entropy" is $-Nk\langle \ln f_1 \rangle$, where f_1 is the *one*-particle probability density. In his 1884 paper Boltzmann formulated both "Gibbs'" microcanonical and canonical ensembles, and described them in terms of the full *many*-body phase space, "γ space", rather than the single-molecule μ space. The manybody formulation leads naturally to "Gibbs' entropy",

$$S_{\text{Gibbs}} \equiv -k\langle \ln f_N \rangle \ .$$

It appears that the many-body *Gibbs'* entropy was actually the independent discovery of both men, though their disparate styles make it hard to be sure. Boltzmann left us a voluminous record documenting his changing views of kinetic theory. Gibbs left us very little trace of his own evolving ideas. We have only his Philadelphia lecture abstract from 1884, his book from 1902, and a handful of notes, mostly without dates, at the Yale University library. Gibbs certainly delivered a lecture at the American Association for the Advancement of Science meeting in Philadelphia in 1884: "On the Fundamental Formula of Statistical Mechanics with Applications to Astronomy and Thermodynamics". According to Klein this is the first appearance of the words "Statistical Mechanics" in print. The published abstract stresses the importance of the incompressible form of Liouville's Theorem, discussed in Section 2.5. Neither in this abstract nor in his book, which appeared eighteen years later, does Gibbs refer to Liouville. But it is quite clear that he had made the connection between the many-body phase-space density and entropy by 1884. He then worked out a detailed exposition over the next 18 years.

Gibbs' reticent style is nicely matched by Onsager's, as a visit to their graves near Yale reveals. Even though Gibbs sent nearly all of his published work to Boltzmann, Bill could find no record of their correspondence beyond a rather formal invitation from Boltzmann, which Gibbs declined, to attend an 1892 meeting in Nuremburg.

Distributions in the six-dimensional single-molecule μ space are evidently much simpler than those in the $6N$-dimensional many-body γ space. It is evident that the many-body phase volume available to the strongly interacting molecules in a condensed phase is greatly reduced by their short-ranged repulsive interactions. We can estimate this effect quantitatively by using the Mayers' theory to analyze van der Waals' picture of hard spheres. Because the centers of no two spheres can come more closely together than the sphere diameter σ, there is a minimum "closest-packing" volume per sphere, $\sigma^3/\sqrt{2}$, at which the many-body partition function must vanish.

The Mayers' virial expansion of the canonical partition function makes it possible to estimate the large-N limit of the phase volume for a dense fluid of N such spheres, at a temperature T and in a volume V:

$$e^{+\langle \mathcal{H}/kT \rangle} \int dq^{3N} \int dp^{3N} e^{-\mathcal{H}/kT}/N! \simeq$$

$$V^N[\,1 - b\rho + 0.1875(b\rho)^2 + 0.0502(b\rho)^3 + \ldots\,]^N (2\pi emkT)^{3N/2}/N! \,;$$

$$b\rho = 2\pi N\sigma^3/(3V)\,;\ e \equiv 2.718281828\,.$$

Though it is evident that Boltzmann correctly *formulated* N-body thermodynamic properties in his 1884 paper, it must be emphasized that his H Theorem explanation of irreversible processes is restricted to low-density gases. The H Theorem is an approximate demonstration—for isolated dilute gases—that the one-body "Boltzmann entropy",

$$S_B(t) \equiv -Nk\langle \ln f_1(r, v, t) \rangle\,,$$

cannot decrease with time. A convincing mechanical proof of the Second Law of Thermodynamics for condensed matter awaited the advances of Green and Kubo.

A dramatic example of apparent Boltzmann-entropy loss occurs during the *inelastic* cold-welding of metallic drops, following their joint collision. Prior to collision each drop can access the entire volume V. Afterward only the single composite particle can do so. A numerical simulation of this process in two dimensions appears in Figure 3.1.

To illustrate the important distinction between Gibbs' and Boltzmann's entropies, Prigogine, Kestemont, and Mareschal simulated an isolated system in which Boltzmann's entropy decreases while Gibbs' does not. They began with an expanded two-dimensional crystal of particles with a short-ranged *purely-repulsive* interaction $\phi(r < \sigma)$. They argued as follows: choose the nearest-neighbor spacing just *exceeding* the range of the potential σ, corresponding to the density ρ and select the initial particle velocities from the Maxwell-Boltzmann equilibrium distribution for the temperature T: $e^{-mv^2/2kT}$. The potential energy vanishes so that the Boltzmann entropy corresponds to that of an ideal gas, $S_{\text{ideal}}(\rho, T)$. The *motion* of the system, at fixed density and energy E, soon leads to a state with a *positive* potential energy $\langle \Phi \rangle \equiv \langle \sum \phi \rangle > 0$ and a correspondingly reduced kinetic energy, $K = E - \Phi$. The Boltzmann entropy then suffers a reduction, while Gibbs' entropy is unchanged.

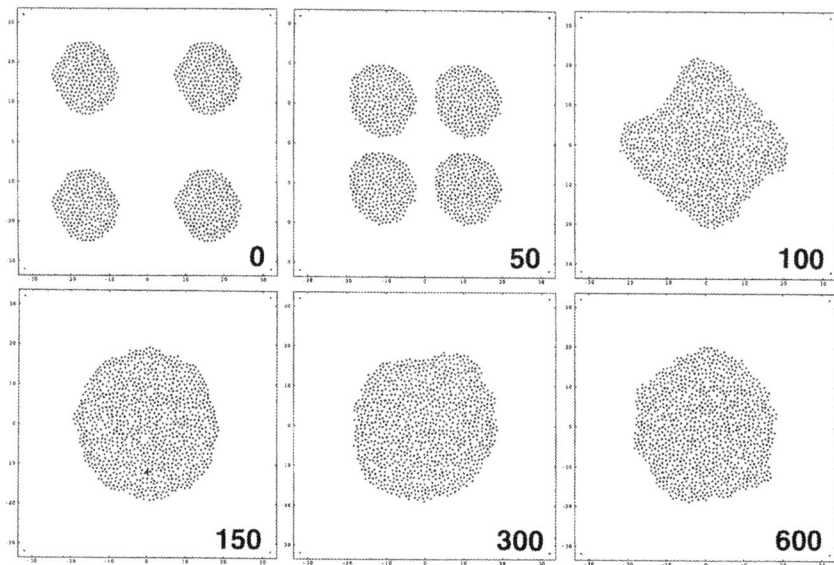

Figure 3.1. Snapshots from a simulation illustrating the cold welding of four colliding two-dimensional metallic drops. In units of the final-state sound traversal time the six snapshots correspond to times of $\{0, 1, 2, 3, 6, 12\}$.

To summarize, Gibbs' entropy is *the* equilibrium entropy, hard to calculate, but soundly founded in thermodynamic measurements. Boltzmann extended the entropy concept to gases far from equilibrium. Extending his ideas further, to dense fluids, is problematic. The fractal nonequilibrium phase-space distributions frustrate analytical approaches. Computer simulation is *the* straightforward approach to understanding nonequilibrium flows. This development needs no nonequilibrium entropy. Instead the focus has shifted toward defining and understanding local thermomechanical variables in situations like shockwaves, where the state variables undergo rapid change in times on the order of the collision time and in distances on the order of the interatomic spacing.

3.8 Number-Dependence and Thermodynamic Fluctuations

A real physical boundary necessarily introduces a host of surface contributions to the energy and the other extensive properties of an enclosed system. These surface contributions are of the order of the surface area $\propto N^{(D-1)/D}$ in D-dimensional systems. They can neatly be avoided by the use of *periodic* boundaries, for which the number-dependent contributions are considerably smaller, typically of order unity. This small systematic number-dependence for periodic systems can be thoroughly and systematically understood for simple models. This is the case for hard disks and spheres, for which the Mayers' cluster-integral expansion has been closely estimated through the first ten terms. See the examples given in Section 3.11.2.

Gibbs' statistical mechanics makes it relatively easy to calculate equilibrium fluctuations. The energy and enthalpy fluctuations, for example, are proportional to the isochoric and isobaric heat capacities:

$$\langle \delta E^2 \rangle = \langle (E - \langle E \rangle)^2 \rangle = C_V k T^2 \; ; \; \langle \delta H^2 \rangle = \langle (H - \langle H \rangle)^2 \rangle = C_P k T^2 \; .$$

The two heat capacities are both "extensive", proportional to N. Thus these examples illustrate the general rule that fluctuations of intensive quantities become negligible in sufficiently-large classical systems:

$$\langle (\delta E/N)^2 \rangle \simeq \langle (\delta H/N)^2 \rangle \propto (kT)^2/N \; .$$

Sufficiently-large means that the parts can be treated as independent, so that the instantaneous state of a simple large system corresponds to many—enough for the Central Limit Theorem to apply—independent small-system samples. The Central Limit Theorem guarantees that large-system equilibrium fluctuations have Gaussian distributions, and become negligibly small in amplitude, $\propto \sqrt{1/N}$, as the system size increases.

3.9 Green and Kubo's Linear-Response Theory

Linear-Response Theory is based on Gibbs' statistical mechanics. It provides a partial explanation of irreversible behavior. Gibbs' "canonical distribution":

$$f_{NVT}(q,p) \propto e^{-\mathcal{H}/kT} \; ,$$

makes it possible to treat ensemble-averaged flows of mass, momentum, and energy as small perturbations. From this point of view equilibrium, represented by a Gibbs ensemble, is characterized by opposing currents, which largely cancel. The residual currents in individual systems, averaged over a large volume with N particles, are of order $\sqrt{1/N}$. The most likely situation, corresponding to the ensemble average, is that the positive and negative contributions to the mass current, $\pm\langle\rho(|v_x|,|v_y|,|v_z|)\rangle/2$, precisely cancel. This delicate balance can be offset by any perturbation (such as a gradient in concentration, velocity, or temperature), giving rise to a net current proportional to the perturbation (the "linear" response).

The simplest illustration of the Green-Kubo theory is a shear flow in the xy plane, with the macroscopic laboratory-frame flow velocity in the x direction and proportional to the y coordinate:

$$\langle \dot{x} + (p_x/m)\rangle = \dot{\epsilon}y \ .$$

Formally, this flow can be generated by using a perturbed [Kewpie] "Doll's-Tensor" Hamiltonian which includes the strain rate $\dot{\epsilon}$:

$$\mathcal{H}_{\text{shear}}(q,p,\dot{\epsilon}) = \mathcal{H}_{\text{equilibrium}}(q,p) + \dot{\epsilon}\sum yp_x \ .$$

Periodic boundary conditions can be constructed in the usual way, provided that the adjacent images in the y direction are displaced horizontally:

$$\{x(y \pm L, t)\} = \{x(y,t) \pm \dot{\epsilon}Lt\} \ .$$

This periodic shear-flow example is worked out in detail in Section 5.8. See page 184. Green and Kubo's theory shows that the limiting small-strain-rate shear viscosity in this example is given by the *equilibrium* integral of the shear stress autocorrelation function (See Section 9.6):

$$\eta = (V/kT)\int_0^\infty \langle P_{xy}(0)P_{xy}(t)\rangle_{\text{eq}}\, dt \equiv (V/kT)\int_0^\infty \langle \sigma_{xy}(0)\sigma_{xy}(t)\rangle_{\text{eq}}\, dt \ .$$

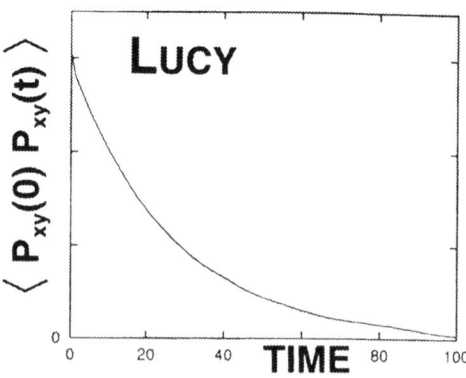

Figure 3.2. A typical dense-fluid Green-Kubo equilibrium shear stress correlation function. The integral gives the shear viscosity coefficient η.

The details of linear-response theory, formulated by Green and Kubo and Onsager, are in agreement with Boltzmann's kinetic theory for gases, but are applicable also to dense fluids and solids as well as to quantum systems. The simple gas-phase case provides a qualitative explanation of irreversible behavior: when the local equilibrium distribution varies, due to the gradients $\{\nabla \rho, \nabla u, \nabla T\}$, the resulting mass, momentum, and energy flows produce perturbations proportional to the products of these gradients and an effective distance between collisions (or "mean free path"). The time required for these perturbations to be established is the time between collisions, while the time required for complete equilibration is much longer, $\propto (L^2/D)$ for a system of size L with a diffusive transport coefficient D with units [meters^2/second]. Although these arguments are not at all rigorous, extensive comparisons of the results of linear-response theory with experiments and with direct nonequilibrium computer simulations confirm their validity.

3.10 An Algorithm for Local Smooth-Particle Averages

In Chapter 2 we formulated a bit-reversible algorithm and applied it to the shockwave/free-expansion problem. In order to compare the results of such atomistic simulations with the predictions of macroscopic continuum theories and corresponding continuum simulations it is necessary to formulate smoothly varying *local* averages. The simplest formulation of such averages results if we distribute each particle's properties smoothly in space according to a finite-range weight function $w(r < h)$. The *range of influence* of each particle on nearby spatial averages is h. Consider the mass and momentum densities, $\rho(r)$ and $\rho(r)u(r)$ as sums of nearby particle contributions. The summed-up contributions *at* the point r are simple sums over such particles:

$$\rho(r) = \sum_{i=1}^{N} m_i w(|r - r_i|) \; ;$$

$$\rho(r)u(r) = \sum_{i=1}^{N} m_i v_i w(|r - r_i|) \longrightarrow u(r) \equiv \frac{\sum m_i v_i w(|r - r_i|)}{\sum m_i w(|r - r_i|)} \; .$$

The weight function $w(r)$ is arbitrary. Typically it resembles a Gaussian distribution with a range sufficient to include about twenty particles. If the weight function has *two* continuous derivatives at its maximum range h then the continuum averages calculated using it will have *two continuous space derivatives*. The twofold differentiability gives averages which connect up with macroscopic phenomenological transport equations like the diffusion equation:

$$(\partial \rho / \partial t) \equiv D \nabla^2 \rho .$$

Such macroscopic evolution equations relate the time derivatives of the mass, momentum, and energy densities to second-order spatial derivatives of these same densities.

If we ask for the *simplest* polynomial weight function which has (i) a spatial integral of unity; (ii) a maximum at $r = 0$, and (iii) two vanishing derivatives at $r = h$, that weight function is Lucy's:

$$w_{1D}(x) = (5/4h)[1 - 6z^2 + 8z^3 - 3z^4] \text{ where } z \equiv (|x|/h) .$$

The integral of this one-dimensional weight function from $x = -h$ to $x = +h$ is unity. In two-dimensional problems Lucy's weight function is

$$z \equiv (r/h) \rightarrow w_{2D} = (5/\pi h^2)[1 - 6z^2 + 8z^3 - 3z^4] \rightarrow \int_0^h 2\pi r w_{2D} dr \equiv 1 .$$

An advantage of the "smooth-particle" averages is that they can be calculated on *any* convenient grid or at *any* point in space.

The following program illustrates the evaluation of the spatially-smoothed density $\rho(r)$ on a square grid, the central 25×25 quarter of which is occupied by points displaced randomly from regular square-lattice positions. This program uses the `rand(intx,inty)` subroutine, introduced in Chapter 2, page 63, to displace particles from the square-grid positions.

```
Call this weight.f, a program computing smoothed density.
implicit double precision (a-h,o-z)
parameter(nx = 100, ny = 100, N = nx*ny)
dimension x(N),y(N),xg(N),yg(N),rho(N)
index = 0
intx = 0
inty = 0
h = 4.0d00
do ix = 1,nx
do iy = 1,ny
index = index + 1
xg(index) = ix - ((nx+1)/2.0d00)
yg(index) = iy - ((ny+1)/2.0d00)
xo(index) = (xg(index) + rand(intx,inty) - 0.5d00)/2.0d00
```

```
yo(index) = (yg(index) + rand(intx,inty) - 0.5d00)/2.0d00
enddo
enddo
do i = 1,N
do j = 1,N
dx = xo(i) - xg(j)
dy = yo(i) - yg(j)
if(dx*dx + dy*dy.lt.h) then
r = dsqrt(dx*dx + dy*dy)
rho(j) = rho(j) + w(r,h)
endif
enddo
enddo
do i = 1,N
write(77,77) xg(i),yg(i),rho(i)
enddo
77 format(3f15.8)
stop
end


function w(r,h)
implicit double precision(a-h,o-z)
pi = 3.141592653589793d00
z = r/h
w = (5.0d00/(pi*h*h))*(1.0d00 + 3.0d00*z)*(1.0d00 - z)**3
return
end
```

The choice of smoothing length h typically requires judgment. Too small a value ($h = 2$ shown in Figure 3.3) results in overly wiggly contours. Too large a value (greater than $h = 8$) gives gradients which are unrealistically small.

102 *Time Reversibility, Computer Simulation, Algorithms, Chaos*

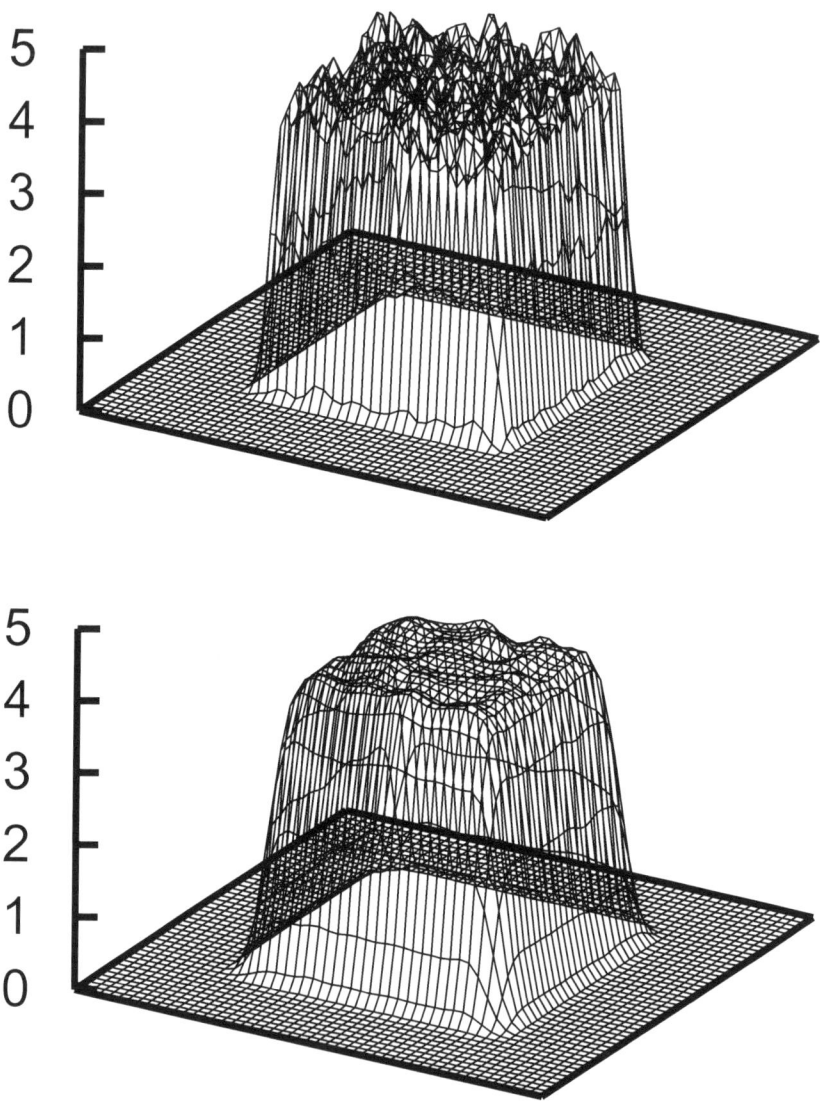

Figure 3.3. Particle density calculated with three different values of the range h: $h = 2$ (top), $h = 4$ (bottom), and $h = 8$ (on the adjacent page). In the central square particle displacements in the range ± 0.25 have been applied to a square lattice with a density of 4.

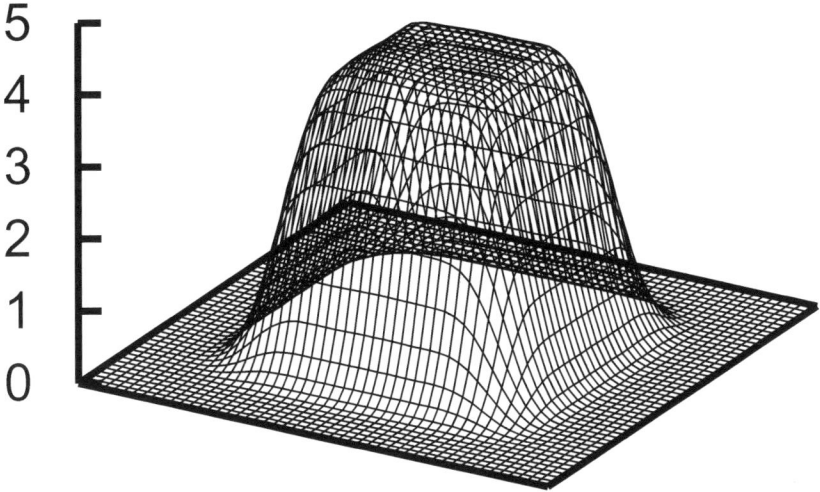

Figure 3.3. (*concluded*). Here $h = 8$.

3.11 Example Problems

The first two example systems described here are prototypes for the discussion of the equilibrium properties of solids and fluids. The "quasiharmonic" approximation to the properties of a crystalline solid, or a glass, replaces the potential part of the Hamiltonian by a (positive) quadratic form describing the interactions of particles through small-displacement force constants. Because such a positive quadratic Hamiltonian can be routinely diagonalized, to give a sum of "normal-mode" Hamiltonians, with a simple product partition function, $Z_{NVT} = \prod z_{\text{mode}}$, this approach to Gibbs' canonical partition function is straightforward.

Hard spheres, particles which cannot overlap, likewise provide a relatively simple approximation to Gibbs' partition function for dense fluids. This approximation proceeds by a systematic evaluation of the Mayers' two-body, three-body,..., corrections to the single-particle ideal-gas partition function. The third and final problem in this Section demonstrates the need for the smoothing algorithm just described in order to understand a classic irreversible flow process, the free expansion of a pressurized fluid into a larger container.

3.11.1 Quasiharmonic Thermodynamics

A static zero-temperature solid can be confined, with periodic boundary conditions, at a pressure given by the derivative of the potential energy with respect to volume,

$$P_0 \equiv -(d\Phi_0/dV) \ .$$

The thermal energy required to heat such a cold solid to a temperature T is of order NkT. The effect of this additional energy on the pressure can conveniently be described by the phenomenological Grüneisen equation of state:

$$P(T) - P_0 \equiv \gamma_{\text{Grüneisen}}(E - \Phi_0)/V \ .$$

Such a picture develops naturally if the energy is considered to be a sum of the minimum possible pair-term energy plus additional thermal contributions to the kinetic and potential energy, $DNkT/2$ each in D dimensions.

A classical crystal is completely motionless at zero temperature, with neighboring particles' relative motion always small compared to the interparticle spacing. Under these conditions a useful approximation for the energy is the "cold" curve, $E(T=0, V) = \Phi_0(V)$, with the *thermal* part of the energy added on. If the thermal motions are relatively small, the dependence of the full N-body potential energy on particle displacements away from the minimum-energy structure, $\{\delta\}$, can be truncated after the quadratic terms:

$$\Phi(\{r\}) = \Phi_0 + (1/2) \sum \delta_i \cdot \kappa_{ij} \cdot \delta_j \ .$$

Evidently the force constants $\{\kappa\}$ in this truncated potential give rise to forces which are *linear* in all the particle displacements, so that the whole problem reduces to a linear one, which can be solved by superposing N independent solutions. The motion governed by such a potential is a linear combination of normal-mode vibrations. The motion is conventionally termed "quasiharmonic", rather than "harmonic", so as to emphasize the fact that both the static energy Φ_0 and the set of force constants $\{\kappa_{ij}\}$ depend upon the volume at which the expansion is carried out. We illustrate the power of Gibbs' statistical mechanics by considering the simplest interesting example of this problem, the thermodynamics of a two-dimensional harmonic crystal at finite temperature.

Imagine a two-dimensional triangular lattice with periodic boundary conditions. For simplicity suppose that only nearest neighbors interact. The vibration of a single particle in this lattice, with all six of its neighbors fixed at their lattice sites, defines the one-particle "Einstein frequency" ω_{Einstein}. If the nearest-neighbor interaction is a purely-harmonic Hooke's-law interaction, with a fixed force constant κ, the Einstein frequency is proportional to $\sqrt{\kappa/m}$:

$$\omega_{\text{Einstein}} = 2\pi\nu_{\text{Einstein}} \equiv \sqrt{3\kappa/m} \; .$$

For the two-dimensional system the "Einstein approximation" to Gibbs' canonical partition function is:

$$Z_{\text{Einstein}} = e^{-\Phi_0/kT} \prod_{2N} z_{\text{Einstein}} = e^{-\Phi_0/kT}(kT/h\nu_{\text{Einstein}})^{2N} \; .$$

An *exact* solution of the small-vibration problem remains relatively simple when *all* the particles move simultaneously, provided that periodic boundaries are used. The zero-pressure solution (adjacent particles are separated, on the average, by the equilibrium separation, where the first derivative of the pair potential vanishes) was first worked out numerically, by evaluating the normal-mode frequency distributions of finite periodic N-particle crystals with $N = 2n^2$ for $2 \leq n \leq 20$. This numerical work made it possible to estimate the large-N "thermodynamic limit" corresponding to an infinite number of particles with five-digit accuracy.

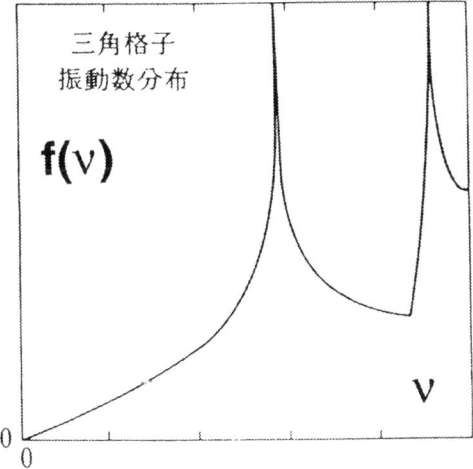

Figure 3.4. Vibrational frequency distribution for the stress-free two-dimensional triangular lattice with harmonic first-neighbor interactions.

Dale Huckaby carried out the *same* calculation analytically. The perfect agreement between the numerical calculation and Huckaby's analysis provided evidence that the usual practice of extrapolating small-system computer results to the large-system limit is valid. The result of this work showed also that the cooperative motion of the particles increased Gibbs' canonical partition function by a numerical factor of $e^{0.27326N}$ relative to the naïve prediction of the Einstein model:

$$Z_{\text{exact}}^{N \to \infty} = e^{0.27326N} e^{-\Phi_0/kT} (kT/h\nu_{\text{Einstein}})^{2N} .$$

The exact oscillation frequencies are typically volume-dependent, justifying the Grüneisen prescription for the pressure, which includes contributions proportional to $(d \ln \nu / d \ln V)$. Figure 3.4 shows the rather intricate form of the exact frequency distribution. For more details see the 1972 paper, "Number Dependence of Small-Crystal Thermodynamic Properties" by Wm. G. Hoover, A. C. Hindmarsh, and B. L. Holian.

3.11.2 *Hard-Disk and Hard-Sphere Thermodynamics*

Hard particles are relatively easy to treat using Gibbs' statistical mechanics. In one dimension the exact partition function can be calculated, just as it can be for a harmonic chain. In two and three dimensions no complete partition functions are available for more than ten particles. But corresponding series expansions *can* be obtained from the Mayers' virial theory. The results for hard disks and spheres in the fluid phase are relatively simple. The Mayers' theory, based on the ideal gas and the perturbation function:

$$f_{\text{Mayers}'} \equiv e^{-\phi/kT} - 1 ,$$

provides a formal density-series expansion for all the thermodynamic properties. For one-dimensional "hard rods" of length σ the *exact* equation of state is

$$PV/NkT = 1/(1 - b\rho) = 1 + (b\rho) + (b\rho)^2 + (b\rho)^3 + (b\rho)^4 + (b\rho)^5 + \dots .$$

The hard-rod second virial coefficient b is equal to the rod length σ.

For hard disks and spheres the first ten terms are accurately known. The "compressibility factor" PV/NkT depends only on density. For hard disks the series begins as follows:

$$PV/NkT = 1 + b\rho + 0.782(b\rho)^2 + 0.532(b\rho)^3 + 0.334(b\rho)^4 + 0.200(b\rho)^5 + \dots .$$

For hard spheres the series begins:

$PV/NkT = 1 + b\rho + 0.625(b\rho)^2 + 0.287(b\rho)^3 + 0.110(b\rho)^4 + 0.037(b\rho)^5 + \ldots$.

In all of these hard-particle expressions b is the "second virial coefficient", half the volume excluded by a stationary "particle" of *radius* σ:

$$b_{\text{rods}} = (2\sigma)/2 = \sigma \; ; \; b_{\text{disks}} = (\pi\sigma^2)/2 \; ; \; b_{\text{spheres}} = (4\pi\sigma^3/3)/2 \;.$$

Successive terms in the series incorporate the additional collisional pressure contributions involving two, three, four, ..., particles.

Figure 3.5 displays the convergence of the fluid-phase virial series for hard disks and hard spheres. Measured pressure results from computer simulations have established that the fluids freeze at about three-fourths the closest-packed density for disks and two-thirds the closest-packed density for spheres. The simulation values for the freezing pressures are indicated by the short horizontal "freezes" lines in the figure.

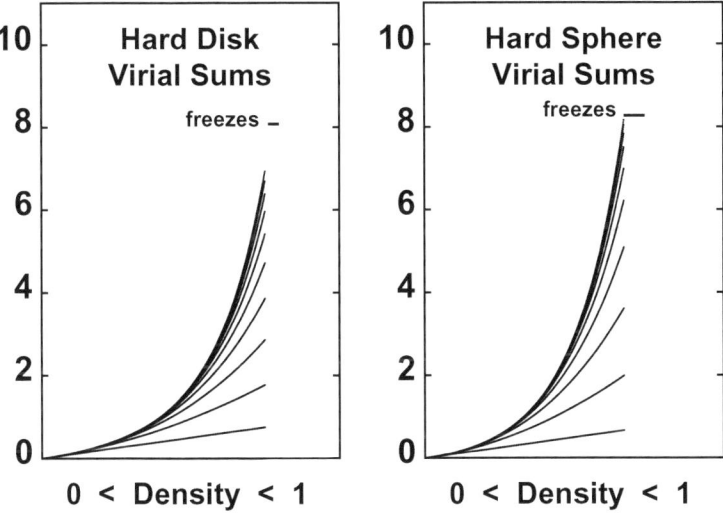

Figure 3.5. The dimensionless pressure PV_0/NkT, where V_0/N is the close-packed volume per particle, $\sqrt{3/4}\sigma^2$ for disks and $\sqrt{1/2}\sigma^3$ for spheres, increases from zero up to a freezing pressure of about 8, as the fluids are compressed. The ten curves in the figure show the summed-up contributions of the first ten virial coefficients to the pressure according to Mayers' virial series. The convergence to the freezing pressure obtained from the computer simulations is somewhat better for spheres than for disks.

It is to be emphasized that the alternative *direct* measurement of the time-averaged pressure, using molecular dynamics to evaluate the tensor form of the virial theorem:

$$PV = \langle \sum_{i<j}(Fr)_{ij} + \sum_i (pp/m)_i \rangle ,$$

is actually much more efficient, and accurate, than is the *tour de force* numerical evaluation of the Mayers' series expansion.

The theory for the hard-disk and hard-sphere *solid* phases is not so well developed as the Mayers' fluid theory—the problem is that the quasi-harmonic expansion of the previous example does not apply to the singular hard-particle interaction. Francis Ree and author Bill worked out the properties of the hard-sphere fluid and solid phases in 1968. They used computer simulations of both phases to locate the transition linking them precisely. The solid-phase pressures, like the fluid-phase pressures, could be evaluated from the virial theorem, applied to a periodic system with an imposed triangular lattice structure (disks) or with a face-centered-cubic solid structure (spheres). For details see the 1968 paper as well as the arχiv papers describing Clisby and McCoy's extensive work on the fluid virial series.

3.11.3   Time-Reversible Confined Free Expansion

With modern work stations many-body molecular dynamics simulations with up to 100,000 particles are today fully as feasible as were the few-body simulations of the 1950s. With parallel computers simulations can treat tens or hundreds of millions of particles. To illustrate a many-body molecular dynamics simulation, let us consider the prototypical macroscopic irreversible process, the confined free expansion of a pressurized many-body system. We begin with a classical two-dimensional dense fluid confined to an $L \times L$ square. The confining boundary is then released, allowing the particles to move within a larger $2L \times 2L$ square container. The equations of motion are simplest if periodic boundaries corresponding to the square container are used:

$$\{\dot{x} = (p_x/m) \ ; \ \dot{y} = (p_y/m) \ ; \ \dot{p}_x = F_x \ ; \ \dot{p}_y = F_y\} ,$$

with continual checks:

$$\{x < -L \rightarrow x = x + 2L \ ; \ x > L \rightarrow x = x - 2L\} ;$$

$$\{y < -L \rightarrow y = y + 2L \ ; \ y > L \rightarrow y = y - 2L\} .$$

With these periodic boundary conditions any particle leaving the confining $2L \times 2L$ square reënters on the opposite side. Equivalently, the motion can be regarded as occurring in an infinite checkerboard array of identical $2L \times 2L$ squares. In the simplest case the forces are negative gradients of a pairwise-additive potential. Figure 3.6 displays several snapshots in the time evolution of 16,384 particles interacting with Lucy's short-ranged purely-repulsive pair potential:

$$\phi_{\text{Lucy}}(r < 6) \equiv w(r < 6)_{2D} = (5/36\pi)[1 + (r/2)][1 - (r/6)]^3 \ .$$

This potential function reduces numerical integration errors because it has two continuous derivatives at the cutoff distance, $r = 6$. Because this potential is also a smooth-particle "weight function", as formulated and detailed in Section 3.10 and further elaborated in Section 7.6, it can be used to compute interpolated variables throughout space.

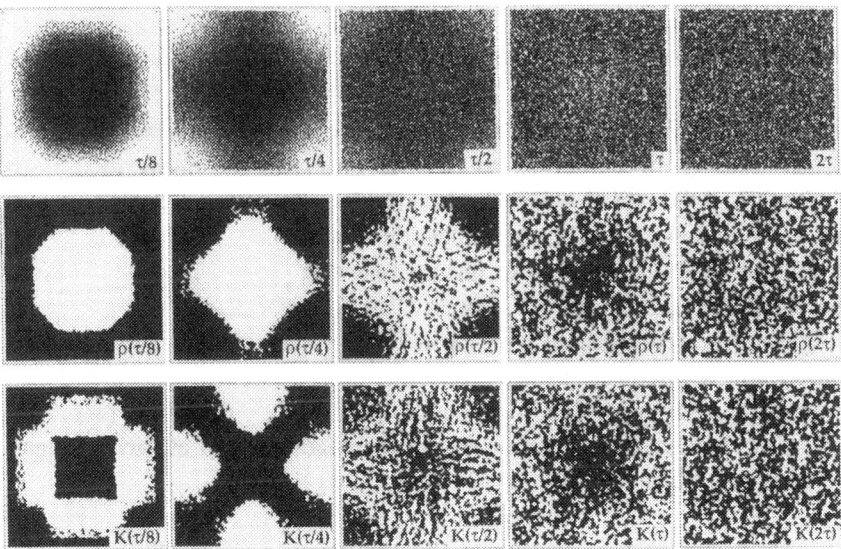

Figure 3.6. Snapshots showing the confined free expansion of 16,384 particles appear in the top row. Corresponding density and kinetic-energy contours, with black indicating below-average values and white above-average, are shown below. The maximum time shown is two sound traversal times. For additional details see the paper by Harald Posch and Bill, in the February 1999 issue of *Physical Review E* **59**, pages 1770–1776.

The many-body particle motions can be made fully reversible by using the *bit-reversible* algorithm of Section 2.11 to advance the coordinates [the simplest example is Levesque and Verlet's bit-reversible Leapfrog algorithm, described in detail as a Fortran implementation in Section 2.11]:

$$\{r_+ = 2r_0 - r_- + [F_0(\Delta t)^2/m]_{\text{integer}}\} .$$

A naïve hydrodynamic analysis of this problem suggests that either viscosity or heat conductivity could dissipate the expansion energy into heat in a time of order (L^2/D), where D is the kinematic viscosity (η/ρ) or thermal diffusivity $(\kappa m/\rho C)$.

Both diffusivities have units [meters^2/second]. If the entropy of the equilibrating system is estimated by summing up local entropies based on local temperatures (velocity fluctuations) and densities the results show instead that nearly all of the thermodynamic entropy increase,

$$\Delta S = Nk \ln[V_{\text{final}}/V_{\text{initial}}] = Nk \ln[4L^2/L^2] = Nk \ln 4 ,$$

actually occurs in a *much* shorter time, less than a single sound-traversal time. Evidently linear hydrodynamics greatly overestimates the decay time. See Figures 3.6 and 3.7. Although the periodic boundaries allow the particles to interpenetrate, simulations of the equilibration of steep sinusoidal density distributions can lead to very similar results, and without any interpenetration. See the 2000 paper by Castillo, Posch, and ourselves.

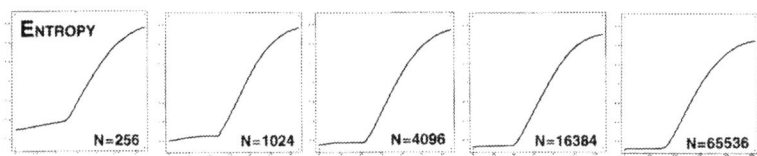

Figure 3.7. Increase of entropy with time for confined free expansions using from 256 to 65,536 particles. The abscissæ correspond to one half sound traversal time. The ordinates correspond to entropy increases of $Nk \ln 4$.

The entropy calculations were carried out by first evaluating a coarse-grained density and temperature at the location of each particle in the system and then summing the corresponding local-equilibrium entropy densities, $\{s(\rho, T)\}$. Unless the *local* velocity fluctuations, in the neighborhood of each particle, are correctly taken into account, using smooth-particle averaging, with

$$2kT \equiv m\langle(v - \langle v \rangle)^2\rangle ,$$

there is no entropy increase at all. Accounting for the fluctuations—missing in the usual hydrodynamic treatment—by measuring both $\langle v \rangle$ and $\langle v^2 \rangle$ separately, the entropy increase agrees well with that calculated using Gibbs' statistical mechanics, $Nk \ln 4$ for N particles.

3.12 Summary

Gibbs' statistical mechanics provides a general phase-space representation for the equilibrium thermodynamic entropy:
$$S = S_{\text{Gibbs}} \equiv -k \langle \ln f_N \rangle \ .$$
It appears that Boltzmann, who invented the one-body "Boltzmann entropy", much earlier,
$$S_{\text{Boltzmann}} \equiv -k \langle \ln f_1 \rangle \ ,$$
arrived at an *identical* formulation of the N-particle "Gibbs'" entropy *independently*, and at about the same time as Gibbs. From the N-body entropy, together with the thermodynamic statement of the First and Second Laws for *reversible* processes, $TdS = dE + PdV$, all of the other thermodynamic quantities can be obtained. Computer simulations—first for hard spheres, but later for a host of different materials—have demonstrated the simplicity and usefulness of this approach. Gibbs' ensemble theory also provides an instructive formulation of fluctuations about the most-likely state, showing that the thermodynamic effects of fluctuations vanish as system size increases, in accord with the Central Limit Theorem.

Gibbs' approach made possible Green and Kubo's exact formulation of the linear response of equilibrium systems to nonequilibrium perturbations. This formulation provides a basis for the algorithms used in *nonequilibrium* simulations, together with an understanding of the link between reversible dynamics and ensemble-averaged dissipative processes. Although Gibbs was careful to avoid a precise description of the coupling required to achieve thermal equilibrium, time-reversible thermostats, providing deterministic couplings, came into common use with the development of fast computers.

3.12.1 Notes and References

Gibbs' *Elementary Principles in Statistical Mechanics* is the basis for the work described here. The Kolmogorov-Arnold-Moser Theorem and the Fermi-Pasta-Ulam problem describe anomalous and unphysical special cases for which Gibbs' methods fail. See the 1972 Tuck-Menzel paper "The Superperiod of the Nonlinear Weighted String (Fermi-Pasta-Ulam) Problem" for some of the details.

The problem of thermostating insufficiently ergodic systems so as to obey Gibbs' *Principles* is described by Kusnezov, Bulgac, and Bauer in "Canonical Ensembles from Chaos", 1990. Another famous approach, valid only for equilibrium systems, is Metropolis, Rosenbluth, Rosenbluth, Teller, and Teller's "Monte Carlo" method, which appears in their 1953 paper, "Equation of State Calculations by Fast Computing Machines". Work at the Los Alamos and Livermore weapons laboratories was continued by Alder, Wainwright, and Wood. See, for instance, the side-by-side 1957 papers "Preliminary Results from a Recalculation of the Monte Carlo Equation of State of Hard Spheres", by Wood and Jacobsen, and Alder and Wainwright's "Phase Transition for a Hard Sphere System".

Vineyard's somewhat more "realistic" work, treating copper rather than hard spheres, is illustrated on the cover of the August 1959 Journal of Applied Physics.

Lucy and Monaghan's "smooth-particle" technique for computing twice-differentiable averages is a particularly useful tool for interpreting atomistic simulations. See Bill's 2006 book, *Smooth Particle Applied Mechanics; the State of the Art*, for more details.

For references to the two-dimensional lattice dynamics of the triangular lattice and Huckaby's analytic work, see Bill's 1982 paper with Ladd, Friesen, and Moran, "Analytic and Numerical Surface Dynamics of the Triangular Lattice". Two of Clisby and McCoy's arχiv papers, "Ninth and Tenth Order Virial Coefficients for Hard Spheres in D Dimensions" (2005) and "New Results for Virial Coefficients of Hard Spheres in D Dimensions" (2004) bring the Hoover-Ree seven-term virial series for disks and spheres up to date.

The free expansion problem is described in our 2000 work with Castillo and Posch, "Computer Simulation of Irreversible Expansions *via* Molecular Dynamics, Smooth Particle Applied Mechanics, Eulerian and Lagrangian Continuum Mechanics."

Chapter 4
Irreversibility in Real Life

Oh come with old Khayyám, and leave the wise
to talk; one thing is certain, that life flies;
One thing is certain, and the rest is lies; the
flower that once has blown forever dies.

Omar Khayyám, as translated by Edward FitzGerald

4.1 Introduction

"Irreversible" means "impossible to annul or to run backward". And life, physics, and nature are all like that. Velocity and temperature differences tend to decay whenever they substantially exceed the equilibrium thermal fluctuations which follow from Gibbs' theory. A simple isolated system—as represented by an isoenergetic numerical simulation of a typical textbook mechanics problem—cannot exhibit this type of irreversibility. The fact that simple isolated systems can just as well run backward as forward is the gist of Loschmidt's *reversibility* objection to Boltzmann's H Theorem. The additional fact that such systems must eventually recur—establishing cyclic rather than monotonic behavior—was Zermélo's *recurrence* objection, also mentioned in Chapter 2. Computer simulations of irreversible behavior, of the kind seen in real life, require either (i) so many degrees of freedom that the decay of the initial state persists for a while, allowing an accurate transient analysis, or (ii) thermal links to the outside world sufficient to lose the extraneous information generated by nonequilibrium decay processes.

Lost information becomes heat when it is transferred to the outside world. When "information" is used in a technical sense, it refers to the precision—in binary bits—with which a calculation is specified. The same information concept can also be used to quantify the precision of experimental measurements. Without *complete* knowledge of the dynamical state the dynamics cannot be *exactly* reversed. *Complete* knowledge is only possible in a space with finite information content, such as the integer-valued coordinate space inhabited by Levesque and Verlet's bit-reversible trajectories. For an "open" system, coupled to the external world, and not isolated, there can be a time-symmetry breaking making a time reversal *impossible in principle*. This is a consequence of the relative stability of processes which destroy, rather than create, information.

In the real world, apart from equilibrium microscopic fluctuations like Brownian motion, heat invariably flows from hot to cold. This is one way, attributed to Clausius, of stating the Second Law of Thermodynamics. Similarly, macroscopic differences in chemical potential or velocity, unless stabilized by external fields or forces, invariably disappear. This inexorable degradation of gradients is quantified by the entropy increase of the macroscopic Second Law. It has nothing to do with the initial conditions, computer algorithms, or the microscopic details describing the past history of the system. It is a simple phenomenological description of inevitable decay, based on observation.

Despite all this natural irreversibility, the fundamental conservation equations describing real continuum flows are neutral regarding reversibility. The *continuity equation* is the basis of this neutrality. It is the partial differential equation expressing the flow of a conserved quantity in the presence of spatial gradients. The usual "conserved quantities" are mass, momentum, and energy. To illustrate the continuity equation, let us apply it first to conservation of mass. The derivation is simplest in a fixed "Eulerian" frame. In this frame the time-rate-of-change of a smoothly-continuous mass density at a fixed location, $(\partial \rho/\partial t)$, must exactly balance the net flow of mass toward that location:

$$(\partial \rho/\partial t) = -\nabla \cdot (\rho u) \ .$$

This is the *Eulerian* form of the continuity equation. An alternative description, in the comoving *Lagrangian* frame, gives an equivalent form of

the differential equation for $\dot{\rho}$, the "Lagrangian" time derivative of density following the flow:
$$\dot{\rho} \equiv (\partial \rho/\partial t) + u \cdot (\nabla \rho) = -\rho \nabla \cdot u \longleftrightarrow (\partial \rho/\partial t) = -\nabla \cdot (\rho u) \;.$$
A straightforward finite-difference algorithm for solving the Eulerian continuum equations is described and applied at the end of this Chapter, with more details. A novel Lagrangian "smooth-particle" method, which models a continuum with *particles* is detailed in Chapter 7, page 251. Evidently reversing time in either version of the continuity equation simply changes the sign of the time derivatives, matching the sign changes of the velocity on the righthand sides of the two versions of the continuity equation. Either version is precisely time-reversible.

The Eulerian and Lagrangian "equations of motion" are derived in a similar way. The equations of motion describe the conservation of momentum. Equate the rate of change of (ρu), the momentum density at a fixed location, to the negative divergence of the corresponding flux. The comoving Lagrangian momentum flux defines (and *is*) the pressure tensor P. The Eulerian momentum flux contains the additional convective contribution, (ρuu). Both P and ρuu are symmetric tensors. The resulting *equations of motion* are:
$$\partial(\rho u)/\partial t = -\nabla \cdot (P + \rho uu) \longleftrightarrow \rho \dot{u} = \rho[(\partial u/\partial t) + u \cdot \nabla u] = -\nabla \cdot P \;.$$
Here the lefthand sides are even functions of time, as can also be the right, provided that the pressure tensor is an even function of the fluid velocity $u(r,t)$. Evidently the time-reversibility of the equations of motion depends solely on the time-reversal properties of the pressure tensor.

The Eulerian and Lagrangian "energy equations" follow similarly, by expressing the energy change in terms of (i) work done and (ii) heat transferred. In the comoving frame the energy equation has the simple form
$$\rho \dot{e} = -\nabla u : P - \nabla \cdot Q \;,$$
where Q is the heat-flux vector—the *conductive* flow of energy in the comoving frame, per unit area and time. The Lagrangian energy equation is the dynamical analog of the First Law of Thermodynamics. In the fixed Eulerian frame, the convective contributions to the energy flux lead to a more complicated, but fully equivalent, Eulerian energy equation:
$$\partial(\rho[e + \tfrac{u^2}{2}])/\partial t = -(\nabla u : P) - \nabla \cdot (\rho u[e + \tfrac{u^2}{2}] + Q) \;.$$
Here the sum $(\nabla u : P)$ includes all terms of the form $(\nabla u)_{ij} P_{ij}$, four of them in two dimensions, and nine of them in three.

In microscopic molecular dynamics, the atomistic equations of motion are generally time-reversible. The instantaneous microscopic pressure tensor, discussed in Chapter 3, is an even function of the particle velocities while the heat flux vector is *odd*. The situation in continuum mechanics is typically very different. There, both nonequilibrium fluxes are generally *irreversible*. The dissipative part of the Newtonian pressure tensor is an *odd* function of velocity and Fourier's heat-flux vector is even. We discuss the linear form of these macroscopic irreversible laws, together with their theoretical bases in Gibbs' theory, in the next two Sections.

4.2 Phenomenology — the Linear Dissipative Laws

In the absence of dissipative constitutive relations, most equations of state would soon lead to unstable continuum behavior. The increase of wave velocity with density eventually leads to catastrophic discontinuous shockwaves unless some offsetting smoothing effect, such as viscosity, heat conductivity, or plasticity, is included. The simplest stable continuum model is a "Newtonian" (linearly viscous) fluid, one exhibiting dissipative (irreversible) behavior.

Newtonian fluids have shear stresses which are *odd* functions of the stream velocity u. The Newtonian pressure tensor for a two-dimensional or three-dimensional continuum fluid is:

$$P_{\text{Newton}} = I[P_{\text{eq}} - \lambda \nabla \cdot u] - \eta[(\nabla u) + (\nabla u)^t]; \; \lambda = \eta_V - (2\eta/D) \; .$$

I is the "unit tensor", with diagonal elements of unity and off-diagonal elements of zero. D is the dimensionality of the system, either 2 or 3. $(\nabla u)^t$ is the "transpose" of (∇u) with

$$(\nabla u)^t_{ij} = (\nabla u)_{ji} \; ,$$

and is included in order to guarantee the symmetry of the pressure tensor, $P_{ij} \equiv P_{ji}$, without which the motion equations are unstable.

The equilibrium pressure P_{eq}, as well as the bulk and shear viscosity coefficients, $\{\eta_V, \eta\}$, are state functions which are independent of the past history of the system and the direction of time. Newtonian viscosity enhances stability by spreading shockwaves over a width of approximately one fluid mean free path. From a macroscopic point of view this width is negligibly small. For this reason a much larger "artificial viscosity", spreading shockwaves over the small-scale numerical resolution length, has to be used to stabilize macroscopic simulations against shockwaves.

In addition to problems involving sound waves and shockwaves, continuum simulations can treat mass and energy flows. These are relatively slow flows, governed by the stabilizing processes of mass and heat diffusion. The diffusive flow of mass can be described by Fick's Law and the flow of heat by Fourier's Law. Fourier's Law *defines* the heat conductivity, κ, in terms of the comoving heat current Q responding to a temperature gradient:

$$Q = -\kappa \nabla T \ .$$

The three phenomenological linear diffusive laws, Fick's, Newton's, and Fourier's, are among the oldest quantitative descriptions of material behavior. All of them provide irreversible solutions to macroscopic initial-value problems. A major accomplishment of Maxwell and Boltzmann's kinetic theory was the demonstration that the *same* linear diffusive laws, found phenomenologically, follow also from an accurate theoretical analysis of low-density gas dynamics. Einstein followed this up with a careful analysis of Brownian Motion, relating two phenomenological irreversible coefficients, the fluid's shear viscosity, and the Brownian particle's diffusion coefficient.

4.3 Microscopic Basis of the Irreversible Linear Laws

Half a century passed before Green and Kubo made a major advance beyond Maxwell and Boltzmann's gas theory. They related the macroscopic irreversible diffusive transport processes to the decay of the microscopic equilibrium fluctuations described by Gibbs' ensemble theory. Green and Kubo evaluated the effect of perturbations to Gibbs' equilibrium ensembles. The perturbations were specially chosen to drive corresponding nonequilibrium mass, momentum, and energy currents. In the linear-response regime, where the effects are proportional to the perturbations, the phenomenological linear diffusive laws of Fick, Newton, and Fourier result. Because Green and Kubo's approach is not at all restricted to gases, but applies equally well to liquids and solids, it provides a *general* understanding of the macroscopic phenomenological laws. These represent, on the one hand, the dissipation found in nonequilibrium steady flows, and on the other, the rate at which equilibrium fluctuations decay. This understanding provides an effective response to Krylov's criticism of Gibbs' equilibrium theory as a basis for treating irreversibility.

Linear-response theory showed that there is no special new idea that is required in order to treat linear macroscopic irreversibility. It was only necessary to develop computer algorithms with appropriate perturbations, or boundary conditions. Such algorithms were developed for use in nonequilibrium molecular dynamics simulations in the 1970s. All three linear laws were reproduced in this way. The transport coefficients could all be expressed as the integrated decays of corresponding equilibrium fluctuations. In the 1970s steady-state thermostats provided an alternative representation of the *same* coefficients, obtained from *energy balance*, in the presence of steady fluxes responding to appropriate driving forces. Despite all this conceptual progress, the early computer simulations, designed to evaluate the microscopic Green-Kubo fluctuation formulas, were disappointing and confusing, containing as they did independent factor-of-two errors in the heat conductivity and bulk viscosity. Dynamical steady-state simulations were carried out soon afterward, and turned out to agree fairly well with the corrected Green-Kubo results.

In 1984 our understanding of nonequilibrium systems was enriched by a further conceptual advance. New thermomechanical motion equations, incorporating fully deterministic and time-reversible "ergostats", "thermostats", and "barostats", were developed. These new motion equations were capable of controlling a wide variety of mechanically driven and thermally-driven nonequilibrium systems. The additional thermostat forces, which were always of the form $\{-\zeta p\}$, came from many disparate sources: Gauss' (Least-Constraint) Principle, Hamilton's (Least-Action) Principle, the Green-Kubo relation for heat flow, and the Nosé and Dettmann Hamiltonians. The main ideas are described in Section 2.8. The older irreversible Langevin approach to temperature control was described in Section 2.9. The new thermostat forces made it possible to simulate transport processes accurately in relatively small systems, with just a few hundred particles.

Now, two generations later, and with our greatly increased computational capability, it is certainly feasible to simulate transport processes in a more "realistic", or at least conventional, way. We *could* simulate a very large isolated Newtonian system incorporating large gradients in its initial conditions. By studying the decay of these gradients we could find the corresponding transport coefficients, including both nonlinear effects and number dependence. "Escape-rate theory", a recently-developed theoretical approach, is a variation of this idea. Escape-rate theory analyzes the decays of equilibrium ensembles, from which mass, momentum, and energy are allowed to escape through an open system boundary. But, given the relative simplicity of steady-state simulations, there is no convincing reason to characterize transport processes through more cumbersome studies of transient decay.

4.4 Solving the Linear Macroscopic Equations

The "equation of motion" for a continuum is just a macroscopic form of Newton's Second Law of Motion. It gives the local acceleration at a point, $\dot{u}(r,t)$, in terms of the local density ρ and the divergence of the pressure tensor (P is the negative of the "stress tensor" σ):

$$\rho\dot{u} = -\nabla \cdot P = \nabla \cdot \sigma \ .$$

Heat flow makes no explicit contribution to the motion equation, though in general the pressure tensor depends on the local temperature. The equation of motion has two linear limits suited to theoretical analysis. Weak long-wavelength disturbances involve negligible heat transfer and viscous dissipation. They are adiabatic and nearly reversible. A systematic expansion of the equation of motion, combined with the continuity equation, $\dot{\rho} = -\rho\nabla \cdot u$, leads to the time-reversible "wave equation":

$$(\partial^2 u/\partial^2 t) \simeq (-1/\rho)(\partial/\partial t)(\nabla \cdot P) \simeq -(\partial P/\partial \rho)_S (1/\rho)(\partial \nabla \rho/\partial t) \longrightarrow$$

$$(\partial^2 u/\partial t^2)_r = c^2 \nabla^2 u; \ c^2 = (\partial P/\partial \rho)_S \ .$$

A natural description of the solution is in terms of lossless density and pressure waves, "sound waves", which propagate at the *sound* velocity c.

If we ignore the density changes leading to such wave propagation, and assume instead an incompressible shear flow with constant shear viscosity, η, constant density, and no velocity divergence, the residual motion, with velocities much less than c, is governed by the *shear* viscosity:

$$\rho\dot{u} = \rho[(\partial u/\partial t) + u \cdot \nabla u] = \eta \nabla^2 u; \ \nabla \cdot u \equiv 0 \ .$$

The nonlinear term $\rho u \cdot \nabla u$ does not necessarily vanish in the incompressible case. In a plane incompressible shear flow, with $u_x \propto y$, $u \cdot \nabla u$ has an x component proportional to the gradient of the "Reynolds' stress" $\rho u_x u_y$. If the nonlinear terms are ignored, "laminar flow" results, governed by a time-irreversible diffusion equation, where the role of diffusion coefficient is played by the "kinematic viscosity", $\nu \equiv (\eta/\rho)$:

$$(\partial u/\partial t) = \nu \nabla^2 u \equiv (\eta/\rho)\nabla^2 u \ .$$

The lefthand side is *even* and the righthand side is *odd* on time reversal. Similar simplifying assumptions for the "energy equation"—negligible velocity and density gradients, with constant heat capacity C and constant heat conductivity κ, lead to another version of the same diffusion equation, but for temperature rather than velocity:

$$(\partial T/\partial t)_r = D_T \nabla^2 T = (m\kappa/\rho C)\nabla^2 T \ .$$

Here the thermal diffusivity $D_T \equiv (m\kappa/\rho C)$ plays the diffusion-coefficient role. This diffusion equation is also intrinsically *time-irreversible*, with the lefthand side *odd*, and the righthand side *even*, in time.

There is no special difficulty in solving diffusion equations for increasing time. Either a truncated Fourier analysis or a grid-based numerical approach can be used. But any attempt to work *backward* in time, to find the past history of a system obeying the diffusion equation, is doomed to failure by a short-wavelength instability. To see this, consider the behavior of a single Fourier component of the temperature, satisfying the diffusion equation. If the wavelength is $\lambda = (2\pi/k)$, the corresponding temperature perturbation decays in a time of order λ^2:

$$\delta T(k) \propto \sin(kx) e^{-t/\tau} \longrightarrow \tau = (\lambda/2\pi)^2/D_T = 1/(k^2 D_T) \ .$$

The decay time τ approaches zero for small wavelengths. The *time-reversed* diffusion equation, which would, if it made sense, determine the past from the present state, exhibits a corresponding exponentially-unstable *divergence* for *any* perturbation:

$$(\partial \delta T/\partial t) = -D_T \nabla^2 \delta T \longrightarrow \delta T(k) \propto \sin(kx) e^{+t/\tau} \ .$$

Thus the diffusion equation cannot be followed *backward* in time without artificially suppressing the short-wavelength components with their unbounded growth rates.

4.5 Nonequilibrium Entropy Changes

Because entropy furnishes the version of equilibrium thermodynamics closest to Gibbs' statistical mechanics, there is a strong motivation to generalize the concept to nonequilibrium systems, as did Onsager in his own linear-response studies. But there is no straightforward way to make the generalization. Changes in the equilibrium entropy, a state function, can only be characterized for processes linking two or more equilibrium states together. Bridgman pointed out that the thermodynamic entropy is undefined for all but the simplest homogeneous fluids or perfect solids. Systems with crystal defects or long-lived chemical complexes cannot generally be created by performing thermodynamically-reversible operations on simple equilibrium states. Bridgman emphasized a subject which he knew well, the irreversible plastic flow of metals. His conclusion was this:

> *The final abiding place of the entropy of irreversibility is in the heat reservoir surrounding the working body ... .*

A *global* form of the Second Law of Thermodynamics states that the total entropy increases with time. Two prototypical systems for the study of steady nonequilibrium flows are shown in Figure 4.1. The steady homogeneous shear of a Newtonian viscous fluid, with strain rate $\dot{\epsilon} = (du_x/dy)$ and shear viscosity η, requires a stress, $\sigma_{xy} = \eta\dot{\epsilon}$ to sustain the flow. The rate at which work is done by this viscous force is

$$\sigma\dot{\epsilon}V = (\eta\dot{\epsilon}^2)V = (\sigma^2/\eta)V \ .$$

Dividing by the temperature gives the corresponding rate of entropy increase of the system. It is significant that the entropy production is *quadratic* in the deviation from equilibrium. If it is assumed that the flow is steady, with heat flowing to the boundary accounting for the work done, then the surrounding heat reservoirs incorporate *all* of the entropy increase, just as Bridgman suggested.

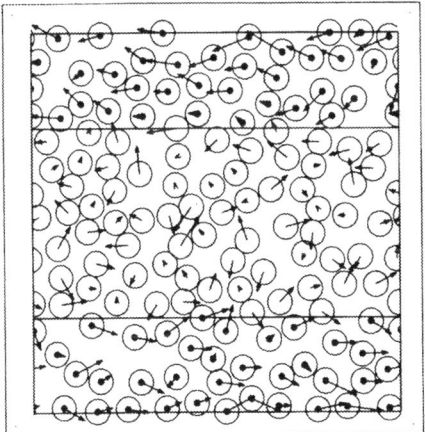 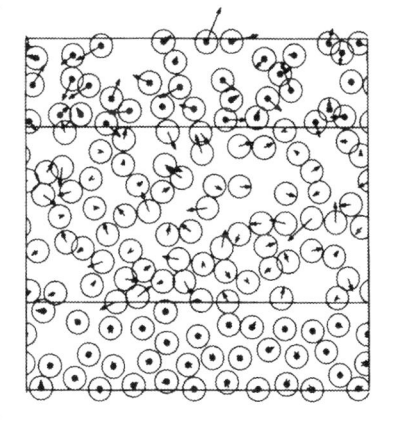

Figure 4.1. Prototypical simulations of two-dimensional shear flow (left) and heat flow (right) driven by two nonequilibrium reservoirs.

In a steady heat flow, with a heat flux Q through a cross-section A between two heat reservoirs separated by a distance L, the total rate of

entropy increase *of the reservoirs* is:
$$(QA/T_{\text{cold}}) - (QA/T_{\text{hot}}) \simeq \kappa V |\nabla \ln T|^2 ,$$
where the total system volume V is AL. Again the entropy production is quadratic in the deviation from equilibrium. If we attempt to apply a similar global reasoning to the *system* entropy, evaluating its change due to heat transfers *from* the hot-boundary heat source and *to* the cold-boundary sink, the result would be a corresponding *decrease* of the system entropy by the same amount, $\dot{S} \simeq -\kappa V |\nabla \ln T|^2$. This decrease is traditionally offset by *fiat*, attributing an additional "entropy production" to the conducting fluid, to ensure that its steady state shows no entropy change whatever. The same general idea can also be applied to shear flow, where the system entropy would likewise *seem* to decrease, due to the loss of dissipated shear work, in the form of heat, at the upper and lower boundaries.

To offset entropy decreases due to both these effects, shear stress and heat flux, the compensating density of "entropy production" in a fluid near equilibrium, with Newtonian shear viscosity η and Fourier heat conductivity κ, is set equal to:
$$\dot{s} \equiv \dot{S}_{\text{prod}}/V = (\eta/T)\dot{\epsilon}^2 + \kappa |\nabla \ln T|^2 .$$
The entropy production needed to keep the steady-state system entropy constant is quadratic in the velocity and temperature gradients, for small deviations from equilibrium. It becomes substantial at strain rates comparable to particle collision frequencies and for energy gradients comparable to the interparticle forces.

How are these prototypical entropy-producing flows related to the Second Law of Thermodynamics? For steady heat flow the *first* of the four informal Second Law statements listed in Section 2.7—"Heat cannot flow from a colder body to a hotter one"—implies that both the conductivity κ, and its contribution to the entropy production must be positive. The latter result is in accord with the *second* statement—"entropy must increase"—as well as the last: $\oint (dQ/T) < 0$. For steady shear flow this last statement implies that heat is produced *within* the sheared system, and flows out. The positivity of the shear viscosity η and its contribution to the entropy production is also required by the *third* statement: "no *cyclic* process can convert heat entirely to work". The "entropy production", within the conducting or shearing fluid, although consistent with the Second Law of Thermodynamics, has no operational significance. Neither entropy, nor its internal production rate, can be measured, for nonequilibrium states. Bridgman and Occam would both suggest we shun nonequilibrium entropies.

4.6 Fluctuations and Nonequilibrium States

In macroscopic continuum mechanics, a nonequilibrium state is easy to recognize. There are gradients, $\{\nabla u, \nabla T\}$ which cause corresponding nonequilibrium fluxes in the stress tensor and heat flux vector. From the microscopic view these nonequilibrium state variables $\{\nabla u, \nabla T, \sigma, Q\}$ all correspond to averages—space averages, time averages, or ensemble averages. To characterize the "nonequilibrium" character of a particular microscopic N-body state is not always so clearcut. It requires significant deviations, over and above the relatively large mechanical and thermal fluctuations $\propto \sqrt{1/N}$. In the laboratory, where the number of particles is of order 10^{24}, sensible nonequilibrium deviations can be many orders of magnitude smaller than those in computer simulations.

Probabilities for equilibrium fluctuations can be calculated on the basis of Gibbs' ensemble theory. At equilibrium the mean values of all the small-system fluctuations, $\{\Delta E, \Delta V, \Delta P, \ldots\}$ vanish, just as do all the *first* derivatives of the entropy with respect to these fluctuations. Because the *second* derivatives vary as $1/N$, macroscopic reservoir contributions are negligible compared to those of a microscopic small system. Straightforward thermodynamic manipulations of the second derivatives lead to the small-system probability density for deviations from equilibrium:

$$(f/f_{\text{eq}}) = e^{-(\Delta P)^2/[2kT(\partial P/\partial V)_S]} e^{-(\Delta S)^2/[2kC_P/m]} .$$

C_P is the constant-pressure heat capacity. Thus, in accord with the Central Limit Theorem, Gibbs' formulation leads to mean-squared pressure and specific entropy fluctuations inversely proportional to system size:

$$\langle (\Delta P)^2 \rangle = \rho c^2 (kT/V) \propto 1/N \propto (kC_P/m)/N^2 = \langle (\Delta S/N)^2 \rangle .$$

Consider a continuum computer simulation of water undergoing steady shear. An $L \times L \times L$ element of volume must be sufficiently large that the pressure fluctuations just estimated,

$$\langle |\Delta P| \rangle = \rho c (kT/mN)^{1/2} \propto L^{-3/2} ,$$

can be ignored relative to the nonequilibrium shear stress, $\eta \dot{\epsilon}$. At the same time, a useful deterministic continuum description requires that a computational volume element not be so large as to exhibit turbulence. Turbulence can occur whenever τ_{shape}, the characteristic time for a volume element to change *shape* $1/\dot{\epsilon}$, is much less than τ_{decay}, the time $L^2/\nu = L^2(\rho/\eta)$ required for the strain rate $\dot{\epsilon}$ to *decay*. The ratio of these two times is "Reynolds' Number":

$$(\tau_{\text{decay}}/\tau_{\text{shape}}) \equiv R \equiv (L^2 \dot{\epsilon}/\nu) .$$

The phenomenological inequality $\dot{\epsilon}L^2 \ll 2000(\eta/\rho)$ describes the range of strain rates for which turbulence can be ignored within a region of sidelength L.

The two restrictions just discussed—sufficiently *large* shear stress with sufficiently *small* strain rate—govern the useful range of cell (or "zone" or "element") sizes for realistic computer simulations of viscous water. Evidently L cannot exceed a meter, and the strain rate cannot lie below 0.001 hertz in a useful description. For a typical zone size of one centimeter, the strain rate would have to lie in the range from 0.03 to 20 hertz.

Atomistic computer simulations are limited by the same formal considerations, but on a much smaller scale. For a "large" micron-sized atomistic simulation, the strain rates for which shear stresses are large enough to measure, but not so large as to induce turbulence, range upward from 3×10^7 hertz, being limited above by atomic vibration frequencies. Thus the regions in which the microscopic and macroscopic approaches can overlap is quite limited. The two approaches are intrinsically different.

A simulation which collectively exchanges work or heat with its surroundings exhibits a *global* deviation from thermomechanical equilibrium. A simulation with a *local* heat flux or shear stress exceeding thermal fluctuations by a wide margin exhibits *local* deviations from equilibrium. An alternative characterization of nonequilibrium states—using local temperature and velocity gradients rather than local heat fluxes and stresses—would require either two-point finite-difference evaluation or an equivalent combination of interpolation and differentiation. Both approaches to describing nonequilibrium states are incorporated in the macroscopic simulation techniques detailed at the end of this chapter and in Chapter 7.

4.7 Deviations from the Phenomenological Linear Laws

The linear laws governing the diffusion of mass, momentum, and energy are based on the observations that the quantities transported are proportional to the gradients driving them. Doubling the magnitude of the velocity gradient $\dot{\epsilon} = (\partial u_x/\partial y)$ in a shear flow doubles the required stress, σ_{xy}:

$$(\dot{\epsilon} \to 2\dot{\epsilon}) \Longrightarrow (\sigma = \eta\dot{\epsilon} \to 2\sigma) .$$

This constitutive relation describes "Newtonian viscosity", which applies over a few orders of magnitude in the strain rate, at fixed system size. Fourier heat conduction is likewise based on the principle that doubling the temperature gradient *doubles* the heat flux:

$$(\nabla T \to 2\nabla T) \Longrightarrow (Q = -\kappa \nabla T \to 2Q) .$$

In both these examples *doubling* the departure from equilibrium, $\dot{\epsilon}$ or ∇T, *quadruples* the dissipation rate and the entropy production.

The validity of the linear laws has occasionally been questioned on formal grounds, based on the observation that Lyapunov instability leads to very large changes in particle trajectories (growing exponentially with time), so that the trajectories for even an infinitesimal perturbation bear no resemblance to their unperturbed cousins. The flaw in this criticism has been reviewed by Ichiyanagi. The exponential growth of perturbations is actually responsible for a *lack* of dependence on initial conditions, *ensuring* the validity of the ensemble approach and *justifying* the linear laws when the nonequilibrium gradients are small. What happens outside the region described by the linear laws?

Shockwaves are the *ultimate* nonlinear irreversible process, converting the kinetic energy of a moving fluid or solid into heat within a distance of a few mean free paths. The conversion process is highly nonlinear. To the extent that transport coefficients can be defined within the shockwave, these coefficients can likewise vary strongly on the microscopic length scale of the mean free path. A generation ago, atomistic simulations of shockwave structure, described in Section 5.7 and in Chapter 6, were compared to corresponding macroscopic simulations. The atomistic simulations used "realistic" short-ranged forces. The macroscopic simulations used linear transport theory, with state-dependent transport coefficients.

The comparison [intended to model the shockwave compression of liquid argon to 10^4 times atmospheric pressure] showed very similar shockwave profiles for density, pressure, and temperature for shockwaves that were not too strong. Evidently Newton's and Fourier's *linear* approximations to transport, though not completely correct, suffice for simple fluids. Stronger shockwaves [10^5 times atmospheric pressure] show substantial deviations from linearity.

More recent simulations, which we carried out with Oyeon Kum as a byproduct of his PhD thesis work, used somewhat "artificial" weaker and longer-ranged repulsive forces. Those simulations produced kinetic-temperature profiles with pronounced *maxima*, corresponding to a negative heat conductivity on the hot side of a steady moderately-strong shockwave! Even more recently we found similar strong temperature maxima within the shock for short-ranged repulsive forces. For more details see Chapter 6.

Because nonlinear effects are often small, the search for nonlinearity in simple fluids *can* be relatively frustrating, and is typically of very little interest from the point of view of engineering design and analysis. Understanding the negative conductivity mentioned above is a challenging problem. Much more complicated constitutive properties *often* arise when the underlying molecular structure is complex. Paint, for instance, displays an interesting coupling between shear and normal stresses—when *stirred* with a rotating rod, it can climb the rod, exhibiting a *vertical* force in response to a horizontal shear deformation. Alloys with a complex microstructure are able to "remember" their original shape after large deformations. But nonlinear effects like these are usually quite small in simple fluids.

4.8 Causes of Irreversibility à la Boltzmann and Lyapunov

There are two relatively simple phase-space explanations for macroscopic irreversibility, one which is more appropriate to isolated systems; the other applying to systems interacting with their surroundings: (i) in *isolated* systems more likely "macrostates" (coarse-grained states, each corresponding to many microstates) evolve from less likely ones; (ii) in *driven* (or "open") systems, interacting at their boundaries, fractal attractors, though of negligible measure at equilibrium, are qualitatively more stable and more probable than are their time-reversed images in phase space, the repellers. The first of these explanations is Boltzmann's idea. It "explains" why equilibrium gas-phase particles are more likely to be found *uniformly* distributed. There are simply many fewer *nonuniform* microstates, such as those states with substantially more particles in the right or in the left sides of their container.

The time reversibility of Newtonian mechanics shows that Boltzmann's state-counting idea can only be correct for times much less than Poincaré recurrence times and for systems which are noticeably disturbed from equilibrium. These difficulties often lead Boltzmann's most enthusiastic supporters to stress the importance of "initial conditions", sometimes as far back as the "Big Bang"!

The second explanation of macroscopic irreversibility, based on modern chaos theory, avoids the reversibility and recurrence objections leveled at Boltzmann's ideas. An understanding of information loss, Lyapunov instability, and computer algorithms are essential to this newer approach. The possibility of information loss, through lack of conservation of phase volume, invariably results in the formation of strange attractor objects with nonvanishing nonequilibrium fluxes. Despite the exploration of many example problems, our understanding of these objects is still imperfect. The relative irreversibility of the fractal attractors can be expressed in terms of the lost information required to recover their initial state. The simplest caricature of this information loss is the "Bernoulli Map", in which binary numbers, $0 < B < 1$, represented as strings of 0s and 1s, are shifted to the left, with the leftmost 0 or 1 discarded, introducing a "bit" of uncertainty and dissipating a bit of "information" each time the shift occurs. For example, the binary number 00010110 can come from the left shift of either of two digit strings (000010110 or 100010110).

To us, *time reversibility* plays an important role in "understanding" irreversibility. Time-reversible equations of motion can lead to just one of two steady-state solution types. On a time-averaged basis it is quite evident— as is discussed at length in Chapter 8—that evolving comoving extensions in phase, volumes in "solution space" (phase space for a system described by $\{q,p\}$ pairs of coordinates and their conjugate momenta), must either (i) stay the same or (ii) decrease. $\langle \dot{\otimes} \rangle \leq 0$. No other possibility (such as a continually increasing volume) is consistent with stability and a nonequilibrium steady state. If the motion equations are both time-reversible and mixing and if dissipation is possible, so that the comoving volume *can* decrease, then any solution of the equations of motion leads to shrinking and collapse, resulting in a strange attractor. The time-reversed repeller, although unarguably a *formal* solution of the same motion equations, is generally both unobservable and unstable.

To make the distinction between microscopic and macroscopic models of nonequilibrium systems clearer, let us turn to developing algorithms suited to simulating nonequilibrium flows of momentum and/or energy, or both, between two planar boundaries. We consider explicitly the details of Rayleigh-Bénard flow. In such flows heat is carried by moving fluid from a hot lower boundary to a cold upper one. A gravitational field, coupled with thermal expansion, provides the mechanism for the convective mass and heat currents. For sufficiently strong flows of heat interesting families of unsteady convection currents, or even chaotic turbulence, can develop.

Here we develop two types of algorithms to simulate such flows. First, we consider the molecular dynamics of microscopic atomistic particles, described by ordinary differential equations; afterward, we will consider the continuum mechanics of an idealized fluid obeying linear transport laws and the partial differential equations of continuum mechanics. In this latter linear-transport case turbulent complexity can still arise—the continuum obeys nonlinear partial differential equations even if the constitutive relations are linear. We begin now with the molecular dynamics.

4.9 Rayleigh-Bénard Algorithm with Atomistic Flow

To apply molecular dynamics to solving atomistic transport and Rayleigh-Bénard flow problems we first develop the *stationary boundary conditions* required to contain a fixed number of purely Newtonian "interior" particles within a square $n_w \times n_w$ "box". This box will be made up of localized "boundary particles". It is straightforward to extend this approach to moving boundaries, simulating shear flow. To describe the stationary boundaries considered here we define four sets of n_w lattice sites at which we tether $4n_w$ "wall particles" at equally-spaced positions along the four walls as shown in the figure.

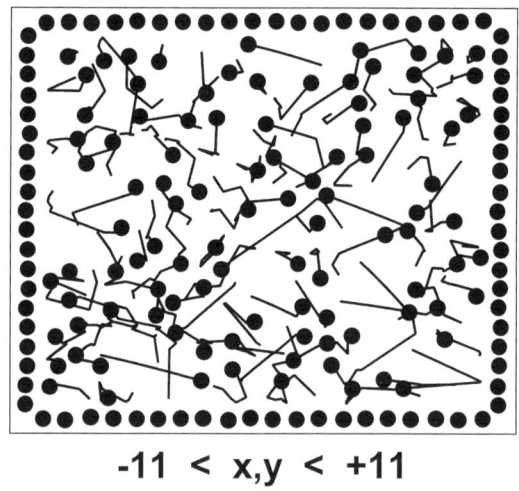

100 + 80 Particles for time = 12.5

-11 < x,y < +11

Figure 4.2. Time exposure of a molecular dynamics simulation of heat flow. There are 100 interior Newtonian particles and 80 boundary particles, with the top and bottom rows thermostated at different temperatures.

The (0,0) origin of our (x,y) coordinate system is at the center of the square container. In the following program the initial coordinates of the boundary particles are specified as {x(i) = x0(i),y(i) = y0(i)}. Then taken in the order of hot (or bottom), cold (or top), left, and right, the wall-particle and tether coordinates are calculated as follows:

```
index = 0
do i = 1,nw
do j = 1,nw
index = index + 1
x(index) = i - 0.5d00*(nw + 1)
y(index) = 1 - 0.75d00 - 0.5d00*(nw + 1)
x(index+nw) = i - 0.5d00*(nw + 1)
y(index+nw) = nw + 0.75d00 - 0.5d00*(nw + 1)
x(index+nw+nw) = 1 - 0.75d00 - 0.5d00*(nw + 1)
y(index+nw+nw) = j - 0.5d00*(nw + 1)
x(index+nw+nw+nw) = nw + 0.75d00 - 0.5d00*(nw + 1)
y(index+nw+nw+nw) = j - 0.5d00*(nw + 1)
enddo
enddo
do i = 1,4*nw
x0(i) = x(i)
y0(i) = y(i)
enddo
```

The offsets of 0.75 in both the x and the y directions have been chosen to avoid high-energy overlaps between the vertical and horizontal particle walls. We place the remaining $(n_w/2)^2$ interior Newtonian particles in a square-lattice arrangement with a spacing of 2, so that the nominal number density of these Newtonian particles is $(1/4)$:

```
index = 4*nw
do i = 1,nw/2
do j = 1,nw/2
index = index + 1
x(index) = 2.0d00*(i - 0.5d00*(1.0d00 + 0.5d00*nw))
y(index) = 2.0d00*(j - 0.5d00*(1.0d00 + 0.5d00*nw))
enddo
enddo
```

The total number of particles in the system is $N = 4n_w + (n_w^2/4)$. The $(0,0)$ origin of our (x, y) coordinate system is at the center of the square container. To impose the thermal boundary conditions at the top and bottom walls, respectively "cold" and "hot", the corresponding sets of boundary particles have their summed-up kinetic energies fixed by velocity rescaling at the end of every timestep. The two sidewalls obey purely Newtonian mechanics. Additional forces are required to prevent any of the interior Newtonian particles from escaping the system by penetrating the four semi-rigid walls of boundary particles.

For this choice of boundaries four kinds of forces appear in the equations of motion. The first of them is a constant vertical gravitational force, $(0, -mg)$, applied to all the interior particles. In addition there are three forces affecting the dynamics of the boundary and the interior particles:

(i): We use *quartic* potentials to tether each one of the $4n_w$ wall particles to its lattice site:

$$\phi_s \equiv 200\delta r^4 .$$

(ii): In addition to these boundary tethers all *pairs* of particles interact with a short-ranged purely repulsive potential which acts between any pair of particles within unit distance of one another:

$$\phi_p(r < 1) \equiv 100(1 - r^2)^4 .$$

(iii): Finally, to prevent particle escapes an additional *one-sided* wall potential,

$$\phi_w = 200\delta r^4 ,$$

acts to return any Newtonian particle toward the interior if it manages to pass through any of the four lines of wall-particle lattice sites. The total energy of the system just described and the forces on the particles can be calculated as follows:

```
calculate tether, gravitational, kinetic energy; initialize.
ekin = 0.0d00
epot = 0.0d00

do i = 1,N
fx(i) = 0.0d00
fy(i) = 0.0d00
```

```
      ekin = ekin + 0.5d00*(vx(i)**2 + vy(i)**2)
      if(i.gt.4*nw) epot = epot + y(i)*gee
      if(i.gt.4*nw) fy(i) = fy(i) - gee
      enddo

      do i = 1,4*nw
      dx = x(i) - x0(i)
      dy = y(i) - y0(i)
      epot = epot + 200*(dx*dx + dy*dy)**2
      fx(i) = fx(i) - 800*dx*(dx*dx + dy*dy)
      fy(i) = fy(i) - 800*dy*(dx*dx + dy*dy)
      enddo

calculate tether, gravitational, kinetic energy; initialize.

calculate pair energies
      do i = 1,N-1
      do j = i+1,N
      dx = x(i) - x(j)
      dy = y(i) - y(j)
      if((dx*dx + dy*dy).lt.1.0d00) then
      rr = (dx*dx + dy*dy)
      epot = epot + 100*(1.0d00 - rr)**4
      fx(i) = fx(i) + 800*dx*(1.0d00 - rr)**3
      fy(i) = fy(i) + 800*dy*(1.0d00 - rr)**3
      endif
      enddo
      enddo
calculate pair energies
calculate outside energies
      do i = 4*nw + 1,N
      if(x(i).lt.-(nw/2)) epot = epot + 200*(-(nw/2) - x(i))**4
      if(x(i).lt.-(nw/2)) fx(i) = fx(i) + 800*(-(nw/2) - x(1))**3
      if(x(i).gt.+(nw/2)) epot = epot + fcons*(+x(i) - (nw/2))**4
      if(x(i).gt.+(nw/2)) fx(i) = fx(i) - 800*(-(nw/2) - x(i))**3
      if(y(i).lt.-(nw/2)) epot = epot + fcons*(-(nw/2) - y(i))**4
      if(y(i).lt.-(nw/2)) fy(i) = fy(i) + 800*(-(nw/2) - y(i))**3
      if(y(i).gt.+(nw/2)) epot = epot + fcons*(+y(i) - (nw/2))**4
      if(y(i).gt.+(nw/2)) fy(i) = fy(i) - 800*(-(nw/2) - y(i))**3
```

```
enddo
calculate outside energies
etot = ekin + epot
write(160,120) time,ekin,epot,etot
```

If the number of particles exceeds 10,000 or so it is worthwhile to speed up the force calculations by keeping an updated list of particle pairs close enough to interact during the next few timesteps. With a maximum relative particle speed of 5 and a timestep of 0.01 such a list, valid for one hundred timesteps, would include all pairs of particles within a distance 1.5. The list is prepared again when necessary.

```
npairs = 0
do i = 1,N-1
do j = i+1,N
dx = x(i) - x(j)
dy = y(i) - y(j)
if((dx*dx + dy*dy).lt.2.25d00) then
npairs = npairs + 1
ipair(npairs) = i
jpair(npairs) = j
endif
enddo
enddo
```

The pair-energy loop (beginning on the previous page) then involves only `npairs` of (`ipair,jpair`) pairs rather than all $N(N-1)/2$, which is *much* slower. For the low density shown in Figure 4.2 and a system with tens of thousands of particles the typical number of pairs in the list is just a few hundred.

For simplicity we impose the lower and upper thermal boundary conditions by rescaling the kinetic temperature of the lower and upper rows of particles. At the end of every timestep the hot and cold boundary particles have their velocities rescaled to reproduce the desired boundary temperature:

```
constrain the hot and cold boundary temperatures
sumhot = 0.0d00
sumcold = 0.0d00
do i = 1,nw
sumhot = sumhot + vx(i)**2 + vy(i)**2
sumcold = sumcold + vx(i+nw)**2 + vy(i+nw)**2
enddo
fachot = dsqrt(2*nw*Thot/sumhot)
faccold = dsqrt(2*nw*Tcold/sumcold)
do i = 1,nw
vx(i) = vx(i)*fachot
vy(i) = vy(i)*fachot
vx(i+nw) = vx(i+nw)*faccold
vy(i+nw) = vy(i+nw)*faccold
enddo
constrain the hot and cold boundary temperatures
```

A typical short trajectory time exposure using this heat flow algorithm, with the particles' end time locations shown as filled circles, appeared in Figure 4.2. In purely conductive heat flow simulations the gravitational acceleration is set equal to zero so that heat conduction, without convection, is measured. Figure 4.3 shows the time development of the boundary fluxes of heat for five hot and cold temperature pairs of $\{1.5, 0.5\}$, $\{1.4, 0.6\}$, $\{1.3, 0.7\}$, $\{1.2, 0.8\}$, and $\{1.1, 0.9\}$.

$$Q_y = [-(dE/dt)_{\text{hot}} \simeq +(dE/dt)_{\text{cold}}]/n_w \equiv -\kappa(dT/dy) > 0$$

$$\longrightarrow \kappa(T \simeq 1) \simeq 1.5 \ .$$

Temperatures of 0.015 and 0.005 show a similar time development and give a smaller heat conductivity $\kappa(T \simeq 0.01) \simeq 0.2$, agreeing quite well with the square-root thermal transport coefficient dependence expected for hard disks or spheres:

$$\eta \propto \sqrt{T}; \ \kappa \propto \sqrt{T} \ .$$

We will apply this atomistic algorithm to the Rayleigh-Bénard problem (in which gravity is added) later in this Chapter, as the second of three example problems. Because we expect that large-system small-gradient molecular dynamics results will approach the continuum description in an appropriate limit, we turn next to an algorithm for obtaining the *continuum* description of Rayleigh-Bénard flows.

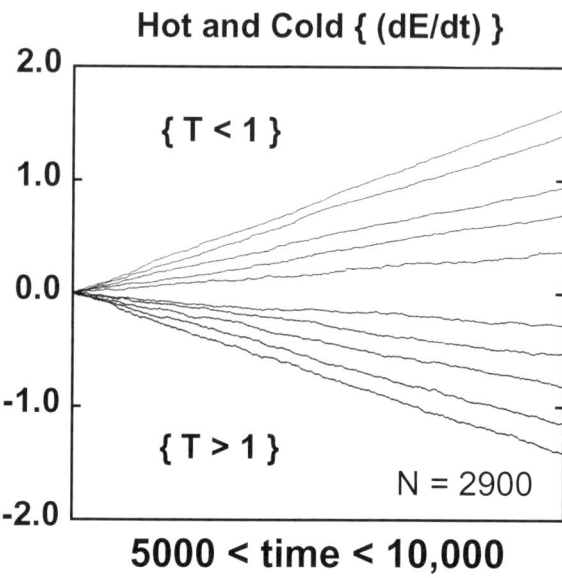

Figure 4.3. Cumulative Newtonian energy change (the negative of the rescaling energy change) of the lower and upper heat-reservoir rows of 100 boundary particles. Gravity is absent. There are 2500 interior Newtonian particles in the 100 × 100 box. The applied temperature gradients for the data shown here span the range from −0.002 to −0.010. All these results can be described by Fourier's law with a heat conductivity $\kappa \simeq 1.5$.

For a two-dimensional problem *molecular* dynamics involves only six variables per particle, $\{x, y, \dot{x}, \dot{y}, \ddot{x}, \ddot{y}\}$. The *continuum* algorithm is more detailed and requires more work. Describing a "simple" fluid with Newtonian viscosities and Fourier conduction requires four state variables, their time derivatives, *and* the pressure tensor and heat flux vector,

$$\{\rho, u, e, P, Q\} \longleftrightarrow \{(\partial \rho/\partial t), (\partial u/\partial t), (\partial e/\partial t)\} \ .$$

The state variables are needed at each node *and* in every cell, making it possible to approximate all the righthand side gradients in the continuity, motion, and energy equations. Thus the total number of variables is about 20 (per cell or per node) in two dimensions and about 30 in three, depending upon the details of the solution algorithm. In the following Sections we describe and apply a computer algorithm which follows such a plan and provides accurate solutions to the continuum Rayleigh-Bénard problem.

4.10 Rayleigh-Bénard Algorithm for a Continuum

To illustrate *continuum* solutions of the Rayleigh-Bénard flow equations we consider again a *square* system confined by rigid walls on which the velocity vanishes and the energy is constant. In continuum mechanics we have to impose our own choices for the constitutive relations and boundary conditions. Simplicity suggests numbers of order unity. Accordingly we choose both the overall density and the average temperature equal to unity. We make up the system geometry from $n_w \times n_w$ square 1×1 cells. We choose a hot temperature of 1.5 at the bottom of our system and a cold temperature of 0.5 at the top. On the vertical walls we fix a linearly-interpolated temperature varying from the hot lower-boundary temperature of 1.5 to the cold upper-boundary temperature of 0.5. For simplicity we consider constant transport coefficients and the monatomic two-dimensional ideal-gas equation of state, $PV = E = NkT$.

To determine a good choice for the number of computational mesh points we can study the mesh-dependence of the single-roll kinetic energy per zone at a Rayleigh number of 6400. See Figure 4.4. Stationary values of the kinetic energy per zone from the computer program described below have a simple mesh dependence. Results for four mesh choices show a number dependence from which the infinite-system truly-continuous result can be estimated with five-figure accuracy: $(K_\infty/n_w^2) = 0.00047287$:

$$20 \times 20 : 0.00049199 \; ; \; 40 \times 40 : 0.00047765 \; ;$$

$$80 \times 80 : 0.00047407 \; ; \; 160 \times 160 : 0.00047317 \; ;$$

$$\longrightarrow K_{n_w} = 0.00047287 n_w^2 + 0.00765 \; .$$

Evidently a simulation with velocities accurate to a fraction of a percent can be based on an array of 80×80 cells, with the velocities and energies computed on a grid of 81×81 nodes. The boundary values of velocity and energy are constant on the 320 outermost nodes.

For convenience we choose `nh*nw` square 1×1 unit cells, with a gravitational field, $(1/80)$, which would correspond to unit density for an ideal gas. A Rayleigh number of 6400, where

$$\mathcal{R} \equiv g(\partial \ln V/\partial T)_P \Delta T L^3 / (\nu D_T) = 6400 = \texttt{nh} * \texttt{nh}/(\texttt{eta} * \texttt{cap}) \; ,$$

results if the shear viscosity and thermal diffusivity are both chosen equal to unity: `eta = cap = 1.0d00`.

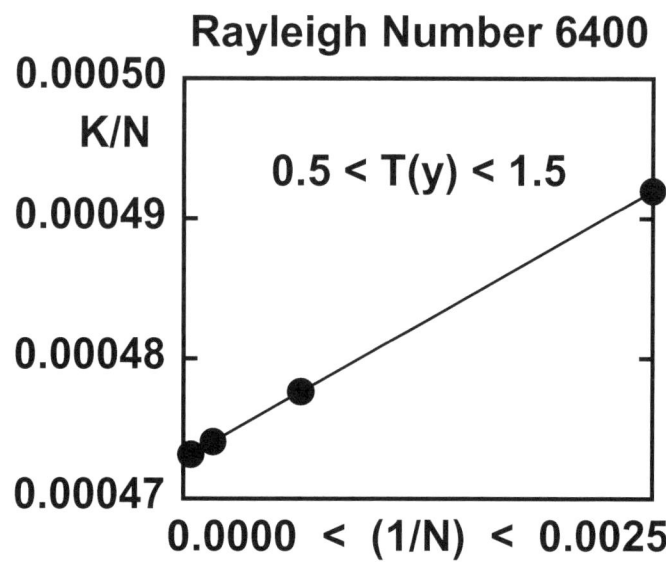

Figure 4.4. Stationary values of the kinetic energy per zone. Problems with mesh sizes of 20x20, 40x40, 80x80, and 160x160 were run to a time of 50,000.

For convenience, we again choose the (0,0) origin at the center of the system. Initial energies at the nodes can be chosen to match Fourier's law,

$$e(y) = T(y) = 1.0 - (y/80); \ [-40 \le y \le +40] \ .$$

Initial cell densities of unity and small random nodal velocities,

$$\{\texttt{rhoc} = 1.0\texttt{d}00\} \ ;$$

$$\{\texttt{uxn}, \texttt{uyn}\} \longleftarrow 0.000001\texttt{d}00 * (\texttt{rand}(\texttt{intx}, \texttt{inty}) - 0.5\texttt{d}00) \ ,$$

complete the initial conditions. Rather than choosing random velocities, we could begin with an assumed roll pattern. The simplest such choice, which turns out to be appropriate to Rayleigh numbers between 5000 and 20,000, is the single-roll structure shown later, in Figure 4.6:

$$\{u_x \propto \cos(kx)\sin(2ky); \ u_y \propto -\sin(2kx)\cos(ky)\}; \ k \equiv (\pi/80) \ .$$

The problem to be solved involves the partial differential equations for the evolution of the mass density, the velocity, and the energy. First is the continuity equation for the conservation of mass. The Eulerian time derivative,

$$(\partial \rho/\partial t) = -\nabla \cdot (\rho u) ,$$

has to be evaluated at each cell. Next, the nodal velocity and energy time derivatives,

$$(\partial u/\partial t) = -u \cdot \nabla u - [(\nabla \cdot P)/\rho]; \; (\partial e/\partial t) = -u \cdot \nabla e - [\nabla u : P + \nabla \cdot Q]/\rho ,$$

have to be evaluated at all of the interior nodes. It is assumed that the boundary values of the velocity and energy are fixed.

The pressure tensor P and heat flux vector Q are next evaluated using Newtonian viscosity (with zero bulk viscosity) and Fourier heat conduction as constitutive relations:

$$P_{xx} = \rho e - \eta[(\partial u_x/\partial x) - (\partial u_y/\partial y)] ;$$

$$P_{xy} \equiv P_{yx} = -\eta[(\partial u_x/\partial y) + (\partial u_y/\partial x)] ;$$

$$P_{yy} = \rho e + \eta[(\partial u_x/\partial x) - (\partial u_y/\partial y)] ;$$

$$Q_x = -D_T(\partial e/\partial x); \; Q_y = -D_T(\partial e/\partial y) .$$

The simplest approach is to evaluate all the gradients on the righthand sides with linear expressions and to integrate the resulting time derivatives, 80^2 equations for $\{(\partial \rho/\partial t)\}$ at the cells, and 81^2 equations for $\{(\partial u_x/\partial t), (\partial u_y/\partial t), (\partial e/\partial t)\}$ at the nodes, with the nodal velocity and energy derivatives along the four boundaries all set equal to zero.

The solution of the $80^2 + 3(81^2) = 26,083$ first-order ordinary differential equations proceeds by following a series of eight steps: (i) Compute the nodal densities and the cell velocities and energies from adjacent cell and nodal quantities. (ii) Compute the velocity and energy gradients in each cell. (iii) Calculate the pressure tensor and heat flux in each cell from the gradients in (ii) above. (iv) Calculate the gradients of the pressure tensor and heat flux in each cell. (v) Evaluate $(\partial \rho/\partial t)$ from the mass fluxes at the cells. (vi) Evaluate the convective parts of $(\partial u_x/\partial t), (\partial u_y/\partial t), (\partial e/\partial t)$ at the nodes. (vii) Evaluate the pressure contributions to $(\partial u_x/\partial t), (\partial u_y/\partial t)$ at the nodes. (viii) Evaluate the pressure and heat flux contributions (work and heat) to $(\partial e/\partial t)$ at the nodes. Once these steps are complete, the numerical values of $\{(\partial \rho/\partial t)_{\text{cell}}, (\partial u/\partial t)_{\text{node}}, (\partial e/\partial t)_{\text{node}}\}$ can be integrated with the fourth-order Runge-Kutta algorithm.

In computing the necessary sums and differences it is convenient first to find the cells neighboring each node and the nodes neighboring each cell with subroutines. For instance, the four nodes at the corners of cell (ix,jy), in counterclockwise order, starting at the bottom left, are returned by the following subroutine:

```
subroutine ijnfromc(ix,jy,i1,i2,i3,i4,j1,j2,j3,j4,nx,ny)
i1 = ix
i2 = ix + 1
i3 = ix + 1
i4 = ix
j1 = jy
j2 = jy
j3 = jy + 1
j4 = jy + 1
return
end
```

For the *first* of the eight steps above, we illustrate a useful approximation, exact for a linear variation of density with coordinate, for the nodal density for the second node, node (2,1), on the bottom row:

```
rhon(2,1) = rhoc(1,1) + rhoc(2,1) - rhon(2,2)
```

After finding all the boundary nodal densities other than those at the four corners, a similar approximation can be applied to find the four corner densities. For example, the density at the lower lefthand corner of the mesh can be calculated in either of two ways:

```
rhon(1,1) = rhon(1,2) + rhon(2,1) - rhon(2,2)   or the alternative
rhon(1,1) = 2.0d00*rhoc(1,1) - rhon(2,2)
```

For the *second* step, getting the velocity and energy gradients, we express the derivatives using the counterclockwise 1234 numbering system, starting at the lower lefthand node. For example, the energy gradient in each cell is worked out as follows (and similarly for the velocity gradients):

```
dedxc(ix,jy)=(-enn(i1,j1)+enn(12,j2)+enn(i3,j3)-(enn(i4,j4))/2
dedyc(ix,jy)=(-enn(i1,j1)-enn(12,j2)+enn(i3,j3)+(enn(i4,j4))/2
```

With the gradients evaluated, the cell pressure tensors and heat fluxes of step *three* follow easily. Step *four*, calculating the cell gradients, $\nabla \cdot P$ and ∇Q follows the scheme used for ∇u and ∇e above. Step *five* requires the righthand side of the continuity equation, $-\nabla \cdot (\rho u)$. The mass flows into and out of each cell can be written as a sum over the four faces. The flux contributions in the x direction are

```
+(rhon(i1,j1) + rhon(i4,j4))*(uxn(i1,j1) + uxn(i4,j4))/4
-(rhon(i2,j2) + rhon(i3,j3))*(uxn(i2,j2) + uxn(i3,j3))/4
```

and those in the y direction are

```
+(rhon(i1,j1) + rhon(i2,j2))*(uyn(i1,j1) + uyn(i2,j2))/4
-(rhon(i3,j3) + rhon(i4,j4))*(uyn(i3,j3) + uyn(i4,j4))/4  .
```

Step *six* requires the "convective" or "advective" contributions, proportional to u, to the nodal values of $(\partial u/\partial t)$ and $(\partial e/\partial t)$. Because the velocity vanishes on the boundary, these derivatives involve only interior nodes. For instance, the energy derivatives are

```
denwrtx=(dendx(i1,j1)+dendx(i2,j2)+dendx(i3,j3)+dendx(i4,j4))/4
denwrty=(dendy(i1,j1)+dendy(i2,j2)+dendy(i3,j3)+dendy(i4,j4))/4
dendt(ix,jy)=dendt(ix,jy)-uxn(ix,jy)*denwrtx-uyn(ix,jy)*denwrty
```

Step *seven* requires the pressure derivatives at the nodes. Because the boundary nodes are fixed only the interior derivatives are required, and those follow from the four neighboring cells. For instance

```
dpxxdx = (+pxxc(i2,j2)+pxxc(i3,j3)-pxxc(i1,j1)-pxxc(i4,j4))/2
dpxydy = (+pxyc(i3,j3)+pxyc(i4,j4)-pxyc(i1,j1)+pxyc(i2,j2))/2
duxdt(ix,jy) = duxdt(ix,jy)-(dpxxdx + dpxydy)/rhon(ix,jy)
```

Finally (step *eight*) $\nabla u : P$ and $\nabla \cdot Q$ need to be evaluated to evolve the nodal energies. Again only the interior nodes are required. The contribution of the heat flux divergence to the energy evolution is typical:

```
xflux = (+qxc(i1,j1)-qxc(i2,j2)-qxc(i3,j3)+qxc(i4,j4))/2
yflux = (+qyc(i1,j1)+qyc(i2,j2)-qyc(i3,j3)-qyc(i4,j4))/2
dendt(ix,jy) = dendt(ix,jy)-((xflux + yflux)/rhon(ix,jy))
```

We apply the atomistic and continuum algorithms just described to Rayleigh-Bénard convection in the following problem Section.

4.11 Three Rayleigh-Bénard Example Problems

The simplest nonequilibrium state is arguably thermal conduction in the absence of convection. Heat flows from a hot boundary to a cold one, as described by Fourier's law, $Q = -\kappa \nabla T$. For Rayleigh numbers up to a few thousand the linear solution, with ∇T constant, results when either molecular dynamics or continuum mechanics is applied.

Heat flow becomes more complicated as the imposed temperature difference increases. Even so, one of the simplest nonequilibrium stationary states is thermal convection. This is because the confining boundaries are motionless. A confined fluid in a gravitational field, with sufficient heating on the bottom and cooling at the top, can be driven into an infinite variety of nonequilibrium states. These include (i) simple motionless conduction, (ii) stationary convection, with a roll or rolls contributing to the heat flow, (iii) time-periodic convection with rolls oscillating or changing in number, and (iv) the wild swirling currents with thermal plumes characteristic of turbulent chaotic heat flow.

A semiquantitative description of the departure of such a nonequilibrium Rayleigh-Bénard system from equilibrium is conventionally characterized by the dimensionless "Rayleigh Number",
$$\mathcal{R} \equiv (\partial \ln V/\partial T)_P \Delta T g L^3/(\nu D_T) ,$$
where g is the gravitational field strength and ΔT is the temperature difference,
$$\Delta T \equiv [T_{\text{bottom}} - T_{\text{top}}] = [T_{\text{hot}} - T_{\text{cold}}] .$$
The kinematic viscosity, $\nu = (\eta/\rho)$, and the thermal diffusivity D_T, both have the same units, those of a diffusion coefficient. For a two-dimensional system of width $2L$ and height L, with vanishing velocity on both horizontal boundaries and with periodic vertical boundaries, two stationary rolls first appear at a Rayleigh number of about 1700. For a square box a single stationary roll appears at a Rayleigh number just below 5000.

For real fluids we don't have the luxury of varying gravity or the transport coefficients. Thus for a fixed temperature difference ΔT the Rayleigh number varies as the *cube* of the system height. For chilly water, at 22° C, with $(\partial \ln V/\partial T)_P = 10^{-4}$/kelvin, cells with $L = 1$ centimeter, 1 meter, and 100 meters, the bottom-to-top temperature differences at this threshold for motion are about 1, 10^{-6}, and 10^{-12} kelvins in the three cases. Only in the last completely-unphysical 100-meter case, with negligible heat flow and hardly distinguishable from an equilibrium situation, would the microscopic equilibrium temperature fluctuations be comparable to the temperature differences in the macroscopic continuum flow. Thus the fluctuations which complicate microscopic descriptions and simulations of transport processes are of little relevance for macroscopic systems.

Even this familiar "Rayleigh-Bénard" convection problem is sufficiently complicated to frustrate analytic work, unless relatively unrealistic approximations are included. Before we take up "realistic" simulations, with many degrees of freedom, let us consider a famous caricature of the heat-flow problem. Lorenz introduced a simplified three-variable model for describing time-dependent heat-driven convection. Lorenz' model was based on Saltzman's more elaborate Fourier analysis of the Navier-Stokes equations.

Lorenz' treatment already incorporates the chaotic fluid motion underlying turbulence. More realistic grid-based simulations, including the effects of compressibility, conduction, and an exact treatment of nonlinear convective effects, can easily be computed with six-figure accuracy by the finite difference methods described in the last Section. Here we describe Lorenz' caricature of thermal convection. We then consider molecular dynamics solutions of the equivalent atomistic problem, by adding heat reservoirs and a gravitational field to what would otherwise have been an equilibrium molecular dynamics simulation.

4.11.1 *Rayleigh-Bénard Flow* via **Lorenz' Attractor**

In the early 1960s computer simulation was still in its infancy, restricted to government laboratories and to a few universities with government contracts. In those days it was natural to express the behavior of complicated flows in terms of simple caricature models. Lorenz' model of Rayleigh-Bénard convection is the best known of these. His approximate three-variable description of a system—where the variables $\{x, y, z\}$ represent respectively the instantaneous values of the rotational velocity and the horizontal and vertical temperature gradients—has the form:

$$\{\dot{x} = -\sigma(x - y) \ ; \ \dot{y} = Rx - y - xz \ ; \ \dot{z} = xy - bz\} \ .$$

Any $\{+x, +y, \pm z\}$ solution has a mirror image twin, $\{-x, -y, \pm z\}$. Notice also that the instantaneous divergence of the (x, y, z)-space "velocity", $(\dot{x}, \dot{y}, \dot{z})$,

$$\nabla \cdot v = (\partial \dot{x}/\partial x) + (\partial \dot{y}/\partial y) + (\partial \dot{z}/\partial z) = (\dot{\otimes}/\otimes) = -\sigma - 1 - b \ ,$$

is constant and negative. Thus any comoving three-dimensional volume, $\otimes = dxdydz$, governed by the Lorenz equations, shrinks rapidly to zero. Lorenz showed that sufficiently large driving (represented by the "reduced Rayleigh Number" $R \equiv \mathcal{R}/\mathcal{R}_c$, where $\mathcal{R}_c$ is the "critical" Rayleigh number, the value at which convection begins) leads to the *exponential* growth of perturbations, "Lyapunov instability". An example trajectory, for the chaotic combination $\{\sigma = 10, R = 28, b = (8/3)\}$, appears in Figure 4.5. For this example the comoving rate of volume change is constant:

$$(\dot{\otimes}/\otimes) = (d \ln \otimes/dt) = -(d \ln f/dt) = -\sigma - 1 - b = -13.667 \ .$$

The constancy of the phase-volume shrinkage rate, $(\dot{\otimes}/\otimes)$, here differs qualitatively from the microscopic manybody situation, as is described by Evans, Morriss, and Cohen in their 1993 Physical Review Letter. They emphasize the strong dependence of the relative probability of fluctuations in the rate of phase-volume change. We briefly explore their idea for the Galton Board problem in Section 9.9.

It is specially significant that Lorenz' model involves three variables. With just one or two ordinary differential equations, and in the absence of external driving, there is no way to obtain chaos. A trajectory confined within a two-dimensional phase space can show no chaos because any deterministic trajectory in a two-dimensional space must either stop or intersect itself, becoming periodic. In three dimensions, where there is no unavoidable need for a trajectory to intersect itself, chaos *can* occur, as it does in Lorenz' model. Lorenz' example retains its pedagogical interest today. It shows that a simple dissipative system, in a three-dimensional state space, can exhibit both chaos and Lyapunov instability.

Lorenz' model is a greatly simplified caricature of real thermal convection. For the usual choice of parameters, corresponding to the solution shown in Figure 4.5, the fluid's rotation undergoes chaotic changes between the clockwise and counterclockwise directions. Today complete two-dimensional or three-dimensional simulations of fluid dynamics are routinely carried out using a variety of computational methods. They show a great variety of solution types, stationary, periodic, and chaotic. Typically, the flow morphology changes with time, with changing numbers of rolls. The timescale of these changes can be slow, on the order of days or weeks.

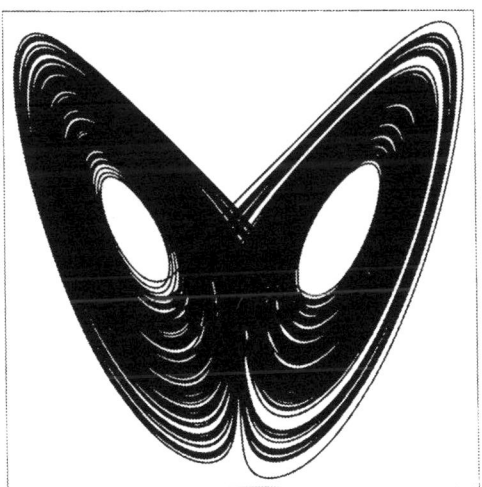

Figure 4.5. Lorenz' Attractor. The calculation shown used 200,000 points generated with Runge-Kutta integration with a timestep of 0.001. Initially $(x, y, z) = (-7, -11, +13)$. The variation of z with x is shown within the ranges $-20 < x < +20$; $0 < z < +50$.

The continuum approach applies over a vast range of Rayleigh numbers. It makes clean predictions, avoiding the troublesome number-dependent fluctuations inherent in molecular dynamics. To illustrate this variety we turn next to a complete solution of the Navier-Stokes-Fourier equations for Rayleigh-Bénard flow. With those results in hand we will then consider this same thermal convection problem from the more fundamental, but necessarily more complex, perspective of atomistic molecular dynamics, following the lead of the simulations pioneered by Kestemont, Mansour, Mareschal, Puhl, and Rapaport in the 1980s.

4.11.2 *Rayleigh-Bénard Flow with Continuum Mechanics*

Simulations with the centered-difference algorithm of the last Section are excellent workstation or laptop projects, generating complete solutions much more rapidly than they can be analyzed. It is relatively easy to answer the simple questions about Rayleigh-Bénard flow: "What is the critical Rayleigh number, above which convection is permanent?"; "Is the flow chaotic?"; "Where is the entropy produced?". The answers to these questions, and many others, follow relatively easily from fully-converged numerical solutions of the continuum equations.

We illustrate the numerical approach by considering again the simplest possible case, a two-dimensional ideal-gas fluid ($PV = NkT = E$), with equal state-independent values of the kinematic viscosity and thermal diffusivity and in the presence of a constant gravitational field g. We will choose the field strength to minimize the density gradient. If the temperature were constant then the gas would follow Gibbs' statistical mechanics and obey the "barometer formula", the equilibrium relationship between pressure and height:

$$P(y) = \rho(y)(kT/m) = P(y=0)e^{-mgy/kT} \longleftrightarrow$$

$$(dP/dy) = -(mg/kT)P(y) \ .$$

Now consider a *nonequilibrium situation* with a vertical temperature gradient. Even *with* this gradient a stationary mechanical equilibrium can be found. A static isochoric (constant density) equilibrium occurs if the gravitational acceleration, ρg *downward*, for unit mass, is selected to balance exactly the *upward* acceleration from pressure, $-(dP/dy)$ for unit mass:

$$\rho g = -(dP/dy) = -\rho k(dT/dy) \longrightarrow g = k[T_{\text{hot}} - T_{\text{cold}}]/L \ .$$

Here L is the box height, the distance between the hot and cold walls. In the end ρ and g are constants so that the pressure varies linearly with y, just measuring the "weight" of the supported gas above.

It is remarkable that this stationary *nonequilibrium* problem has such a simple analytic solution. Both the pressure and the temperature can be calculated (or measured) using the usual equilibrium relations from Gibbs' statistical mechanics:

$$P = \rho\langle(v_x^2 + v_y^2)\rangle/2 = \rho e = \rho(kT_{xx} + kT_{yy})/2 \ .$$

Even so, the presence of a nonequilibrium heat flow implies that the atomistic velocity distribution *cannot* be exactly Maxwell-Boltzmann.

An exact calculation of the perturbation induced by gravity depends on the assumed form of the interatomic force law. Solving Boltzmann's kinetic-theory equation for the coordinate-dependent velocity distribution, $f(v,y)$ results in a perturbation, proportional to the temperature gradient (or to g) and nearly quadratic in the velocity. In D (equal to two or three) space dimensions one finds, in the small-gradient approximation:

$$f(v) \simeq f_{\rm eq}(v)\left[1 - (d\ln T/dy)v_y\tau[(mv^2/2kT) - (D/2) - 1]\right] \ ,$$

where τ is a mean collision time. The perturbation is consistent with constant pressure and density (so that there is no macroscopic mass flow) but does give rise to a heat current (so that the perturbation can be used to compute the value of the heat conductivity).

For a given set of boundary conditions flow begins for a sufficiently large temperature gradient. For simplicity we choose a square $L \times L$ bounding box for the flow. All along the box perimeter, $(x \text{ or } y) = \pm L/2$, the velocity vanishes and the temperature is fixed, proportional to the y coordinate:

$$u_x = u_y = 0 \ ; \ kT = e = [T_{\rm hot} + T_{\rm cold}]/2 + (y/L)[T_{\rm cold} - T_{\rm hot}]$$

[on the Boundaries] .

With the isochoric choice $g = k[T_{\rm hot} - T_{\rm cold}]/L$, the Rayleigh number,

$$\mathcal{R} = (\Delta T/T)^2 n_w^2/(\nu D_T/T) \ ,$$

for a fixed number of cells, n_w^2, can be controlled by adjusting the transport coefficients or the *size* of the particles. Comparison of a simulation at a Rayleigh number of 6400 with the analytical Fourier mode (Figure 4.6) shows remarkable agreement. Three choices of initial velocities, random, single-roll, and double-roll, all lead to the same stationary single roll solution.

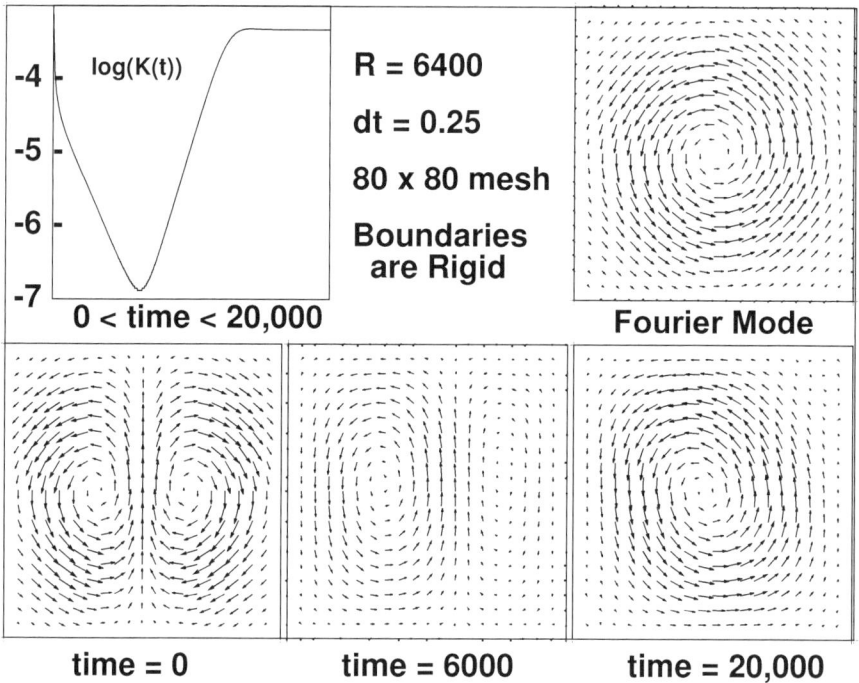

Figure 4.6. Prototypical one-roll continuum flow velocity field. A fully-converged continuum simulation, shown on the bottom right, is not very different to the analytic form shown at the top right:

$$\{u_x \propto \cos(kx)\sin(2ky);\ u_y \propto -\sin(2kx)\cos(ky)\};\ k \equiv (\pi/L).$$

The double roll initial condition changes to a single roll at approximately the minimum in the kinetic energy (shown in the top left) at time 6000.

Choosing an 80x80 grid with a temperature difference of +1 nicely reproduces the linear variation of the flow's kinetic energy with the inverse square root of the Rayleigh number. See Figure 4.7. The critical Rayleigh number, below which there is no convection, is about 5000.

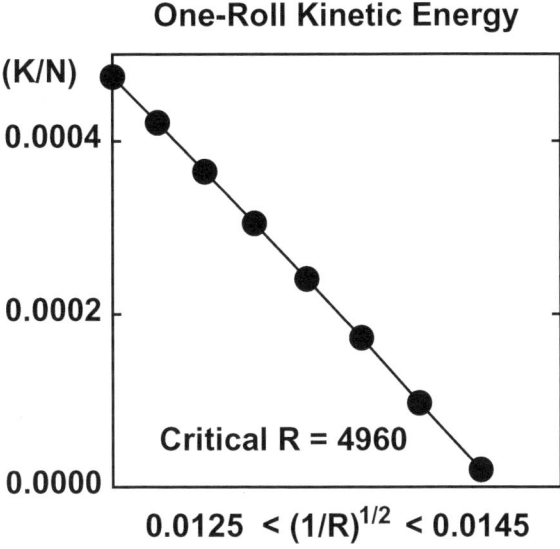

Figure 4.7. Dependence of the kinetic energy per zone on the inverse square root of the Rayleigh number for an ideal gas with constant transport coefficients. The bottom-to-top temperature difference is $\Delta T = 1.5 - 0.5$. The gravitational field strength is $g = \Delta T/n_w$. The sidelength of the square box is $n_w = 80$ with the boundary temperature varying linearly along the sides of the box.

The simple finite-difference algorithm described earlier converges nicely for Rayleigh numbers as high as $800,000$. For $\mathcal{R} = 160,000$ there is a time-periodic solution, going through the series of roll morphologies abstracted in Figures 4.8 and 4.9. Notice that Figure 4.9 is a mirror image of Figure 4.8 with the points equally spaced in time as in the kinetic energy time history shown in Figure 4.10. For $\mathcal{R} = 800,000$ the flow is chaotic, and never exactly repeats. The chaotic nature of the roll patterns is clearly evident in the morphologies of Figures 4.12 and 4.13. Typical kinetic energy histories for the two Rayleigh numbers are compared in Figure 4.11.

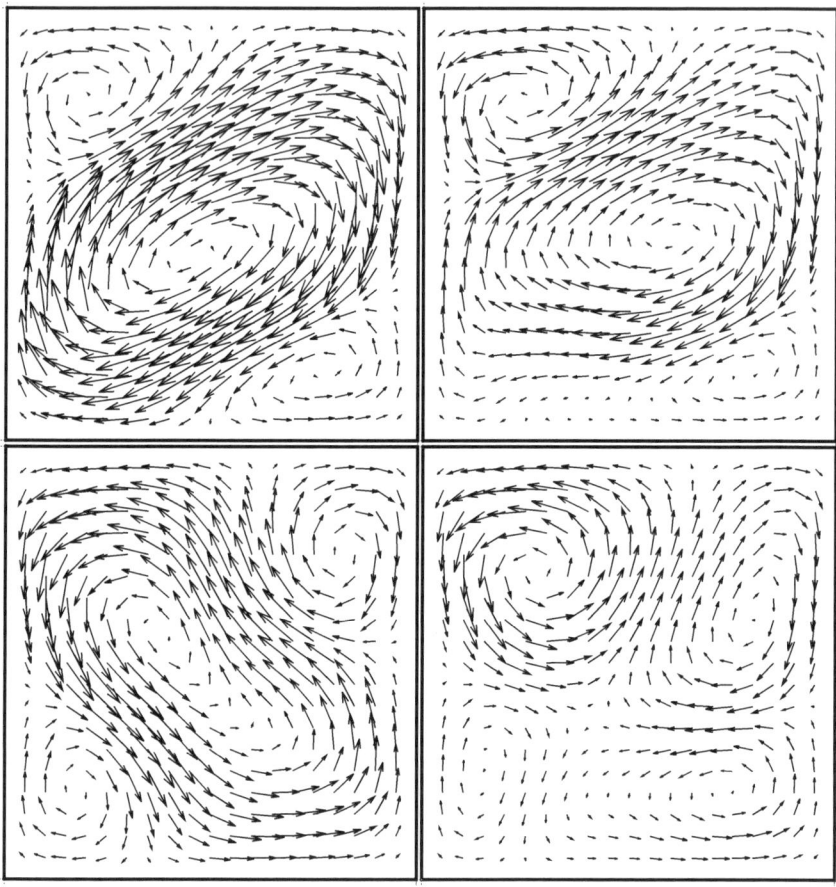

Figure 4.8. Morphology for a typical continuum solution using 80×80 computational cells and $\mathcal{R} = 160,000$. The time sequence of these plots starts at the upper left and continues clockwise to the lower left. The sequence corresponds to the points 1–4 on the kinetic energy time history in Figure 4.10. Figure 4.9 shows the later points on the kinetic energy curve. Notice that the kinetic energy is periodic and that Figures 4.8 and 4.9 are mirror images of each other.

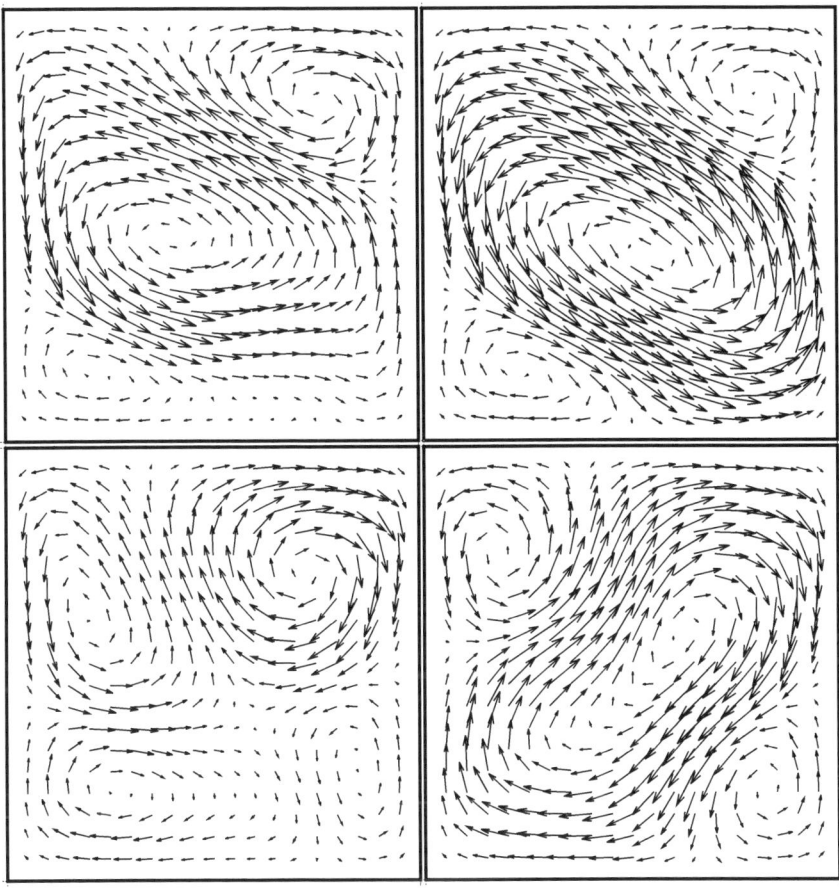

Figure 4.9. Morphology for a typical continuum solution using 80 × 80 computational cells and $\mathcal{R} = 160,000$. The time sequence of these plots starts at the upper right and continues counterclockwise to the lower right. The sequence corresponds to the points 5–8 on the kinetic energy plot in Figure 4.10. Figure 4.8 shows the earlier points on the kinetic energy time history. Notice that the kinetic energy is periodic and that Figures 4.8 and 4.9 are mirror images of each other.

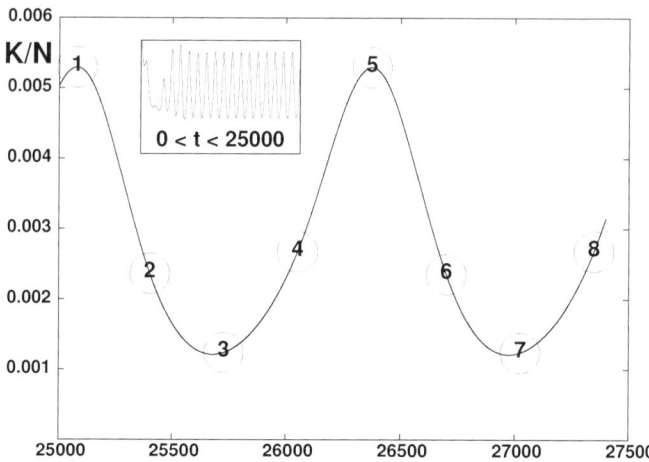

Figure 4.10. Kinetic energy per zone for $\mathcal{R} = 160,000$ (inset). The kinetic energy is periodic ($\tau = 2592.2$). The numbered points over one cycle correspond to the roll patterns shown in Figures 4.8 and 4.9.

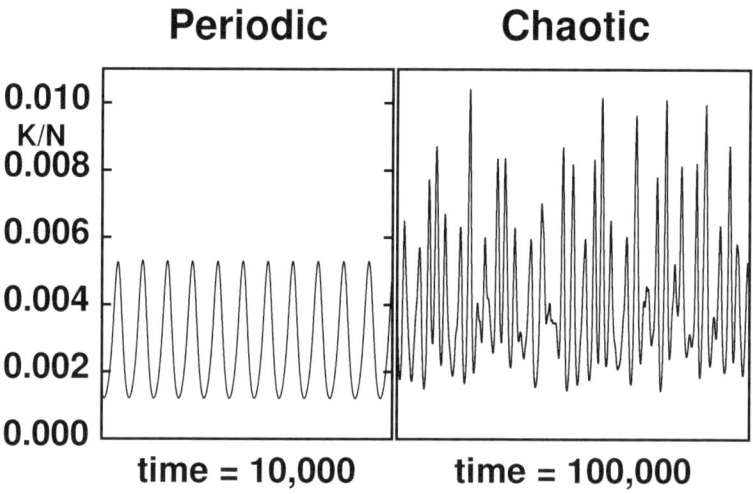

Figure 4.11. Kinetic energy per zone showing periodic (left, $\mathcal{R} = 160,000$) and chaotic behavior (right, $\mathcal{R} = 800,000$). Morphology for the periodic case is shown in Figures 4.8 and 4.9 and for the chaotic case in Figures 4.12 and 4.13.

Irreversibility in Real Life 151

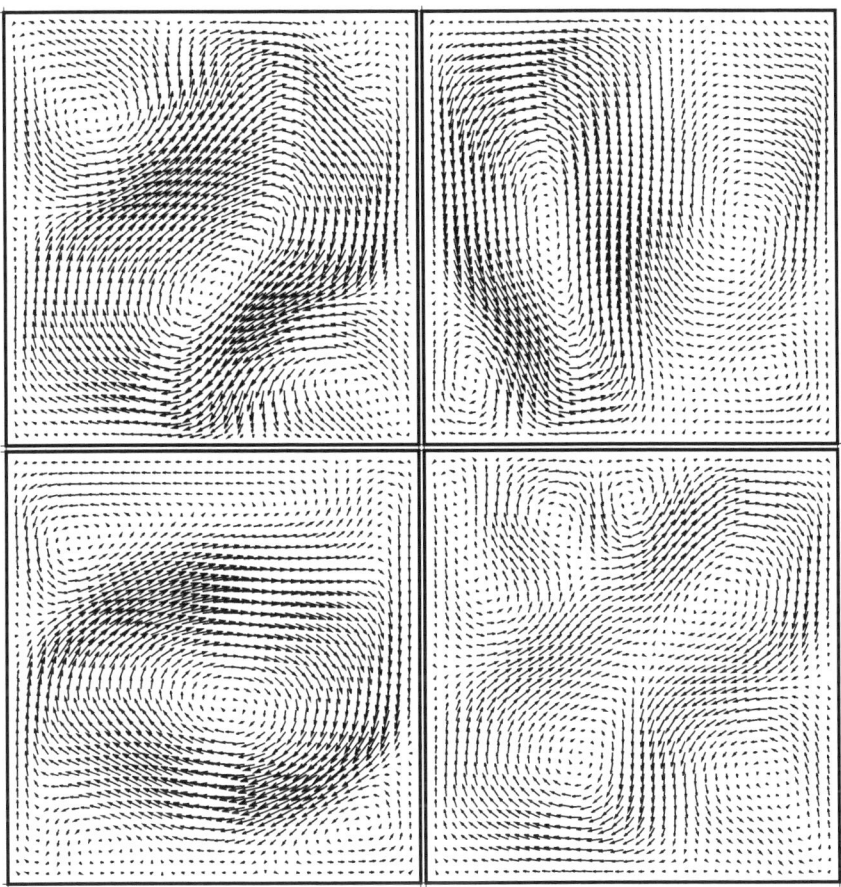

Figure 4.12. Morphology for a chaotic continuum solution using 160 × 160 computational cells and $\mathcal{R} = 800,000$. These patterns begin at the upper left and continue clockwise to the lower left, corresponding to the points labeled 1 to 4 on the kinetic energy time history shown in Figure 4.14.

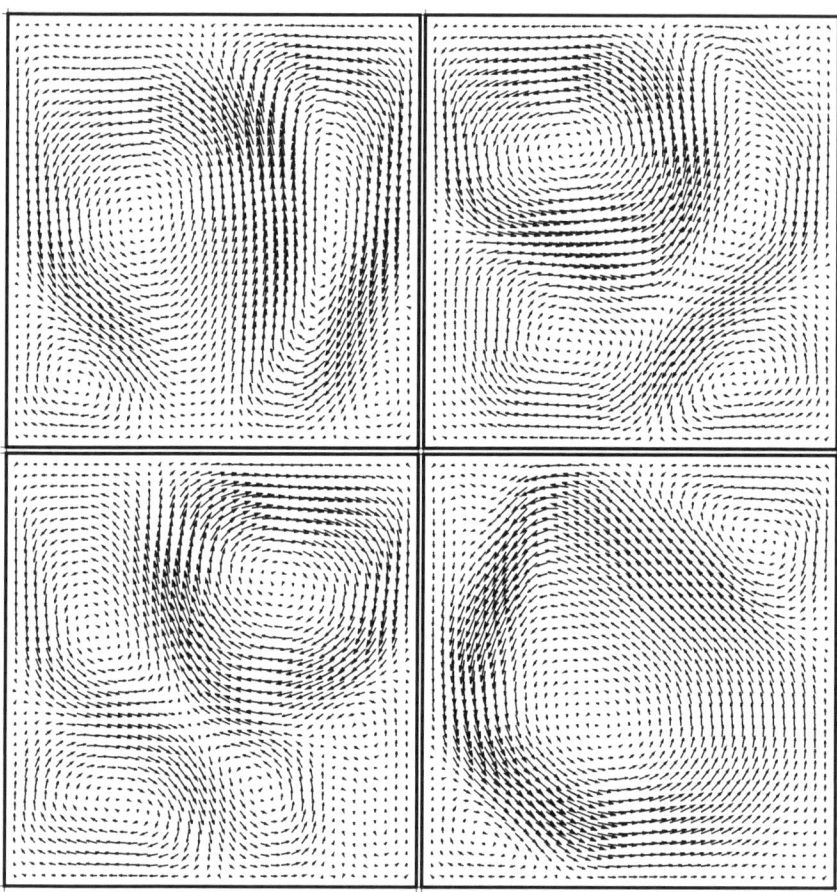

Figure 4.13. Morphology for a chaotic continuum solution using 160×160 computational cells and $\mathcal{R} = 800,000$. These patterns begin at the upper left and continue clockwise to the lower left, corresponding to the points labeled 5 to 8 on the kinetic energy time history shown in Figure 4.14.

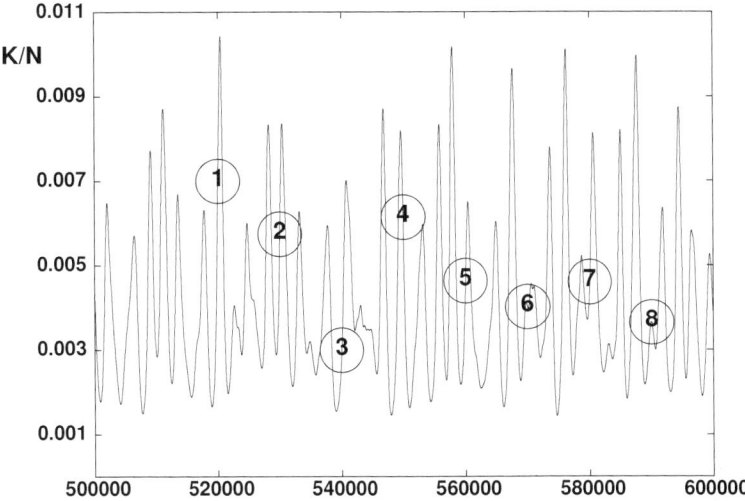

Figure 4.14. Kinetic energy per zone for $\mathcal{R} = 800,000$. The kinetic energy is chaotic. The numbered points correspond to the morphology shown in Figures 4.12 and 4.13.

The chaotic nature of the flow, reminiscent of Lorenz' model, can be checked by following the growth of the separation $\sqrt{\sum[\delta\rho^2 + \delta u^2 + \delta e^2]}$ between two nearly identical simulations. See Figure 4.15 for a typical time history of a small separation.

The Rayleigh-Bénard problems shown here exemplify the lack of uniqueness of the solutions of the nonlinear Navier-Stokes-Fourier equations. Although it is sometimes argued that entropy-based criteria can predict the relative stability of such complex flows, the multiplicity of stable solutions to the Rayleigh-Bénard problem shows that such a "maximum-entropy" idea is a flawed oversimplification.

We certainly anticipate that molecular dynamics, which carries with it the chaos of thermal motions, will agree with the results of the continuum theory, including the lack of uniqueness, for sufficiently large systems. To appreciate the degree of correspondence between the microscopic and macroscopic approaches, and its difficulties, we turn next to this application of molecular dynamics.

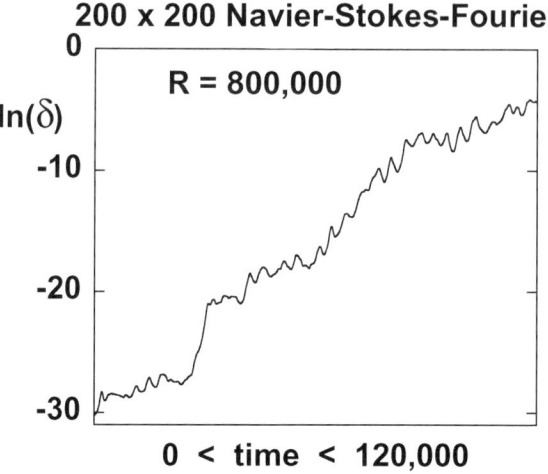

Figure 4.15. Phase-space perturbation growth, $\delta \propto e^{+\lambda t}$ for a Lyapunov unstable Rayleigh-Bénard flow. The components of the offset vector δ are differences in 201×201 nodal values, $\{\delta u_x, \delta u_y, \delta e\}$, and in 200×200 cell values of $\{\delta \rho\}$ for two very similar solutions of the Navier-Stokes-Fourier equations. The Rayleigh number is $800,000$. The corresponding largest Lyapunov exponent, $\lambda \simeq 0.0002$, has an e-folding time of 5000.

4.11.3 *Rayleigh-Bénard Flow with Molecular Dynamics*

We have seen that numerical solutions of two- (or three-) dimensional continuum problems involve *dozens* of variables at each mesh point and in each cell. The differential equations involve a long list of state and constitutive variables as well as their gradients and the underlying time derivatives:

$$\{\rho, u, e \ \to \ \nabla u, \nabla T, P, Q\} \longleftrightarrow \{(\partial \rho/\partial t),(\partial u/\partial t),(\partial e/\partial t)\} \ .$$

Despite the relative complexity of the continuum mechanics it holds a clear advantage over molecular dynamics because many fewer degrees of freedom are required. Only a few hundred nodes and cells are required for an accurate solution of stationary Rayleigh-Bénard flows.

Microscopic molecular dynamics requires just two (or three) coordinates, momenta, and force components for each point mass. It is therefore considerably simpler to program a solution for the *ordinary* differential equations of molecular dynamics, four (or six) first-order equations per particle. For maximum simplicity we consider throughout *two*-dimensional example problems, reducing the number of variables in molecular dynamics by one third, as well as greatly reducing the required number of meshpoints for an accurate solution. The main remaining difficulties lie in (i) defining "averaged flows" and (ii) developing the special thermal boundaries required to enclose the particles and enforce an overall temperature gradient on the system.

Averages can be *defined* on a coarse grid by dividing the system up into rectangular zones. Such averages are necessarily discontinuous in space and in time. They are poor choices if one would like to understand the correspondence of microscopic dynamics and macroscopic continuum mechanics We prefer defining averages which have two continuous derivatives in both space and time. Such an approach facilitates comparisons with continuum mechanics, where the nonequilibrium diffusion of mass, momentum, and energy are expressed in terms of first and second spatial derivatives. Our continuous approach is based on the smooth-particle algorithm detailed in Chapter 6. Our continuously-varying density and velocity *at* the field point r are described by smoothed spatial averages over all those particles within a range h about r:

$$\langle \rho(r,t) \rangle \equiv \sum_N m_i w(|r - r_i(t)|) \; ;$$

$$\langle u(r,t) \rangle \equiv \sum_N v_i(t) w(|r - r_i(t)|) / \sum_N w(|r - r_i(t)|) \; ;$$

$$w(|r| < h) = (5/\pi h^2)[1 + 3(|r|/h)][1 - (|r|/h)]^3 \longrightarrow \int_0^h 2\pi r w(r) dr \equiv 1 \; .$$

Here w is a short-ranged very smooth "weight function", normalized to unity. The range h can be chosen so as to optimize the correspondence between molecular dynamics and continuum mechanics. Defining the averages in this way guarantees the validity of the continuum continuity equation, $\dot{\rho} = -\rho \nabla \cdot u$. For a sufficiently large range h the velocity fluctuation, $\langle (v - \langle v \rangle)^2 \rangle$ can likewise be calculated so as to define local values of the temperature.

Like averages, boundary temperatures can be defined and imposed in a variety of ways. For Raylcigh-Bénard flow the simplest thermal boundary condition can be imposed by introducing a row or column made up of n_w tethered "wall" particles. For an *isothermal* wall the total kinetic energy of the wall particles is constrained to reproduce the desired value of the boundary's kinetic temperature:

$$2n_w kT \equiv \sum_{n_w} m(v_x^2 + v_y^2) \; .$$

In our Rayleigh-Bénard flow examples here the bottom row of particles has a constant kinetic energy $n_w k T_{\text{hot}}$ and the top row has kinetic energy $n_w k T_{\text{cold}}$.

In order that a recognizable convective flow can occur, it is necessary that the system size be sufficiently large. For large enough systems, the spatial and temporal thermal fluctuations should become insignificant, with the atomistic flows closely resembling the corresponding continuum idealizations from the Navier-Stokes-Fourier equations. Continuum flows become "interesting" for Rayleigh numbers of several thousand and chaotic for Rayleigh numbers on the order of a million.

To estimate how many atomistic particles these problems involve we first simplify our considerations by choosing the gravitational constant consistent with constant density, $mgL = k\Delta T$ where σ is the "collision diameter" of the particles and $N(\sigma^2/L^2)$ is of order unity. In this case the (two-dimensional) diffusive transport coefficients, ν and D_T are both of order $\sqrt{kT/m}(L^2/N\sigma)$ so that we can estimate the number of particles required for a Rayleigh number of order 10,000:

$$\mathcal{R} \simeq 10,000 = g\Delta T (\partial \ln V/\partial T)_P L^3/(\nu D_T) \simeq$$

$$k(\Delta T/mL)(\Delta T/T)L^3(m/kT)(N^2\sigma^2/L^4) = (\Delta T/T)^2 N \simeq N \ .$$

Early molecular dynamics simulations confirmed this estimate, showing that about 5000 particles are required, in two dimensions, for systematic convection to occur. In three dimensions the same argument gives $10,000 \simeq N(\sigma/L) \simeq N^{2/3}$ so that interesting three-dimensional molecular dynamics simulations require *millions* of particles, not just tens of thousands.

The number and types of rolls observed both depend upon the aspect ratio of the system and on the boundary conditions. In three dimensions hexagonal convection cells with complex spiral particle trajectories are typical. In two dimensions various roll-like morphologies are the rule. As the temperature gradient is increased, the single two-dimensional stationary roll pattern becomes unstable. New roll patterns can appear, as we will detail for square-box boundary conditions. With a wider box, giving rise to two rolls at a critical Rayleigh number, vertical oscillations of the rolls occur as the Rayleigh number is increased. At still higher Rayleigh numbers, chaotic plumes, moving from side to side, can be seen.

In the 1980s Rapaport was able to model the vertical roll oscillations by using 57,600 hard-disk particles in a microscopic molecular dynamics simulation. More recently, with parallel computation, he studied a relatively thin *three*-dimensional Rayleigh-Bénard flow with eight roughly-hexagonal cells and 3,507,170 soft-sphere particles. By way of contrast, a few hundred "smooth particles" are sufficient to simulate two-dimensional convective flow and only tens of thousands are needed in three dimensions, using the macroscopic continuum equations, as is described in detail in Section 7.7 and Oyeon Kum's 1995 PhD thesis (University of California at Davis/Livermore).

Rapaport made detailed movies (the individual frames were averaged over both space and time) showing the results of his hard-disk and soft-sphere simulations. He found convective flows in semiquantitative agreement with analogs from numerical continuum simulations. He could distinguish three different flow regimes: (i) conductive flow without convection; (ii) stationary circulation of two counter-rotating vortices, as in Figure 4.6 at time 0 ; and (iii) periodic vertical oscillations of two rotating vortices. In the three-dimensional work (with periodic lateral boundaries) he observed merging of the convective columns as their numbers decreased from about thirty to a final number of eight. The convective plumes which both experiments and continuum simulations reveal at higher values of the driving temperature gradient are not so easily accessible to atomistic simulations.

Suppose, as with hard disks and hard spheres, that the pressure is proportional to temperature. We have seen that the resulting Rayleigh number is of order N in two dimensions. The convective plumes, which become prominent at Rayleigh numbers of order 2×10^5 and higher, require atomistic simulations with hundreds of thousands of particles, just feasible for modern work stations. On the other hand, continuum simulations on work stations can easily reach Rayleigh numbers of order 10^8. See Vic Castillo's 1997 papers and his 1999 PhD thesis (University of California at Davis/Livermore).

To illustrate an atomistic Rayleigh-Bénard simulation we use the pair potential along with the single-particle boundary tethering potential described in Section 4.9, page 128:

$$\psi(r < 1) = 100[1 - r^2]^4 \; ; \; \phi(\delta r) = 200(\delta r)^4 \; .$$

Our exploratory simulations were discouraging, in that the atomistic flows are intrinsically dominated by thermal fluctuations. Although the effect of spatial fluctuations can be minimized by choosing appropriate weight functions, the additional temporal fluctuations, lasting on the order of a few collision times, require a further time-averaging step. With a timestep of 0.0025 the flow stabilizes at a time of order 100,000. Visualizing the underlying flows requires not only spatial, but also temporal, averaging, over time intervals of at least several collision times. Figure 4.16 compares a longtime average (eight million timesteps) to an instantaneous snapshot for the molecular dynamics of $N = 150 \times 150 + 4 \times 300 = 23,700$ particles. 1200 boundary particles define the 300×300 square box. As usual the gravitational constant was chosen to promote constant density, $g = (\Delta T/300)$. The hot and cold temperatures are 0.015 and 0.005 so that, using kinetic-theory estimates of the viscosity and conductivity, the Rayleigh number is of order

$$\mathcal{R} = g\Delta T (\partial \ln V/\partial T)_P L^3/(\nu D_T) \simeq$$

$$(0.01/300)(0.01/0.01)300^3/(0.2 \times 0.1) = 900/0.02 = 45,000 ,$$

which should be within the range of single-roll stability.

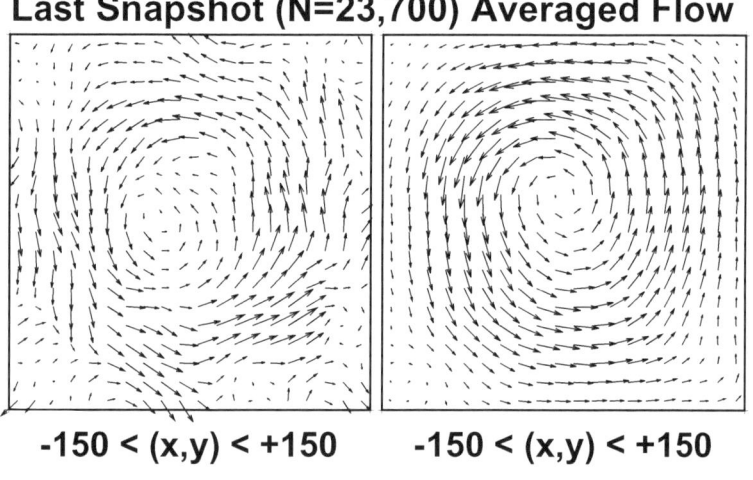

Figure 4.16. Rayleigh-Bénard flow velocity from molecular dynamics, using Lucy's weight function with $h = 50$, where the system width is 300. The longtime average shows a well-formed single roll, similar to the continuum solution of Figure 4.6. The snapshot, from the last timestep, is spoiled by thermal fluctuations. The Rayleigh number for this 23,700 particle simulation is approximately 45,000.

Evidently the correspondence between continuum mechanics and molecular dynamics is relatively poor for this problem. Perhaps this reflects the fundamentally different time-dependences of the two approaches. In continuum mechanics the fluctuations in stress and strain rate, like fluctuations in heat flux and temperature gradient, are in phase with each other,

$$\delta\sigma_{xy} \propto \delta\dot{\epsilon}_{xy} \ ; \ \delta Q \propto \delta \nabla T \ .$$

In atomistic mechanics these fluctuations are necessarily *out of phase*. The atomistic strain rate and heat flux change sign with time reversal while stress and temperature do not. This is opposite to the continuum situation.

Despite the convergence problems, an overall thermodynamic analysis of these Rayleigh-Bénard flows follows relatively easily from our choice of deterministic boundary conditions. As described in Sections 2.8 and 3.4, the instantaneous change in the phase-space probability density $f(\{q,p\},t)$ and the extension in phase $\otimes$ depend upon the instantaneous heat flows at the "cold" top and "hot" bottom of the system. We count Q as positive in the upward direction:

$$(d \ln f/dt) = -(d \ln \otimes /dt) = n_w[(Q_y/kT)_{\text{cold}} - (Q_y/kT)_{\text{hot}}].$$

In the long-time limit the entering and leaving flows of heat must balance one another, giving the simpler result:

$$\langle (d \ln f/dt) \rangle = -\langle (d \ln \otimes /dt) \rangle = n_w \langle Q_y \rangle [(1/kT)_{\text{cold}} - (1/kT)_{\text{hot}}] \ .$$

Because the combination of temperature terms in the square brackets is necessarily positive, it is necessary that Q be positive too, corresponding to the *increase* of f with time and the *decrease* of the comoving extension in phase $\otimes$ which is characteristic of a strange attractor. This ability to predict the direction of the heat flow—from hot to cold—as well as the rate of collapse of the probability density to the strange attractor, is an important advantage of the deterministic thermostats.

4.12 Summary

How is the irreversibility seen in real life, and exemplified here in our atomistic and continuum simulations, to be understood? In most cases interactions with the *surroundings* are involved. Without the cat, and the floor, the cup would not shatter. Without the small boy and the rock, or at least the wind, or frogs, fish, or raccoon, there would be no waves dissipating at the water's edge. The *surroundings* act as sources and sinks for momentum and energy, and the "information" generated by chaos. The simple Rayleigh-Bénard problem is already sufficiently complex to display these features of irreversible chaos.

Fick's, Newton's, and Fourier's phenomenological macroscopic transport laws describe the simplest nonequilibrium states, the linear mass, momentum, and energy flows which result when sufficiently small gradients are present. Any system, coupled to its surroundings, is subject to perturbing influences which are then amplified by Lyapunov instability. Chaotic expansion and contraction in phase space creates and destroys information, making it impossible in principle to retrace the past or to know the future. Forward evolution invariably seeks out relatively stable attractors while the inaccessible reversed record of this evolution corresponds to an unstable unobservable repeller.

Evidently irreversibility is a consequence of nonlinearity, chaos, mixing, and bifurcations, all leading to the overall creation and destruction of information, with the destruction dominant. With nonlinearity, small perturbations can only die, oscillate forever, without change, or diverge. Without chaos and mixing there is no possibility of the memory loss (which gives an "insensitivity to initial conditions") which makes a physical deterministic description, with observations and simulations, possible. Bifurcations, of the "heads *versus* tails" type, are a simplified model for the time-reversible Lyapunov instability which creates and destroys information and directs systems toward multifractal attractors obeying the Second Law of Thermodynamics. Although bifurcations are a common feature of fine-grained molecular dynamics they are not necessarily always present in the continuous solutions of discretized continuum mechanics. The chaos responsible for irreversible behavior in continuum mechanics has a more global, as opposed to local, character.

4.12.1 *Notes and References*

Pars' *Treatise on Analytical Mechanics* and Sommerfeld's *Mechanics* texts both discuss Gauss' Principle. Landau and Lifshitz' *Fluid Mechanics* provides a succinct introduction to continuum concepts. Bridgman discusses some of the difficulties in treating metal plasticity with thermodynamics in "The Thermodynamics of Plastic Deformation and Generalized Entropy".

A relatively formal entropy-based approach to nonlinear transport has been promoted by Jou, Casas-Vasquéz, and Lebon, in their book, *Extended Irreversible Thermodynamics*.

The microscopic details of nonequilibrium fluctuations have given rise to a voluminous and sometimes contentious literature. The macroscopic approach, as in Landau and Lifshitz' *Statistical Physics*, is well accepted. Evans, Cohen, and Morriss introduced an approximation, widely applicable and exact at long times, for nonequilibrium flucutations in "Probability of Second Law Violations in Shearing Steady States". Chris Jarzynski has written a comprehensive review and guide to the corresponding research literature, "Equalities and Inequalities: Irreversibility and the Second Law of Thermodynamics at the Nanoscale".

The ultimate nonequilibrium steady state, a stationary shockwave, can be modeled accurately with molecular dynamics. See Holian, Hoover, Moran, and Straub's "Shockwave Structure *via* Nonequilibrium Molecular Dynamics and Navier-Stokes Continuum Mechanics" as well as our seven arχiv contributions on this subject listed at the end of Chapter 6.

The Rayleigh-Bénard problem has been a staple of experimental and computational work. The rapid evolution from Mareschal, Mansour, Puhl, and Kestemont's "Molecular Dynamics *versus* Hydrodynamics in a Two-Dimensional Rayleigh-Bénard System" to Rapaport's more recent [three-dimensional] work, "Hexagonal Convection Patterns in Atomistically Simulated Fluids", Physical Review E **73**, 025301 (2006), makes fascinating detail available in highly-complex chaotic flows. The smaller-scale caricature of this problem, developed by Lorenz, has been the subject of a book, Sparrow's *The Lorenz equations: Bifurcations, Chaos, and Strange Attractors*, while Lorenz' basic model has itself been simplified by Sprott, in "Simplifications of the Lorenz Attractor".

Chapter 5
Microscopic Computer Simulation

*Faith is a fine invention
for gentlemen who see;
But microscopes are prudent
in an emergency.*

Emily Dickinson

5.1 Introduction

Microscopic flows begin with individual particle coordinates, velocities, and accelerations. They take place on atomistic space and time scales. To simulate such microscopic flows, using fast computers, involves first of all, a specification of the interparticle forces. The forces are ordinarily derived from a potential function: $\{F \equiv -\nabla\Phi(\{q\})\}$ which depends upon the coordinates $\{q\}$. The potential energy Φ could be a sum of simple pair terms (hard spheres or Lennard-Jones', for example), and could also include complicated many-body interactions chosen to represent anisotropy, defects, and surface properties of metals. Over a relatively limited range of thermodynamic states, this force-law approach can lead to useful correlations of data for real materials.

In modeling and understanding real materials it is necessary to estimate the quantum effects which both cause and complicate interparticle forces inferred from experiments. These adjustments to purely classical behavior are both linear (λ_h/σ), where σ is an effective collision diameter, and quadratic ($\lambda_h^2 \nabla^2 \Phi$), in the quantum de Broglie wavelength, $\lambda_h \equiv h/\sqrt{2\pi mkT}$, with h Planck's constant. A simpler wholly-classical approach views simulation, for given forces, as a data source alternative to laboratory experiments and theoretical explanations. It is convenient to imagine that the chosen force law defines a corresponding "material", whether or not any similar material exists in the real world. With the forces chosen, an algorithm for solving the corresponding motion equations, with or without the influence of boundaries, is required. Finally, an analysis of the results is needed. Because, as kinetic theory indicated, even the simplest interparticle forces, either impulsive or smooth, are quite sufficient for an understanding of fundamental physics, we will not here touch on the complications which result from more elaborate models incorporating detailed molecular structure and long-ranged Coulomb forces. Instead, we emphasize here the *simplest* choices of integration algorithms and boundary conditions, as well as methods of analysis for treating the results gleaned from computer simulations.

5.2 Integrating the Motion Equations

Suppose that we have a representation of the forces in terms of all the current particle coordinates: $\{F(\{r_0\})\}$. Newton's motion equations can then be discretized and solved with the "leapfrog algorithm",

$$\{r_+ - 2r_0 + r_- \equiv \Delta t^2 (F_0/m)\},$$

where the subscripts indicate three successive times, separated by the timestep Δt. Reasonable accuracy requires that Δt be considerably smaller than the shortest vibrational period or collision time. The long-time coordinate error for this algorithm is formally of order $t(\Delta t)^2$ but can actually increase *exponentially* with time when Lyapunov instability is present (see Chapter 8). If increased accuracy is desired, the fourth-order Runge-Kutta algorithm is an excellent choice. For a set of coupled first-order ordinary differential equations, $\dot{x} = y(t, x)$ this algorithm takes the form:

$$x_1 = x_0 + (\Delta t/2) y(0, x_0) \ ;$$

$$x_2 = x_0 + (\Delta t/2) y(\Delta t/2, x_1) \ ;$$

$$x_3 = x_0 + (\Delta t) y(\Delta t/2, x_2) \ ;$$

$$x_{\Delta t} \equiv x_0 + (\Delta t/6)[y(0, x_0) + 2y(\Delta t/2, x_1) + 2y(\Delta t/2, x_2) + y(\Delta t, x_3)] \ .$$

Because the Runge-Kutta algorithm corresponds to a truncated series expansion of the motion equations' solution, it generally does not reproduce the time-reversibility of the underlying differential equations, This lack of consistency can be used to estimate the integration error. Because the formal one-step integration error is $\dddot{x}\,(\Delta t)^5/5!$, it is quite feasible to estimate the timestep required to reduce the error to the level of computational roundoff, usually about one part in 10^{16}.

The corresponding timestep Δt is of order $(\tau/1000)$ where τ is a typical vibration period. Practical calculations can use a timestep ten times larger. It is an advantage of the Runge-Kutta algorithm for first-order ordinary differential equations that it needs no special modifications for nonequilibrium simulations with velocity-dependent forces. It is also possible to generalize the leapfrog approach to deal with these cases. The control variables required to thermostat nonequilibrium steady states are more easily included in a Runge-Kutta integration of the motion equations. In such a case the extent of irreversible heating must first be quantified, and then extracted from the system using the thermostat forces described in Chapter 2. In Section 5.8 we illustrate the algorithm for steady-state isoenergetic shear flow. The special feature the algorithm requires is a treatment of periodic boundary conditions that is consistent with steady shear.

5.3 Interpretation of Results

In order to draw parallels linking microscopic simulations to macroscopic analyses—thermodynamics and hydrodynamics—mechanical analogs of the pressure tensor, heat flux vector, and temperature are all required. Temperature is basic to nonequilibrium simulations, for any steady-state simulation converts work, or stored energy, to heat. In Chapter 2 we saw that the ideal-gas interpretation of temperature,

$$kT_{xx} = \langle p_x^2/m \rangle;\ kT_{yy} = \langle p_y^2/m \rangle\ ,$$

has both a *dynamical* basis in kinetic theory and a *thermodynamic* basis in Gibbs' statistical mechanics. Unlike the microcanonical equilibrium definition $T \equiv (\partial E/\partial S)_V$ discussed in Sections 2.8 and 3.5 the ideal-gas kinetic temperature is useful *far from equilibrium* too, where T_{xx} and T_{yy} can differ. We adopt this ideal-gas definition of temperature under any and all conditions.

The pressure tensor and heat flux vector likewise have two dynamical bases. They can be derived by considering "balance equations" for the mo-

mentum and energy within a small volume, or, equivalently, by expressing particle contributions to locally-defined fluxes of momentum and energy, using the forces and energies governing the microscopic equations of motion. The volume-averaged pressure tensor P within the volume V is given by:
$$PV = \sum_k [(pp/m)_k + (rF)_k] \ .$$
Each of the tensor components in the first term, a sum over the individual particles $\{k\}$, gives the rate at which the corresponding x or y momentum component is transported by that particle in the x or y direction. $P_{xy}V$, the summed-up flow of x momentum in the y direction, includes a contribution $(p_x p_y/m)$ for each particle in the volume V contributing to the flux. The second term is the dyadic product of the individual particle coordinates and the total forces on the same particle. For pair forces, from a pair potential ϕ,
$$\Phi = \sum_{i<j} \phi_{ij} \ ,$$
the single-particle (rF) sum can be written as a sum over all distinct particle pairs, with each pair contributing the dyadic $(r_i - r_j)F_{ij}$ to the sum, where $F_{ij} = -F_{ji}$ is the force on particle i resulting from its interaction with particle j.

The heat flux vector Q is likewise composed of two terms, a convective energy transport and a contribution due to changes in interparticle interactions. The two terms taken together give the complete heat flux. For pair forces the instantaneous, but volume-averaged, heat flux Q within the volume V is given by:
$$QV = \sum_i (pe/m)_i + \sum_{i<j} r_{ij}[F_{ij} \cdot (p_i + p_j)/(2m)] \ ,$$
where the individual particle energies $\{e_i\}$ include the comoving kinetic energy as well as half of each interaction energy ϕ_{ij} shared by particle i with another particle j. Evaluating the shear stress or the heat flux for appropriate nonequilibrium boundary conditions is the simplest route to the transport coefficients. The shear stress algorithm illustrated in Section 5.8 is a good choice for determining shear viscosity. The tethered-boundary heat flow algorithm of Chapter 4 fills the need for a thermal conductivity algorithm.

For completeness, we sketch the derivations of the pair-force expressions for the pressure tensor and the heat flux vector. More details can be found in Section 5.5 of *Computational Statistical Mechanics*. We start by writing the time derivatives of Particle i's momentum and energy and multiplying those derivatives by the particle coordinate r_i:

$$r_i \dot{p}_i = r_i \sum_j F_{ij}; \ r_i \dot{e}_i = -r_i \sum_j F_{ij} \cdot (p_i - p_j)/(2m) + r_i \sum_j (p_i/m) \cdot F_{ij} \ .$$

The last term comes from the time derivative of particle i's kinetic energy, $(p_i^2/2m)$. We have omitted here, on the righthand sides, the *boundary* contributions to the system momentum and energy. We will include them below. Here we assign half the potential energy ϕ_{ij} of any (ij) pair to Particle i and the other half to Particle j. Next, we consider two longtime averages:

$$\langle (d/dt)(rp)_i \rangle = \langle (\dot{r}p)_i + (r\dot{p})_i \rangle \equiv 0; \ \langle (d/dt)(re)_i \rangle = \langle (\dot{r}e)_i + (r\dot{e})_i \rangle \equiv 0 \ .$$

The averages vanish. This is because the longtime averaged time derivative of any bounded quantity is necessarily zero.

We can use these averages to compute the summed time averages of $(\dot{r}p)_i$ and $(\dot{r}e)_i$:

$$\langle \sum_i (\dot{r}p)_i \rangle = -\sum_i r_i \sum_j F_{ij} + PV \rangle \ ;$$

$$\langle \sum_i (\dot{r}e)_i \rangle = -\sum_i r_i \sum_j F_{ij}(p_i + p_j)/(2m) + QV \rangle \ .$$

Notice that the time-averaged momentum transfer at the x boundaries, $\pm L_x$, multiplied by the cross-sectional area L_y (in two dimensions), and similarly for y, gives the pressure volume product $-PV$ and that the time-averaged heat transfer at the boundaries likewise gives the heat flux vector volume product $-QV$. Combining the i and j sums into a pair sum and introducing $r_{ij} \equiv r_i - r_j$ gives the final forms used for computation:

$$PV = \sum_k [(pp/m)_k + (rF)_k] \ ;$$

$$QV = \sum_i (pe/m)_i + \sum_{i<j} r_{ij} [F_{ij} \cdot (p_i + p_j)/(2m)] \ .$$

These are respectively the "virial theorem" and the "heat theorem".

These instantaneous expressions, without the time-averaging brackets, can also be derived by considering the local momentum and energy fluxes as sums of convective and action-at-a-distance contributions. We can put them to use in comparing simulation results, from molecular dynamics, with solutions of the continuum equations.

In such cases it is desirable to define *local* values of temperature, pressure, heat flux, and the like. A particularly useful description of such variables, with the advantage of two continuous space derivatives, relies on the smooth-particle interpolation method illustrated in Section 3.10 and applied to the Rayleigh-Bénard problem in Section 7.7. Briefly, local averages of *any one*-particle property X_i, at the location r, are calculated using "weighting functions" $w(r - r_i)$ with two continuous derivatives,

$$\langle X(r) \rangle \equiv \sum_i X_i w(|r - r_i|) / \sum_i w(|r - r_i|) \ .$$

The sums include all particles $\{i\}$ within the range of the weighting function. This link between particle and continuum properties is also the basis for a simple numerical method for simulating *continuum* dynamics with *particles*, "Smooth Particle Applied Mechanics", discussed in Chapter 7. In what follows here we will illustrate the solution and analysis of nonequilibrium motion equations with a simple one-dimensional example, with two-dimensional examples of momentum and energy flows, and with a shockwave simulation. To conclude this Chapter we will work out three nonequilibrium example problems, the Galton Board, and heat flow in both one-body(!) and many-body systems.

5.4 Control of a Falling Particle

An isolated nonrelativistic particle in a gravitational field continues to accelerate for all time, never reaching a steady state. A particle falling in a *viscous* medium *can* reach a steady nonequilibrium state in which the frictional viscous force just balances the gravitational acceleration. A particle falling in a vacuum, or through an array of fixed scatterers, can also reach a steady state, provided that the motion is controlled, or thermostated, by artificial thermostat forces of the type described in Chapter 2.

As an example of artificial control, consider a falling particle from the standpoint of Dettmann's Hamiltonian, the simplest approach to isothermal Nosé-Hoover dynamics, as discussed in Section 2.8. If we set the constants defining the temperature, viscous relaxation time, gravitational field strength, and mass, $\{kT, \tau, g, m\}$, all equal to unity, Dettmann's Hamiltonian is:

$$\mathcal{H}_{\text{Dettmann}}(q, p, s, p_s) = (p^2/2s) + s[-q + \ln s + (p_s^2/2)] \equiv 0 \ .$$

We will see that the momentum p_s conjugate to s corresponds to the usual time-reversible frictional control variable ζ.

For Dettmann's special choice, $\mathcal{H} \equiv 0$, Hamilton's equations of motion are as follows:

$$\{\dot{q} = +(\partial\mathcal{H}/\partial p); \, \dot{p} = -(\partial\mathcal{H}/\partial q); \, \dot{s} = +(\partial\mathcal{H}/\partial p_s); \, \dot{p}_s = -(\partial\mathcal{H}/\partial s)\}$$

$$\longrightarrow \{\dot{q} = (p/s); \, \dot{p} = s; \, \dot{s} = sp_s; \, \dot{p}_s = (p/s)^2 - 1\} \, .$$

If we express $\ddot{q}$ in terms of $\dot{q} \equiv v$ and set $p_s \equiv \zeta$, then Hamilton's equations for this falling-body problem take on the usual Nosé-Hoover form, with s absent:

$$\{\dot{q} = v; \, \ddot{q} = \dot{v} = 1 - \zeta v; \, \dot{\zeta} = v^2 - 1\} \, .$$

Evidently the stationary dissipative value of the velocity, $v = +1$, corresponds to a friction coefficient $\zeta = 1$, so that a particular solution of the equations is:

$$\{q = +t; \, v = +1; \, p = +e^{+t}/e; \, s = e^{+t}/e; \, p_s = \zeta = +1\}$$

$$\longrightarrow \mathcal{H}_{\text{Dettmann}} \equiv 0 \, .$$

From the physical standpoint, gravitational field energy is converted into heat by the motion, $[-\dot{q} = -\zeta v = -1]$, with the corresponding heat extracted by the control force $-\zeta v$ representing the interaction of the falling particles with a heat reservoir. The divergence of the flow velocity, in (q, v, ζ) space, is negative:

$$(\partial \dot{q}/\partial q) + (\partial \dot{v}/\partial v) + (\partial \dot{\zeta}/\partial \zeta) = 0 - 1 + 0 = -1 \, ,$$

so that the solution near the stable fixed point $(v, \zeta) = (+1, +1)$ is "attractive", with the infinitesimal comoving three-dimensional volume element contracting toward zero, $\otimes \simeq e^{-t}$.

This dissipative solution, like the original set of Hamilton's equations for the motion of $\{q, p, s, p_s\}$ is also *time-reversible*, with the variables $\{\dot{q} = v, \, p, \, p_s = \zeta\}$ all changing sign in the reversed motion:

$$\{q = -t; \, v = -1; \, p = -e^{-t}/e; \, s = e^{-t}/e; \, p_s = \zeta = -1\} \longrightarrow \mathcal{H} \equiv 0 \, .$$

The divergence of the reversed flow velocity in (q, v, ζ) space is *positive*:

$$(\partial \dot{q}/\partial q) + (\partial \dot{v}/\partial v) + (\partial \dot{\zeta}/\partial \zeta) = 0 + 1 + 0 = +1 \, ,$$

or "repulsive", and corresponds to exponential instability characterizing the *unstable* "repulsive" fixed "point": $(v, \zeta) = (-1, -1)$, with $\otimes \simeq e^{+t}$.

This simplest of deterministic time-reversible thermostated problems illustrates the relative stability of a dissipative process converting potential energy, or work, into heat. The evolving flow can be thought of as linking the unstable repeller "source" to the stable attractor "sink". See Figure 5.1. Projected into (v, ζ) space, the ultimate steady-state attractor is the single point $(v, \zeta) = (+1, +1)$. More complicated chaotic problems generate strange attractor-repeller pairs in phase space, with the attractors stable relative to their time-reversed repeller counterparts.

If we had applied a stochastic heat reservoir with a constant friction coefficient ζ to this same falling-particle problem we would have obtained the Langevin equation of motion,

$$\ddot{x} = \dot{v} = 1 - \zeta v + \text{"noise"} ,$$

with the noise amplitude determining the kinetic temperature of the thermostated system. In the simplest case, with the bath temperature much less than that of the falling particle, the Langevin noise can simply be ignored and the solution becomes the same as that just obtained from Dettmann's Hamiltonian:

$$\{\dot{x} = v;\ \dot{v} = 1 - v\} \longrightarrow \{x = t;\ v = 1\} .$$

These Langevin equations of motion, and their solution, are *irreversible* (because the friction coefficient is constant), but they can still be "inverted", to calculate the past history of the falling ball. The inverted solution, like that calculated with Dettmann's time-reversible friction, is then unstable. If we represent by Δ a small velocity perturbation at time zero, the inverted perturbation *grows* exponentially in the time:

$$\{v, t\} \to \{-v, -t\} \to \{\delta \dot{x} = \delta v;\ \delta \dot{v} = \delta v\} \to \{\delta x = \Delta e^{+t};\ \delta v = \Delta e^{+t}\} .$$

The reversible friction-coefficient approach to simulation has three advantages over the stochastic Langevin alternative: (i) the equations have the useful analytic property of time reversibility, with the friction coefficient(s) $\{\zeta\}$ changing sign in the time-reversed solution; (ii) the results are reproducible, independent of any random number generators or computer idiosyncrasies; (iii) the summed-up friction coefficients are directly related to the rate of phase-space volume collapse—onto a strange attractor—through the compressible version of Liouville's Theorem, detailed in the following Section:

$$-\dot{\otimes}/\otimes \equiv \dot{f}/f \equiv -\sum[(\partial \dot{q}/\partial q) + (\partial \dot{p}/\partial p)] = \sum_p \zeta = \sum_\zeta \#_\zeta \zeta ,$$

where $\#_\zeta$ is the number of degrees of freedom thermostated by the friction coefficient ζ. In addition, if the attractor is independent of the initial conditions, which is typical, then the attractor-repeller pairs characteristic of time-reversible motion equations are necessarily ergodic. This means that *any* possible initial condition will recur in the future.

Figure 5.1 shows streamlines in (v, ζ) space which correspond to choosing different initial conditions for Dettmann's time-reversible equations. The attractive fixed point at $(+1, +1)$ is always the stable *sink*, while the unstable fixed point at $(-1, -1)$ acts as a repeller *source* for the flow.

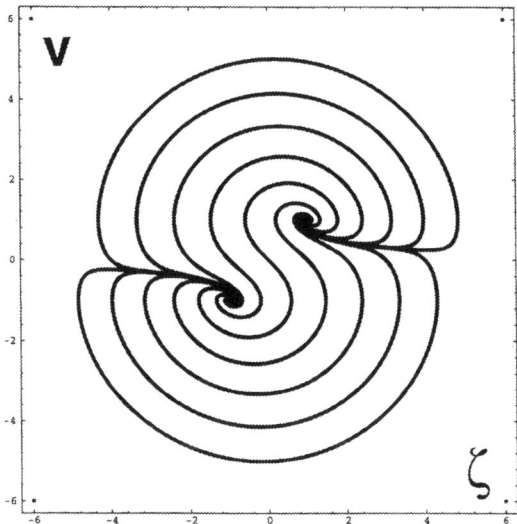

Figure 5.1. Typical trajectories in the (v, ζ) space for the falling particle. The trajectories link an unstable fixed point $(-1, -1)$ to the stable "attractive" fixed point at $(+1, +1)$.

Microscopic simulations can only provide suggestive indications of macroscopic behavior, because, short of Hurculean efforts, processor speeds and capacities limit them to tens or hundreds of millions of degrees of freedom. Without these speed and size limitations there would be no need for continuum mechanics. Of course continuum mechanics is likewise limited in the duration and number of equations solved, so that the range of time and space scales which can be studied is strictly limited in two and three-dimensional Eulerian or Lagrangian continuum simulations.

In Section 2.5 we pointed out that Liouville's Incompressible Theorem for the time-rate-of-change of phase volume ⊗ is a direct consequence of Hamilton's equations of motion for an isolated system:
$$\mathcal{H} \Longrightarrow \{\dot{f} = 0 \longleftrightarrow \dot{\otimes} = 0\} \ .$$
Neither the phase-space probability density f nor the comoving phase volume ⊗ can change in the purely Hamiltonian case. An extremely useful consequence of this probability-conservation property is a geometric interpretation of isoenergetic thermodynamic equilibrium. Evidently any *equilibrium* phase-space distribution, being time independent, should correspond to a distribution with a fixed probability density and with a fixed "extension in phase". The simplest such distribution is Gibbs' microcanonical distribution, which contains all accessible states of equal energy, all of them with equal weights.

The intuitive reasoning linking an unchanging distribution to equilibrium could likewise be applied to nonequilibrium steady states, which evidently must also be unchanged. Liouville's Theorems can be misleading for that situation because the nonequilibrium distributions are typically fractals, with an overall perpetually-contracting attractive flow. Liouville's Theorems only apply to differentiable distributions. Applied to nonequilibrium situations, for which fractals are appropriate, Liouville's compressible theorem predicts that the fine-grained comoving distribution continually changes. This conclusion would be true for any continuous solution. The rub is that no such solutions exist! The long-time limit is instead invariably *fractal*, as is discussed in more detail in Chapter 8.

For the *non*Hamiltonian thermostated equations of motion discussed here, both the density f and the *infinitesimal* comoving volume element ⊗ can change with time. This follows from the *compressible* form of Liouville's Theorem:
$$\dot{f} = (\partial f/\partial t) + v \cdot \nabla f = -f \nabla \cdot v = -f \dot{\otimes}/\otimes \ .$$
In the simplest case, using Nosé-Hoover thermostats, with the friction coefficient(s) included as new independent variables, the phase-space velocity is $\{\dot{q}, \dot{p}, \dot{\zeta}\}$. The divergence of the flow is then just the sum of the friction coefficients:
$$\dot{f} = -f \nabla \cdot v = -f \sum_p \partial(-\zeta p)/\partial p = f \sum_p \zeta \ .$$
where the sums are over all thermostated degrees of freedom. As a corollary, the instantaneous "local" identity (where ⊗ is infinitesimal),
$$\nabla \cdot v \equiv d \ln \otimes /dt \equiv -d \ln f/dt = -\sum_p \zeta \ ,$$

shows that positive friction causes a loss of phase volume, while negative friction corresponds to increasing volume. From this simple observation it is clear that any stable stationary state must correspond to *positive* friction, overall, and to the *shrinkage* of phase volume. This conclusion, the mechanical analog of the Second Law of Thermodynamics, became apparent in 1986, on looking at computer-generated phase-space distributions for simple nonequilibrium systems. The distributions revealed the fractal character which we discuss more fully in Chapter 8. Here we will consider applications of this thermostat idea to both shear flows and heat flows.

The phase-space distribution typically "collapses" onto an ergodic multifractal attractor—$(f \to \infty \, ; \, \otimes \to 0)$—like the dissipative attractor shown on page 71 resulting from the time-reversible map considered in Section 2.12.1. Figures 5.8-5.10, shown with the Galton-Board example problem, at the end of this Chapter, likewise illustrate typical multifractal phase-space distributions.

The Galton Board, or "Lorentz Gas" was the first such system to be investigated. It consists of a thermostated mass point driven through a regular periodic lattice of hard elastic disks by a constant external field. The *global* time average of the local collapse rate, $\langle d \ln \otimes / dt \rangle < 0$, can be expressed in terms of the time-averaged values of the reservoirs' entropy change $\dot{S}$:

$$\sum \langle \dot{S}_{\text{external}}/k \rangle \equiv -\langle d \ln \otimes /dt \rangle \equiv \langle d \ln f /dt \rangle \equiv -\sum \lambda \equiv \sum \langle \zeta \rangle > 0 \, ,$$

where the long-time-averaged Lyapunov exponents, $\{\lambda\}$, can be calculated as is described in Section 8.4.

It seems paradoxical that this unidirectional collapse, onto a fractal attractor, is achieved with *time-reversible* motion equations. The details of the unidirectional behavior found with thermostats can depend upon the precise form the thermostats take. This sensitivity to boundary conditions has macroscopic analogs. It corresponds macroscopically to the dependence of drag coefficients on surface friction. This effect is specially noticeable in the frictional force a flowing fluid exerts on the inner walls of pipes with various finishes and surface roughnesses.

An inverted movie of a far-from-equilibrium numerical solution, with the movie projected backward, would certainly show a violation of the Second Law of Thermodynamics. In such an inverted movie heat would "naturally" flow from cold to hot in the presence of temperature gradients; heat would likewise be converted into velocity *differences* in inverted viscous flows. These artificially-*reversed* phenomena, like unburning papers, or divers reëmerging dry from a pool, make no sense because they contradict our everyday experience. It is noteworthy that such artificially-inverted movies rely on *stored* information. They can be continued backward only to the initial state of the motion—the first frame. Any attempt to integrate the equations of motion further backward in time, beyond the initial condition, by replacing $+\Delta t$ with $-\Delta t$, produces a solution indistinguishable, on the average, from the solution forward in time.

Why is the reversed trajectory obtained by changing the signs of all the velocities so bizarre, despite the time-reversibility of the equations? From the physical standpoint, the flux directions are counter to our experience. From the mathematical standpoint, a more detailed answer involves the Lyapunov spectrum and Lyapunov instability of any such time-reversed trajectory. Figure 5.2 shows a typical situation in the "extended" phase space, where this phase space incorporates also any additional friction coefficients required to control the flow. As time proceeds, an infinitesimal comoving hypersphere centered on a trajectory deforms, to a hyperellipsoid (shown at the center of the Figure), which *rotates* as it moves. The logarithmic rates of growth $\{\dot{\delta}/\delta\}$ of the offset vectors $\{\delta\}$ (proportional to the corotating principal axes of the infinitesimal hyperellipsoid), define the local Lyapunov exponents. Eventually deformation beyond the linear range of the Lyapunov exponents causes the hyperellipsoid to bend—as shown in the third view of Figure 5.2, and later to fibrillate. On a global time-averaged basis, the sum of the linear local Lyapunov exponents is necessarily negative (for attractive "stability", as opposed to repulsive divergence) and gives the time-averaged rate of collapse of a small comoving hyperellipsoid onto the ultimate multifractal strange attractor.

Now consider the time-reversed motion. The central trajectory runs *backward* and its comoving corotating hyperellipsoid *grows*, on the average, rather than shrinking. The long-time-averaged Lyapunov sum in this reversed motion is necessarily positive, rather than negative. The sign change corresponds to the sign change from the reversed equations of motion:

$$+\langle d\ln f/dt\rangle_{\text{forward}} \equiv -\langle d\ln \otimes/dt\rangle_{\text{forward}} = -\sum \lambda_{\text{forward}} =$$

$$-\langle d\ln f/dt\rangle_{\text{backward}} \equiv +\langle d\ln \otimes/dt\rangle_{\text{backward}} = +\sum \lambda_{\text{backward}}\ .$$

A comoving phase volume which grows on the average, equivalent to a time-averaged positive Lyapunov spectrum sum, would be *unstable*. The extension in phase would diverge and could not be contained in any bounded region of phase space. In fact, the instability of such a reversed motion would cause that motion to leave the repeller and to seek out the attractor. Any attempt to reverse a dissipative trajectory, short of storing and playing it back precisely, must necessarily fail, with the trajectory soon again seeking out the same attractor which is pursued in the forward direction of time. The severe fractal instability of cold-to-hot heat flow is responsible for time-averaged flows' *invariably* following the Second Law of Thermodynamics, as is discussed at length in Section 8.8.

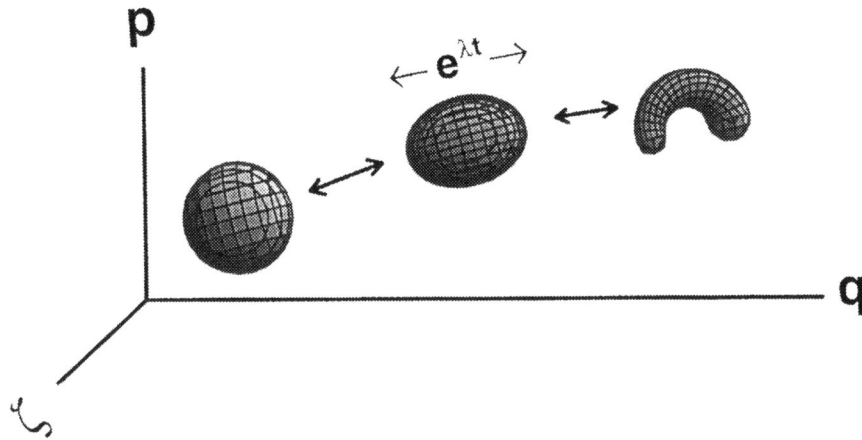

Figure 5.2. Three stages in the schematic time development of a small comoving phase-space hypersphere. The second view shows the rotation and deformation of the hypersphere to form an evolving hyperellipsoid. The third stage shows the nonlinear bending which, along with shrinking, ultimately converts the local hyperellipsoid to a global fractal attractor. The Lyapunov exponents describe the linear, but global, long-time-averaged growth and decay rates parallel to the corotating principal axes of the hypervolume $\otimes$.

This instability necessarily arises because the phase-space strange attractor is far more stable than is its time-reversed image, the strange repeller, upon which f would be a *decreasing* function of time so that the comoving phase volume $\otimes$ would have to increase. The observed and natural asymmetry, with *some* information discarded and *lost* going forward in time, but with additional information absolutely required in order to go *backward* in time, explains the inability of nonequilibrium simulations, or experiments, to recapture the past. Any attempt to do so simply generates another dissipative trajectory, just like those going forward in time. Only by keeping track of the entire trajectory is it possible to regenerate the information-starved reversed trajectory. The lack of symmetry between the forward and backward directions of time, despite formal time reversibility, provides a basis for the analysis of phase-space distributions through unstable periodic orbits.

The introduction of similar constraints, analogous to thermostats, into quantum mechanics can readily be implemented through the use of Lagrange multipliers in the Schrödinger equation. Rather than using an external force to drive the current, a constant current can likewise be maintained by imposing an appropriate constraint on the wave function. Kusnezov discusses an alternative approach, representing quantum thermostats with Gaussian random matrices, in his very readable 1999 paper.

5.5 Second Law of Thermodynamics

For thermostated systems there can generally be no analog of the time symmetric *bit-reversible* leapfrog algorithm. Whenever the comoving phase volume undergoes local changes, in a space with integer coordinates, a faithful mapping must be locally either one-to-many or many-to-one. Though neither mapping can be exactly reversed, the two possibilities are logically different. A one-to-many mapping requires continual decisions *creating* information while a mindless many-to-one mapping *destroys* it. This destruction of information, making it impossible to recapture the past, is typical of computer simulations of irreversible processes. Except in very special circumstances—good random-number generators, for instance—any timeperiodic mapping, reversible or not, cannot be simultaneously ergodic.

The isokinetic Galton Board problem corresponds to a thermodynamic system which can do work, through a particle's falling in a gravitational field, and generate heat, which is then removed by the isokinetic thermostat forces. This dissipative conversion of work to heat is made possible by an essentially infinite energy reservoir in the coordinate-dependent gravitational potential.

Heat flow by itself is also dissipative. Consider a many-body nonequilibrium steady state without any macroscopic motion a system confined between two reservoirs, one hot and one cold. Heat flow, from the hot reservoir to the cold one, leads to an overall increase in the summed-up reservoir entropies:

$$\Delta S_{\text{Reservoirs}} = -(\Delta Q/T_{\text{hot}}) + (\Delta Q/T_{\text{cold}}) > 0 \ .$$

Both the heat-flow example and the thermostated Galton Board are consistent with Clausius' formulation of the Second Law, [heat flows from hot to cold] with the net change of the reservoir entropies necessarily positive.

5.6 Simulating Shear Flow and Heat Flow

We have already considered the macroscopic description of steady nonequilibrium flows, with fixed velocity and temperature gradients. Take a homogeneous "Newtonian" fluid, confined to a volume V (which could be a small volume element), and endowed with a Newtonian shear viscosity η and Fourier's heat conductivity κ. In the presence of a shear strain rate $\dot{\epsilon}$ and a temperature gradient ∇T, the fluid dissipates energy into heat at the rate given by its total "entropy production" $\dot{S}$:

$$\dot{S} = [(\eta/T)\dot{\epsilon}^2 + \kappa |\nabla \ln T|^2] V \ .$$

Steady flow simulations *defining* the transport coefficients η and κ, can be based directly on observations of the shear stress or heat flux induced in a Newtonian simulation region bounded by two thermostated reservoir regions. If the velocity differences driving a shear flow are sufficiently large the generated heat produces an (approximately) parabolic temperature profile, from which the heat conductivity can simultaneously be determined. See Figure 5.3.

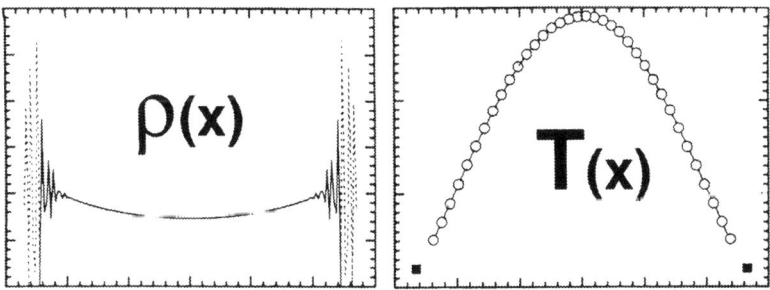

Figure 5.3. Density and Temperature profiles found by Liem, Brown, and Clarke in a stationary simulation of plane Couette flow with thermostated boundaries. See *Physical Review A* **45**, 3706 (1992) for the details.

The first atomistic transport simulations, based on the definitions of η and κ and stabilized with thermostating forces, were carried out about forty years ago. They are nicely described in Bill Ashurst's 1974 PhD thesis (University of California at Davis/Livermore). The most straightforward type of implementation in his work used a three-part system, like that of Figure 4.1 on page 121. There are two reservoirs driving the flow through a central Newtonian region. There are *periodic* boundary conditions in the x direction, perpendicular to the systematic nonequilibrium flow of momentum or energy in the y direction. Both the reservoir particles and the Newtonian particles are confined to one of the three subsystems by elastic hard-wall potentials, corresponding to impulsive contact forces acting in the y direction. The equations of motion in the two reservoir regions include "friction coefficients" designed to maintain the flow velocity and temperature within the reservoirs. With the velocities thermostated in both the x and the y directions, as is the usual choice, the reservoir equations of motion are:

$$\{\dot{x} = \dot{x}_0 + (p_x/m);\ \dot{y} = (p_y/m)\}$$

$$\{\dot{p}_x = F_x - \langle F_x \rangle - \zeta(T)p_x;\ \dot{p}_y = F_y - \zeta(T)p_y\};$$

$$\zeta = \sum (F - \langle F \rangle) \cdot p / \sum p^2/m\ .$$

The differences $\{F_x - \langle F_x \rangle\}$ ensure that the mean flow velocity $\dot{x}_0$ is unchanged. The friction coefficient ζ is chosen to maintain the desired reservoir temperature, $2kT \equiv \langle (p_x^2 + p_y^2)/m \rangle$. For the simulation of heat flow exactly the same equations can be used, but with the two reservoirs' mean velocities $\{\dot{x}_0\}$ vanishing.

It is important to note that both the Newtonian and the reservoir equations of motion are time-reversible, with the overall strain rate, and the variables $\{\dot{x}_0\}, \{p\}, \{\zeta\}$ all changing signs along the time-reversed trajectory. Thus any solution of these motion equations, played backward with these sign changes, would satisfy exactly the same equations, to the same numerical accuracy as did the corresponding forward solution. The *stability* of this hypothetical reversed trajectory is totally different. The reversed trajectory corresponds to an *unstable unobservable repeller flow*, as will be considered in detail in Chapters 8 and 9.

Consider the example problem shown in Figure 5.4. It is a computational model of a laboratory viscometer. A many-body fluid, accelerated downward by a gravitational field and slowed by the viscous resistance of a heat-conducting boundary wall, can reach a steady state in which the mechanical acceleration is balanced by viscous drag. Then the heat transfer from the fluid to the wall offsets the energy gain from the gravitational field which is dissipated by viscosity. This flow is particularly instructive because it incorporates nonlinear mass, momentum, and energy fluxes which are respectively first-order, second-order, and third-order in the particle velocities. An instantaneous time reversal of the flow, with the gravitational field direction unchanged, reverses the direction of the vertical mass flux ρu_y, the horizontal conductive heat flux $Q_r = -\kappa \nabla_r T$, and the vertical *convective* energy flux $\rho u_y [e + (u_y^2/2)]$, while leaving the shear stress σ_{ry} unchanged. In the time-reversed situation both the "apparent viscosity" $\eta = (\sigma_{ry}/\dot{\epsilon}_{ry})$ and the "effective heat conductivity" $\kappa = -Q_y/(\nabla T)_y$ would be negative. Heat would flow *from* the walls into the fluid, providing the energy for the upward mass flux. This unstable flow would violate the Second Law of Thermodynamics in *two* ways, by allowing heat to flow from cold to hot while simultaneously converting heat into work.

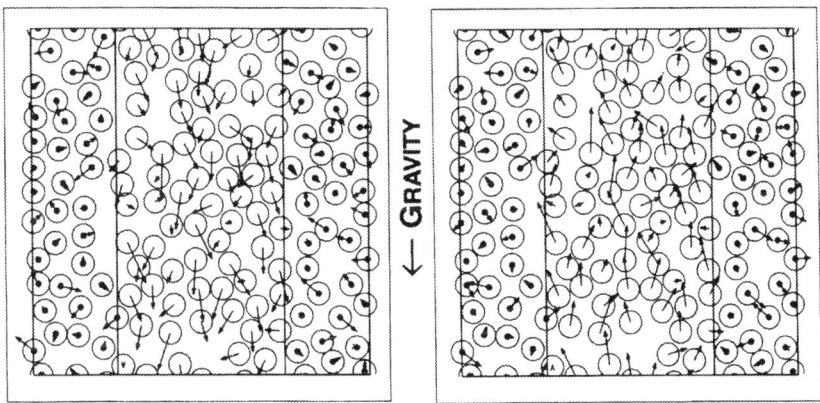

Figure 5.4. Dissipative flow in a model viscometer. Both the *forward* downward flow (at the left), with the signs of (σ_{ry}, Q_r, J_y) respectively $(+, +, -)$, and the instantaneously *reversed* unstable backward flow (at the right), with the last two signs reversed, $(+, -, +)$, are shown. Gravity is downward, in the negative y direction. Thermostated particles, identified here with a central spot, are confined to the two vertical reservoir regions.

Relatively small simulations, with 100 particles in two space dimensions, are sufficient to establish convincing linear velocity or temperature profiles, with transport coefficients determined with errors of order one percent. Spatially-homogeneous modifications of the more complicated three-part systems have also been designed. These simpler systems eliminate the role of the boundaries, and reduce the dependence of the results on system size. Such homogeneous simulations have been applied to the irreversible plastic flow of solids as well as to the shear and dilatation of fluids. The fluid viscosities and heat conductivities so found were quite consistent with those from Green and Kubo's linear-response theory based on Gibbs' ensembles.

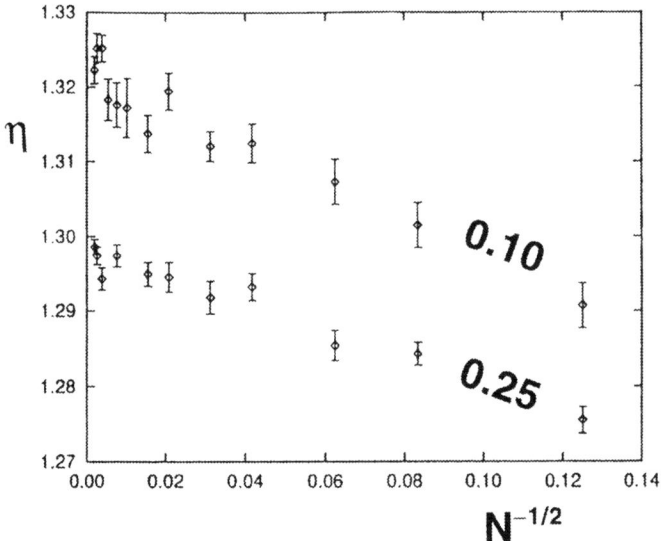

Figure 5.5. Number dependence of shear viscosity in stationary periodic two-dimensional molecular dynamics simulations with a short-ranged repulsive pair potential. The data, all for the same energy per particle, and the same density, but with two different strain rates, indicate the existence of a finite rate-dependent large-system limiting viscosity, with small-system deviations of order $\sqrt{1/N}$. The two-dimensional large-system limiting viscosity is thought to have a logarithmic dependence on the strain rate, here $\eta \simeq 1.27 - 0.02 \ln \dot\epsilon$. See author Bill's 1995 paper with Posch for details.

Thus the equilibrium Green-Kubo shear viscosity,

$$\eta = (V/kT) \int_0^\infty \langle P_{xy}(0) P_{xy}(t) \rangle_{\text{eq}} dt \ ,$$

agrees with that obtained by carrying out direct simulations of the corresponding nonequilibrium systems. The alternative spatially-homogeneous simulations, using periodic boundaries in both directions, have been developed for shear flows and for heat flows stabilized by artificial energy-sensitive fields. This work shows that there are no significant disagreements among these various routes to the simple linear transport coefficients.

It is noteworthy that simulations over a wide range of system sizes indicate that the homogeneously-thermostated two-dimensional systems undergoing shear have no special instabilities. At fixed strain rate, density, and energy the shear stress extrapolates smoothly to a well-behaved large-system limit, as is shown in Figure 5.5.

5.7 Shockwaves

A one-dimensional shockwave is arguably the simplest far-from-equilibrium steady state. Here "one-dimensional" means that the macroscopic averages (stress, energy, heat flux, and the like), as well as the microscopic long-time-averaged distribution functions, depend solely on the spatial coordinate in the direction parallel to the shockwave propagation. One-dimensional shockwaves link two equilibrium states—a "cold" pre-shocked state and a "hot" post-shocked state—with a relatively-localized and strongly-nonequilibrium transition region linking them. Because the velocity and temperature gradients can be quite large, with orders of magnitude changes occurring in just a few interatomic spacings, we might expect considerable deviations from the simple Newtonian and Fourier flow laws that apply with small gradients. The possibility of characterizing these deviations, together with the relatively simple geometric requirements, has motivated many shockwave simulations.

The simplest approach to shockwave simulation takes in cold material at one end of the simulation and expels hot material at the opposite end. A proper balance of the two boundary mass currents leads to a stationary position for the shockwave linking the two materials. Entropy *always* increases in the shock process. Typically pressure, density, and temperature increase too, but with the time-averaged mass flux, $\langle \rho v \rangle$, constant. By measuring the velocity and "temperature" and their gradients—where both $(m/k)\langle (v_x - \langle v_x \rangle)^2 \rangle$ and $(m/2k)\langle (v_x - \langle v_x \rangle)^2 + v_y^2 \rangle$ are possible temperature definitions—the corresponding momentum and energy fluxes, σ_{xx} and Q_x, as well as the effective transport coefficients within the shockwave, can be computed. These phenomenological nonlinear transport coefficients can then be compared to the linear transport coefficients obtained in homogeneous systems. For weak shockwaves the agreement is excellent. For typical pair potentials, even in the very strongest shockwaves, with widths of only a few interatomic spacings, the effective nonlinear transport coefficients lie within thirty percent of their linear counterparts.

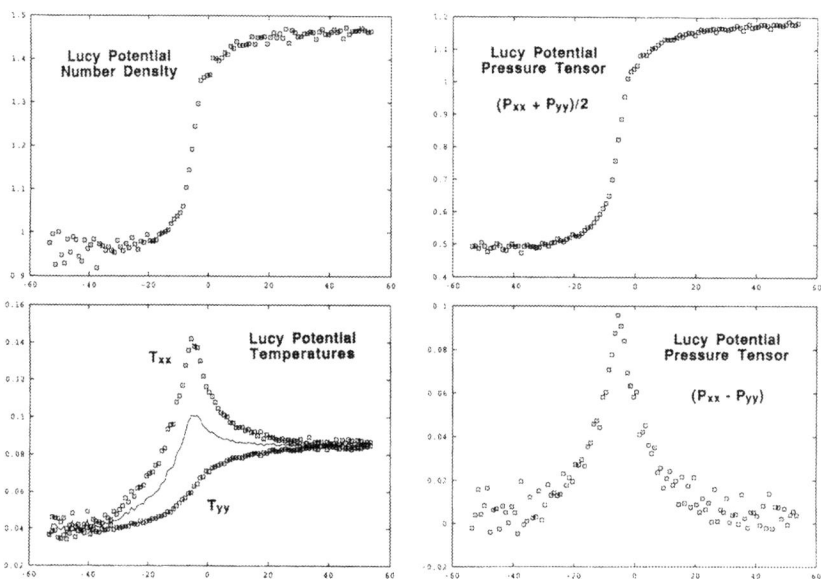

Figure 5.6. Density, temperature, and pressure variations within a dense fluid shockwave, the Lucy potential was used. Although the heat flux flows from the hot boundary toward the cold boundary throughout, both the transverse and the average temperatures, T_{yy} and $(T_{xx} + T_{yy})/2$ reach pronounced maxima within the shock wave, while the heat flux does not change sign, an aspect of the flow which is inconsistent with Fourier's Law.

In shockwaves the deviations from the linear transport laws are typically not very large. Figure 5.6 illustrates an exception to this rule—interesting kinetic temperature data for a two-dimensional strong shockwave using the relatively-weak and long-ranged repulsive forces derived from Lucy's potential,

$$\phi(r < 3) = [5/(9\pi)][1 + r][1 - (r/3)]^3 .$$

In this case it is noteworthy that the "temperature" reaches a maximum within the shockwave. If Fourier's Law held, even with a nonlinear conductivity, such a temperature profile would imply that the direction of the heat flux changes sign within the shockwave. Thus the heat flux and the temperature gradient in this particular simulation are in qualitative disagreement with Fourier's law. This work is described in detail in our 1997 paper with Oyeon Kum.

The evident irreversibility of the shockwave compression process is in qualitative contrast with the strict time-reversibility of an underlying molecular dynamics simulation of that same process. Stress (the negative of the pressure tensor) in molecular dynamics is an *even* function of the momenta and coordinates while strain rate [the averaged value of $(\partial v_x/\partial x)$ for shock propagation in the x direction] is *odd*, changing sign in a reversed movie of the motion. This different time symmetry is completely at odds with the Newtonian viscosity law, stress $\propto$ strain rate. A similar inconsistency governs heat flow. In molecular dynamics the heat flux is an odd function of velocity—reversing the motion reverses the direction of the heat flow. But the temperature gradient (quadratic in the velocities) is unaffected by reversal. Again the macroscopic law, Fourier's heat-flow law in this case, is inconsistent with microscopic time-reversible dynamics.

In an attempt better to understand these contradictions we have looked at the details of the (i) stress and strain rate profiles as well as the (ii) heat flux and temperature gradient profiles. In both cases we found a noticeable delay, of the order of the atomistic collision time. In a strong one-dimensional shockwave the strain rate *precedes* stress. Likewise, the temperature gradient *precedes* heat flux.

A formal treatment of these effects can be based on the Maxwell-Cattaneo relaxation model. For stress, the relation

$$\sigma + \tau_\sigma \dot{\sigma} = \eta \dot{\epsilon} ,$$

introduces the relaxation time τ_σ to describe the timelag between the deformation (strain rate) and the response (stress). Similarly the Maxwell-Cattaneo modification of Fourier's law introduces a relaxation time τ_Q forcing the heat flux to lag behind the temperature gradient:

$$Q + \tau_Q \dot{Q} = -\kappa \nabla T .$$

Much of the literature describing these relaxation effects is confusing due to the use of partial, rather than comoving, derivatives with respect to time. Instead, the derivatives $\dot{\sigma}$ and $\dot{Q}$ *following the motion* are required in order to interpret stationary nonequilibrium states such as the steady shockwave.

Although the stress relaxation equation by itself can be solved for arbitrarily large values of the time delay τ, the shockwave problem is unstable if τ is too large. Some special cases have analytic solutions. For example:

$$\sigma + \dot\sigma = \dot\epsilon = \frac{1}{e^{+t} + e^{-t}} \longrightarrow \sigma = e^{-t}\ln(\sqrt{1+e^{+2t}}) \ .$$

5.8 Algorithm for Periodic Shear Flow with Doll's Tensor

A clever way to avoid the relatively slow convergence caused by surface effects is to *omit* the surface entirely, by using *periodic* boundaries. Early hard-disk and hard-sphere equilibrium simulations showed, as the Mayers' virial theory suggested, that deviations from large-system results are typically of order $(1/N)$ with periodic boundaries. Surface effects with other sorts of boundaries are of the order of the surface-to-volume ratio, $(1/N)^{1/D}$ in D dimensions. As usual in simulation, the boundary conditions are a crucial aspect of the modeling. The periodic boundary conditions that reduce surface effects can be used away from equilibrium too.

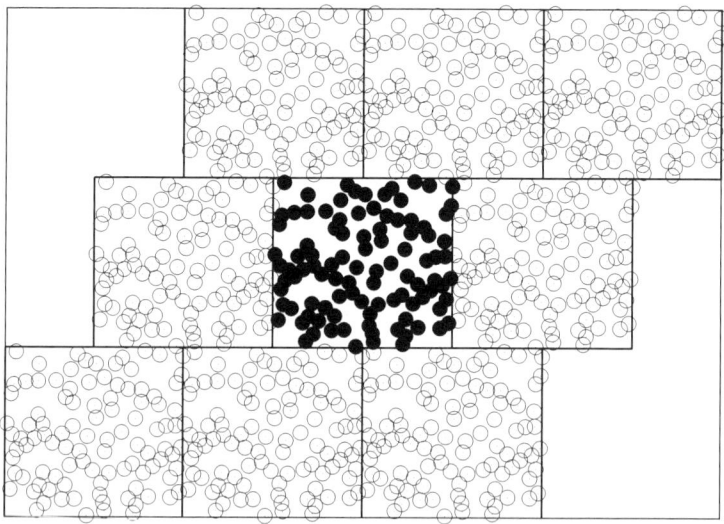

Figure 5.7. Periodic shear flow. Eight images of the central periodic cell are shown, three above, three below, and one each to the left and right. $L = 20$; $(N/V) = 0.25$.

Periodic shear flow, illustrated in Figure 5.7, can be modeled by using as boundaries *moving* images of the system undergoing shear. The strain rate $\dot\epsilon$ is fixed. Thus an "image system" above or below the system moves at the fixed relative speed of $\pm L\dot\epsilon$ relative to the $L \times L$ square centered on the coordinate origin $(0,0)$. In Figure 5.7 eight image systems are shown, three above, three below, and one each to the left and right. The usual nearest-image approach to computing pair separation distances needs a corresponding modification, taking the shear into account. If the integer part of the shear `istrain` is subtracted, then the remaining relative displacement of the images `dx` is just the fractional remaining part of the strain multiplied by the system sidelength `el`:

```
istrain = (edot*time)
dx = (edot*time - istrain)*el
```

In Figure 5.7 `dx` is `el/2`. Taking the shear into account, the "nearest image" periodic boundary conditions for particle pairs can be applied as follows:

```
i = ipair(ip)
j = jpair(ip)
xij = x(i) - x(j)
yij = y(i) - y(j)
if(yij.lt.-(el/2)) then
yij = yij + el
xij = xij + dx
endif
if(yij.gt.+(el/2)) then
yij = yij - el
xij = xij - dx
endif
if(xij.lt.-(el/2)) xij = xij + el
if(xij.gt.+(el/2)) xij = xij - el
if(xij.lt.-(el/2)) xij = xij + ol
if(xij.gt.+(el/2)) xij = xij - el
```

Because the x displacement can change by as much as `el` in the periodic-boundary calculation the x displacement must occasionally be corrected by *more* than `el`, as is done by the last two lines above.

Similarly, it is convenient, at the end of each timestep, to replace any particles which have left the central cell:

```
do i = 1,n
x(i) = yy(i)
y(i) = yy(i+n)
px(i) = yy(i+n+n)
py(i) = yy(i+n+n+n)
if(y(i).lt.-(el/2)) then
y(i) = y(i) + el
x(i) = x(i) + dx
endif
if(y(i).gt.+(el/2)) then
y(i) = y(i) - el
x(i) = x(i) - dx
endif
if(x(i).lt.-(el/2)) x(i) = x(i) + el
if(x(i).gt.+(el/2)) x(i) = x(i) - el
if(x(i).lt.-(el/2)) x(i) = x(i) + el
if(x(i).gt.+(el/2)) x(i) = x(i) - el
yy(i) = x(i)
yy(i+n) = y(i)
enddo
```

We measure particle momenta $\{p_x, p_y\}$ relative to the local mean flow:

$$\{\dot{x} \equiv (p_x/m) + \dot{\epsilon}y;\ \dot{y} \equiv (p_y/m)\}\ ,$$

and calculate the pressure tensor using these same comoving momenta. After computing the kinetic part of the pressure tensor,

```
do i = 1,n
pvxxk = pvxxk + px(i)*px(i)
pvyyk = pvyyk + py(i)*py(i)
pvxyk = pvxyk + px(i)*py(i)
enddo
```

a sum over all `npairs` particle pairs, again taking into account the displacement of the system images, $\pm$dx completes the pressure calculation:

```
do ip = 1,npairs
i = ipair(ip)
j = jpair(ip)
xij = x(i) - x(j)
yij = y(i) - y(j)
if(yij.lt.-(el/2)) then
yij = yij + el
xij = xij + dx
endif
if(yij.gt.+(el/2)) then
yij = yij - el
xij = xij - dx
endif
if(xij.lt.-(el/2)) xij = xij + el
if(xij.gt.+(el/2)) xij = xij - el
if(xij.lt.-(el/2)) xij = xij + el
if(xij.gt.+(el/2)) xij = xij - el
r = dsqrt(xij*xij + yij*yij)
if(r.lt.1.0d00) then
call force(r,f,phi)
pvxxp = pvxxp + f*xij*xij/r
pvyyp = pvyyp + f*yij*yij/r
pvxyp = pvxyp + f*xij*yij/r
endif
enddo
```

Then the equations of motion for $\{\dot{x}, \dot{y}, \dot{p}_x, \dot{p}_y\}$ follow from a special Doll's-Tensor Hamiltonian (named by Bill for the "Kewpie" $\simeq qp$ doll). This Hamiltonian provides equations of motion with the energy conservation in the form of the First Law of Thermodynamics—the rate of change of internal energy is equal to the rate at which work is done by the shear stress $-P_{xy}$:

$$\mathcal{H} - \mathcal{H}_0 + \dot{\epsilon}\sum_i (yp_x)_i \longrightarrow 0 \equiv (d/dt)\mathcal{H} = (d/dt)\mathcal{H}_0 + (d/dt)\dot{\epsilon}\sum_i (yp_x)_i =$$

$$\dot{E}_0 + \dot{\epsilon}\sum_i \dot{y}(p_x/m) + \dot{\epsilon}\sum_{i<j}(xyF/r)_{ij} \equiv \dot{E}_0 + \dot{\epsilon}P_{xy}V .$$

$$\longrightarrow \dot{E}_0 \equiv -\dot{\epsilon}P_{xy}V .$$

The corresponding equations of motion are:
$$\{\dot{x} = (p_x/m) + \dot{\epsilon}y;\ \dot{y} = (p_y/m);\ \dot{p}_x = F_x;\ \dot{p}_y = F_y - \dot{\epsilon}p_x\}\ .$$

The Rayleigh-Bénard simulations of Section 4.11.3 used average temperatures of 1.00 and 0.01. We can estimate the shear viscosity η with this Hamiltonian method using an additional friction coefficient ζ to keep the internal energy constant as the shear progresses:
$$\{\dot{p}_x = F_x - \zeta p_x;\ \dot{p}_y = F_y - \dot{\epsilon}p_x - \zeta p_y\}$$
$$\zeta \equiv -\dot{\epsilon}P_{xy}V/\sum_i [p_x^2 + p_y^2]/m\ .$$

Simulations with these isoenergetic motion equations, carried out for a total shear of 200 gave the following samples of viscosity:

$\eta = 0.034$ for $\{(E/N) = 0.01;\ (N/V) = 0.25;\ \dot{\epsilon} = 0.01\}$;

$\eta = 0.018$ for $\{(E/N) = 0.01;\ (N/V) = 0.25;\ \dot{\epsilon} = 0.10\}$;

$\eta = 0.32$ for $\{(E/N) = 1.00;\ (N/V) = 0.25;\ \dot{\epsilon} = 0.10\}$
with uncertainities of order one percent .

5.9 Example Problems

We consider three problems which illustrate the use of deterministic thermostats. First, a single particle falling through an array of fixed scatterers (the isokinetic "Galton Board"); second, a single particle oscillating in a confining potential in the presence of a temperature gradient. These first two problems, involving a single particle driven away from equilibrium, provide multifractal phase-space distributions in few-dimensional spaces, making visualization and analysis of the fractal extensions in phase feasible. Finally, we apply the same thermostat techniques to the simulation of a *many*-body nonequilibrium heat flow in a many-dimensional phase space. Although the many-body heat flow considered here takes place in only two space dimensions, so as to simplify visualizing the particle trajectories, exactly the same equations of motion, supplemented by those for motion in a third dimension, can be solved with equal ease. The three problems all share the characteristics of time-reversible deterministic dynamical equations, with an overall dissipative conversion of work to heat.

5.9.1 Isokinetic Nonequilibrium Galton Board

In 1873 Sir Francis Galton used balls falling through a regular lattice of pegs to demonstrate the binomial and Gaussian distributions. An up-to-date review of many aspects of this problem includes the contributions of Carl Dettmann and Harald Posch given in the Bibliography. In the periodic field-free case, worked out in Chapter 1, such a ball moves without friction, and eventually comes arbitrarily close to all of its accessible states. The motion is ergodic. An ensemble of such balls, initialized at the center of a large board, eventually spreads out as is described by the *irreversible* diffusion equation. This prototypical illustration of diffusion is simultaneously interesting and manageable, so that the problem has attracted considerable well-deserved attention.

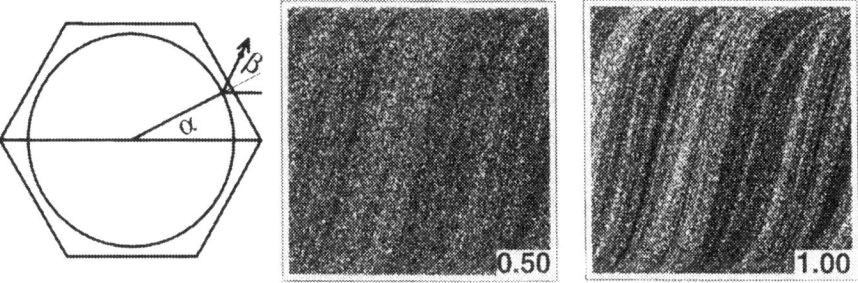

Figure 5.8. 100,000 points $\{\alpha, \sin(\beta)\}$ for the isokinetic Galton Board with two moderately-strong fields. Compare the thermostated results on the left to the Newtonian results for the same field strength, shown in Figure 1.5, on page 28. Definitions of the angles $[\alpha, \sin(\beta)]$ appear on the left. Here the field strengths are $\{mE\sigma/p^2\} = \{0.50, 1.00\}$, where a particle with mass m and speed $|p|$ is scattered by particles of diameter σ. The density of the scatterers is $(4/5)$ the close-packed density. See Bill's 1987 paper with Bill Moran.

A frictionless Galton Board would not function well, for the gravitational energy would eventually cause the falling ball to heat, resulting in a thermal barometer-formula distribution rather than the desired Gaussian distribution at the base. The simplest way to thermostat the falling ball is to constrain it to move at constant speed. The least force required to impose this constraint is the drag force, $-\zeta p$, where ζ is $F \cdot p/2K$, K is the (fixed) kinetic energy, and E is the strength of the accelerating field.

Figure 5.9. 100,000 Attractor (left) and repeller (right) collisions for the isokinetic Galton Board with a relatively-strong field, $(mE\sigma/p^2) = 2.00$. The (unstable and unobservable) repeller states, with $\langle \dot{\otimes} \rangle > 0$, are obtained from the attractor by changing the sign of the velocity and reversing the time-ordering of the points. The abscissa is $0 < \alpha < \pi$ and the ordinate is $-1 < \sin(\beta) < +1$.

The corresponding isokinetic equations of motion are still time-reversible! Both ζ and p change sign in the reversed motion. Similar results, but in higher-dimensional phase spaces, can be obtained with irreversible motion equations by using a "coefficient of restitution" (reducing the energy of the moving particle at each collision) or by using a linear drag force $-p/\tau$ between collisions.

It has been argued that time-reversible equations can show no behavior different in the future from that which they displayed in the past. This argument is reasonable, even true, for isolated systems. But it is certainly *false* for systems driven away from equilibrium, or for systems in the presence of external driving fields. Consider, for simplicity, a ball in a gravitational field. Independent of the initial conditions the ball will eventually fall. In the distant past the ball must have been rising. Thus the future state (falling) and the past state (rising) are quite different. This distinction between future and past presupposes a psychological "arrow of time" making possible the definition of velocity, $v \equiv \dot{r}$.

Figure 5.9 shows also the mirror-image repeller for this same flow, corresponding to the time-reversed attractor trajectory with unstable upward flow, and with the order of the points reversed in time. The inevitable future motion is stable, dynamically, in that the flow collapses onto a strange attractor with a negative Lyapunov sum, $\sum \lambda = -(\dot{S}/k) < 0$.

The relative stabilities of the forward and backward motions can differ dramatically if the dynamics is *constrained*, as it is in the Galton Board problem. Consider an ensemble of moving particles with each particle constrained to move through its own Board at constant kinetic energy. On the average, the future flow is *downward*, and the distribution of collisions is given by the familiar strange-attractor Poincaré sections shown in Figures 5.8 and 5.9.

The conductivity of the Galton Board, $\langle p_y/E \rangle$ varies relatively irregularly with the field strength for low fields. At reduced field strengths above $(mE\sigma/p^2) = 3.69$ interesting limit cycles occur, in which the Lyapunov unstable scattering of the moving particle is overwhelmed by the ordering effect of the field plus thermostat, giving rise to stable quasiperiodic motions.

Because the isokinetic Galton-Board motion equations,

$$\{\dot{x} = (p_x/m);\ \dot{y} = (p_y/m);\ \dot{p}_x = F_x - \zeta p_x;\ \dot{p}_y = F_y - E_y - \zeta p_y\},$$

$$\zeta = -E_y p_y / 2K,$$

are time-reversible, with

$$(+t, +p_x, +p_y, +\zeta) \longleftrightarrow (-t, -p_x, -p_y, -\zeta)$$

in the time-reversed motion, it is easy to generate hypothetical, but wholly unrealistic, states lying in the past. The steps required to generate the past are exactly the same as those detailed in the oscillator example of Section 2.2. A long trajectory segment, obtained by integrating a reversed initial condition "forward" becomes a fictitious unobservable repeller evolution as the result of two operations: (i) reversing the velocities, and (ii) reversing the time-ordering of the points. These steps produced the repeller shown above in Figure 5.9. The unlikely past states at the left—reversed to give the repeller at the right—correspond to trajectories which move upward in the Board, against the field, and with overall negative entropy production. This reversed past upward motion is dynamically unstable, corresponding to phase-space expansion and to a positive Lyapunov sum. Time-reversible driven systems *typically* display such a lack in symmetry between future and past, with future states likely, obeying the Second Law, and past time-reversed states unlikely, violating it. The Lyapunov spectrum is a key diagnostic quantifying the lack of symmetry between the very unlikely, unstable, and unobservable reversed "past" relative to the more-certain relatively-stable inevitable future.

In principle, the attractors generated by computer simulation must actually be periodic orbits because any finite-precision computation must eventually repeat. For most problems, with roughly 10^{16N} state points for an N-dimensional solution space, the typical orbital period is much too long for an accurate determination. For the two-dimensional Poincaré sections of the Galton Board Figure 5.10 shows the resolution attained with 10^{10}, 10^{14}, and 10^{18} points, corresponding to 5, 7, and 9 digit accuracy in the abscissa and ordinate. The 5-digit simulation is an 11,951-collision periodic orbit or "limit cycle". The singularities separating the various S-shaped regions in the Poincaré sections correspond to boundaries separating scattering collisions with different neighbors of the preceding scatterer.

Figure 5.10. Convergence of resolution of typical stationary states for the Galton Board, using five-, seven-, and nine-figure accuracy (left to right) for the abscissa and ordinate values. The field strength is $(mE\sigma/p^2) = 3.00$ and each of the simulations includes 300,000 successive collisions.

5.9.2 Heat-Conducting One-Dimensional Oscillator

The essential singularities associated with the Galton Board problem, and the multifractal phase-space distribution which it generates, complicate the numerical analysis somewhat, and might also complicate theoretical approaches. For that reason we describe a system which is simpler, from the analytic standpoint, a one-dimensional oscillator in a temperature gradient. The motion takes place in four-dimensional phase space, $\{q, p, \zeta, \xi\}$. The temperature (kinetic energy) of the oscillator is governed by a generalized Nosé-Hoover control of the second velocity moment, $\langle p^2 \rangle$, using *two* time-reversible friction coefficients, ζ and ξ. The extra coefficient is required to ensure ergodicity in the limiting equilibrium isothermal case.

For simplicity we set the oscillator mass and force constant, as well as both thermostat relaxation times, all equal to unity. The thermostat variables, ζ and ξ, control the oscillator's kinetic temperature $\langle p^2 \rangle = kT$ as follows:

$$\{ \dot{q} = p;\ \dot{p} = -q - \zeta p;\ \dot{\zeta} = p^2 - kT - \xi\zeta;\ \dot{\xi} = \zeta^2 - kT \} .$$

In the equilibrium case (where the temperature T is constant) these motion equations provide ergodicity with a Gaussian distribution function:

$$kT(q) = 1 \longrightarrow f = (2\pi)^{-2} e^{-(q^2 + p^2 + \zeta^2 + \xi^2)/2kT} .$$

Ergodicity is most simply checked by propagating a grid of points taken from the surface of a unit hypersphere $q^2 + p^2 + \zeta^2 + \xi^2 = 1$ forward in time, and choosing those points with the most disparate time averages for long-time study. The stationary Gaussian distribution follows directly from the Liouville equation for $(\partial f / \partial t)$, using the Gaussian distribution given above:

$$(\partial f/\partial t) = -\nabla \cdot (fv) = -f\nabla \cdot v - v \cdot \nabla f = -f[-\zeta - \xi] - v \cdot \nabla f =$$

$$f[\zeta + \xi] + p(qf) + (-q - \zeta p)(pf) + (p^2 - 1 - \xi\zeta)(\zeta f) + (\zeta^2 - 1)(\xi f) \equiv 0 .$$

A nonequilibrium *dissipative* time-reversible variation of this isothermal conducting oscillator model results if the temperature varies smoothly between the limits $(1 \pm \epsilon)$:

$$kT(q) \equiv 1 + \epsilon \tanh(q) .$$

As would be expected, so long as the perturbation ϵ is small, the long-time-averaged distribution function is independent of the initial conditions (again, as judged by the long-time-averaged values of moments), making it possible to assess the nonequilibrium dissipation and entropy production.

The problem is a nice illustration of the difficulties involved in analyzing nonlinear dissipation. The "entropy production" is simply the mean value of the heat extracted from the oscillator, divided by the corresponding temperature:

$$\langle \dot{S}_{\text{external}}/k \rangle = \langle (\zeta p^2 / mkT(q) \rangle \simeq \langle \zeta \rangle + \langle \xi \rangle .$$

Because the temperature fluctuates, the entropy production is not simply related to the rate at which a four-dimensional comoving hypervolume $\otimes \equiv dqdpd\zeta d\xi$ collapses to a strange attractor:

$$\langle d\ln \otimes /dt \rangle = \langle \nabla \cdot v \rangle = \langle -\zeta - \xi \rangle \ .$$

On the other hand, if we consider a "generalized" energy $\tilde{E}$:

$$\tilde{E} \equiv (q^2 + p^2 + \zeta^2 + \xi^2)/2 \ ,$$

the corresponding entropy change, $(d\tilde{E}/dt)/T$ exactly reproduces the comoving time-rate-of-change of (the logarithm of the) phase volume:

$$(1/T(q))(d\tilde{E}/dt) \equiv (q\dot{q} + p\dot{p} + \zeta\dot{\zeta} + \xi\dot{\xi})/T(q) \equiv$$

$$-\zeta - \xi \equiv d\ln \otimes /dt = -d\ln f/dt \ .$$

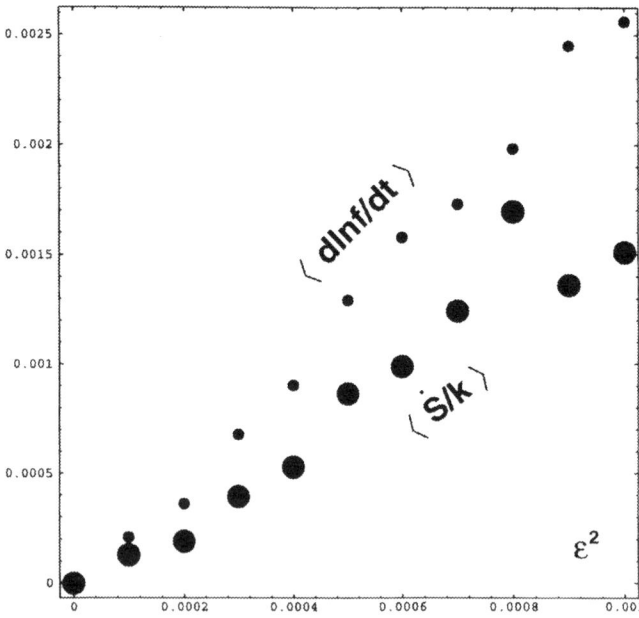

Figure 5.11. $\langle d\ln f/dt \rangle = -\langle \nabla \cdot v \rangle$ (above) and $\langle (\dot{S}/k) \rangle$ (below) as functions of the *square* of the maximum temperature gradient ϵ for a one-dimensional oscillator. The data indicate quadratic relationships, as would be expected from linear-response theory.

For larger values of the maximum temperature gradient ϵ the motion collapses to a limit cycle, a one-dimensional periodic trajectory in the four-dimensional phase space. Numerical solutions for this problem, for a series of equally-spaced values $\{\epsilon^2\}$, show increases in both $\langle \dot{S}/k \rangle$ and $\langle d\ln f/dt \rangle$ which are close to quadratic, but not equal, for small ϵ^2. A limit cycle, corresponding to a strictly-periodic oscillator trajectory, becomes stable at maximum gradient values larger than those shown in Figure 5.11. At equilibrium, with $\epsilon = 0$, the largest of the four Lyapunov exponents is approximately 0.066 for this problem. In Section 8.10 we will consider additional details of nonequilibrium oscillator problems in both three- and four-dimensional phase spaces.

5.9.3 Many-Body Heat Flow

To capture the flavor of a serious many-body simulation, while maintaining the graphic advantages of two space dimensions, consider a doubly-periodic four-chamber system (an area in two dimensions rather than a volume) with a hot chamber and a cold chamber separated by two Newtonian chambers. Each of the four chambers contains 324 identical particles at an overall number density of unity. The hot-to-cold temperature ratio is three. Pairs of particles interact with a very-smooth short-ranged Lucy potential,

$$\phi_{\text{Lucy}}(r<3) = [5/(9\pi)][1+r][1-(r/3)]^3 .$$

The particle motion in the two thermostated regions is moderated by friction coefficients depending upon the specified equilibrium temperatures of these regions. If we express these equilibrium temperatures by specifying average reservoir kinetic energies, $\{K_0(T)\}$, using Nosé-Hoover temperature control, the corresponding reservoir equations of motion are:

$$\{\dot{x} = (p_x/m);\ \dot{y} = (p_y/m);\ \dot{p}_x = F_x - \zeta(T)p_x;\ \dot{p}_y = F_y - \zeta(T)p_y\} .$$

$$\zeta(T) = \int_0^\infty \left[\frac{K(t')}{K_0(T)} - 1\right] \frac{dt'}{\tau^2} .$$

The Newtonian motion equations have the same form, but without any frictional terms. Figure 5.12 shows a typical snapshot of the dynamics and a time-averaged temperature profile. Analysis of this problem has established that the hot-to-cold heat flow causes the collapse of the corresponding probability density onto a strange attractor.

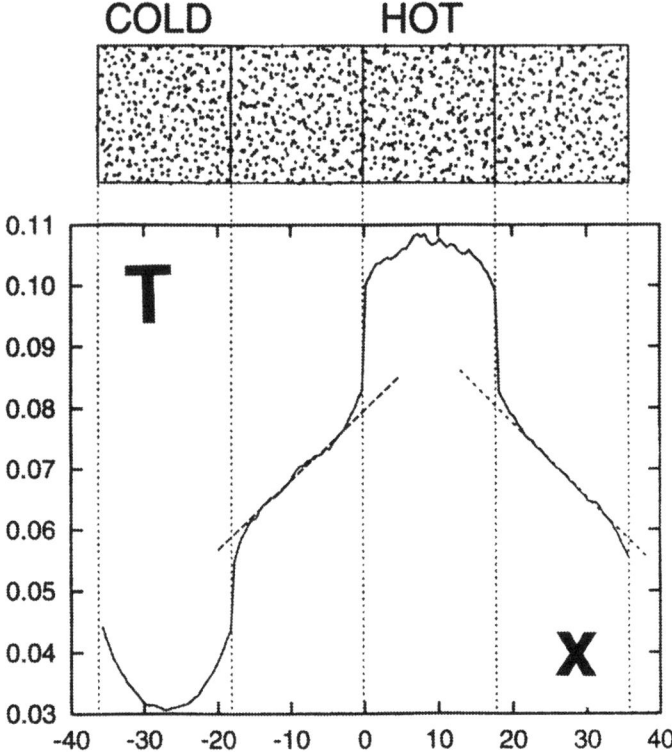

Figure 5.12. Snapshot of the four-chamber 1296-particle simulation of heat flow between two reservoirs and time-averaged temperature profile. For details see "Numerical Heat Conductivity in Smooth Particle Applied Mechanics", by Harald Posch and author Bill, *Physical Review E* **54**, 5142–5145 (1996).

5.10 Summary

The deterministic time reversibility of Newtonian, Lagrangian, and Hamiltonian mechanics for isolated systems can be carried over to the simulation of thermostated nonequilibrium atomistic systems. The nonequilibrium flows which result exhibit a subtle time asymmetry, which was first seen in simulations of the Galton Board problem. Despite the deterministic and time-reversible nature of the equations of motion, the resulting trajectories generate a probability density which is singular everywhere (ergodic and multifractal), a *strange attractor*. The long-time-averaged time-rate-of-change of the comoving phase-space volume $\langle \dot{\otimes} \rangle$ is invariably negative, corresponding to contraction and collapse, while nearby trajectories invariably diverge from one another exponentially, with time. The underlying strange attractor is typically *ergodic*, filling all of the accessible phase space, and corresponds to a dissipative state, as described by the Second Law of Thermodynamics.

The time reversibility of the equations of motion guarantees the existence of a topologically similar repeller structure, on which the Second Law is violated. The repeller is less stable than the attractor, and unobservable, in a sense which can be made precise through an analysis of the Lyapunov spectrum, as is explained in Chapter 8. These generic properties of the Galton Board problem were pointed out in Bill's 1986 work with Bill Moran. There is a closely-related paper, describing thermostated motion in a one-dimensional *sinusoidal* potential, which Harald Posch, Brad Holian, Mike Gillan, Michel Mareschal, Carlo Massobrio, and author Bill published at about the same time. The generic features which this group found in the computer simulation of the Galton Board and sinusoidal work, were validated theoretically seven years later, by Chernov, Sinai, Eyink, and Lebowitz.

The time-reversibility of the underlying motion equations is more apparent than real. In the forward direction of time more information is destroyed than is created. The Second Law of Thermodynamics is necessarily satisfied, $\langle \dot{S} \rangle > 0$. A multifractal phase-space distribution is formed by the dynamics on a zero-measure phase-space set. Such a nonequilibrium flow can only be "reversed" by storing the entire trajectory and playing it backward.

5.10.1 *Notes and References*

It is a pleasure to look back on the development of nonequilibrium atomistic simulation techniques, appreciating the resulting understanding we have reached. The earliest days are nicely chronicled in the special volume of Physica devoted to Howard Hanley's 1982 Boulder meeting, where people interested in theory, experiments, and simulations all gathered together to discuss *Nonlinear Fluid Behavior*. Ever-faster computers fuel the explosive growth of our field, evidenced by the frequent workshops and meetings which continue today. The contagious excitement of simulation motivated Bill to write *Molecular Dynamics* and *Computational Statistical Mechanics* as well as many summary articles, like the reminiscent "Nonequilibrium Molecular Dynamics: the First 25 Years". Denis Evans and Gary Morriss chronicled their views of the progress in hundreds of publications, as well as a useful book, available at their websites, *Statistical Mechanics of Nonequilibrium Liquids*. By 1984 Nosé's work made it possible to link the algorithmic development directly to Hamiltonian mechanics. Dettmann and Morriss made a related valuable connection in 1996 in "Hamiltonian Formulation of the Gaussian Isokinetic Thermostat".

A generation earlier Green and Kubo had linked transport to the decay of equilibrium fluctuations. So we can agree with Hanley that *nonlinear effects* are *the* legitimate goal of nonequilibrium simulations. In small systems nonlinearity is obscured by boundary effects. In larger systems distance is the barrier to nonlinear propagation. Liem, Brown, and Clarke showed this with their 43,110-particle effort, "Investigation of the Homogeneous-Shear Nonequilibrium-Molecular-Dynamics Method". Our own investigation, with Janka Petrovic, compared many homogeneous and boundary-driven shear algorithms, and showed that the nonlinear effects, though real, are small. Malek Mansour's "Temperature Profile of a Dilute Gas Undergoing a Plane Poiseuille Flow" and Puhl, Mansour, and Mareschal's "Quantitative Comparison of Molecular Dynamics with Hydrodynamics in Rayleigh-Bénard Convection" contrast the microscopic and macroscopic treatments of nonlinear flows.

An excellent way to avoid complicated surface effects is to study shockwaves, far from equilibrium and far from annoying boundaries. Shockwave studies date back to 1967. The Cold War weapons programs motivated parallel efforts in Russia and the United States.

Simple models are another route to understanding, particularly attractive to mathematicians. Hopf's Baker Map, Lorenz' Attractor, the Lorentz Gas/Galton Board problems, and their many-body generalizations have stimulated thousands of investigations. The Galton Board illustrates three different kinds of periodic orbits, each with infinitely many representatives: unstable orbits, characterizing Lyapunov instability, periodic orbits stabilized against instability by a driving field, and the coarse-grained periodic orbits induced by finite-precision arithmetic. In our "Nonequilibrium Fluctuations in a Gaussian Galton Board (or Periodic Lorentz Gas) Using Long Periodic Orbits" we analyze a 12-digit periodic orbit composed of 793,951,594-collisions.

Chapter 6

Shockwaves Revisited

*It is better to ask Some of the Questions
than to know All the Answers.*

James Thurber

6.1 Introduction

In the early days of computers, just after the Second World War, shockwave simulations mainly consisted of grid-based solutions of the atomistic Boltzmann Equation and the Navier-Stokes-Fourier continuum equations. An important experimental application of shockwaves is equation-of-state measurements at pressures comparable to those generated by nuclear weapons. Relatively simple velocity measurements, coupled with conservation of energy, can provide pressure-energy-volume $P(E, V)$ points at pressures up to 100 Megabars. An assumption necessary to this analysis is that the shockwave is a steady one-dimensional wave, with the hot fluid reaching an equilibrium state during the measurement. The early computer simulations of shockwaves, in the 1970s, provided welcome checks of this assumption.

The shock process is particularly interesting because it requires no special boundary conditions. Experimentally all that is required is a high-pressure source (compressed gas, or a chemical or nuclear explosive) to provide acceleration, and accurate velocity measurements. Supporting simulations, using molecular dynamics or continuum mechanics, can then generate models of the acceleration process through measurements of simulation values of the pressure and energy. The simulation needs to provide an acceleration or deceleration process corresponding to the velocity change in the corresponding laboratory experiment.

In their *Fluid Mechanics* text Landau and Lifshitz analyzed the shockwave compression of a viscous fluid and showed that the width of the compression wave is related to the viscosity and the pressure. Because the strain rate in the shock is of order (u/λ), where u is the velocity change in the shock process and λ is the shock width, the pressure within the shock (which is primarily viscous) can be used to estimate the shock width:
$$P \simeq \eta(u/\lambda) \longrightarrow \lambda \simeq (\eta u/P) \ .$$
We will illustrate the simulation of a continuum shock profile for a variety of constitutive models with simple finite-difference algorithms, in Section 6.6. We also describe shockwave stability and profile results from molecular dynamics simulations. Their analysis, using the smooth-particle averaging technique of Section 3.10, provides transport information far from the linear regime of Newtonian viscosity and Fourier conductivity. The dynamics simulations are particularly valuable in suggesting improvements to the continuum constitutive relations.

In the 1970s it became possible to simulate dense-fluid shockwaves with molecular dynamics and to compare these simulations with corresponding continuum solutions. Today a laptop computer is not only sufficient for both these approaches, but is also capable of providing much more detailed analyses. We decided that it was time to revisit the dense-fluid shockwave problem. Details of the work are included in a series of seven papers published in the Los Alamos arχiv. Two of them are joint work carried out in collaboration with Paco Uribe. We addressed several interesting questions.

The first has to do with the planar *stability* of shockwaves. Are the waves really one-dimensional, as assumed in most analyses? This question has been addressed experimentally by machining sinusoidal surfaces on metal blocks. Accelerating these blocks with high explosives and observing the subsequent decay of the sine wave toward planarity provides experimental estimates for metals' viscosities and plastic yield strengths at high pressure. The same idea can be pursued with computer simulations. Here we will simulate the decay of sinusoidal shockfronts with molecular dynamics. Another way to address the stability question is to use smooth-particle averaging to assess the flatness of the propagating waves.

The somewhat uncertain location of a *wave* within an *atomistic* simulation brings up a second set of interesting research questions: how are the *local values* to be defined? We have seen that smooth-particle averaging provides local variables and we will demonstrate here that this technique can be usefully applied to shockwaves. With local measurements of velocity, we can determine the tensor nature of the comoving kinetic temperature and stress, as well as the heat flux, and the strain rate. Such measurements make it possible to assess the accuracy and utility of Newton's viscosity formulation and Fourier's heat conduction law.

A third question arises next: how is temperature to be *defined* in a shockwave? Ideas from Gibbs' statistical mechanics and Maxwell and Boltzmann's kinetic theory suggest several useful models for thermometry. With these Fourier's law of heat conduction can be checked, and validated or improved. An outcome of these thermal investigations has been a clarification of the tensor nature of temperature, not just for gases, but for dense fluids as well.

With the profile variables defined we sometimes see significant delay times between "cause" and "effect" in shockwaves. How important are these delays? How can they be modeled, so as to incorporate them into continuum simulations? To what extent can the well-known classical models be generalized to describe the new data? More detailed microscopic questions arise and impinge on chaos and nonlinear dynamics: How does the *irreversible* shockwave stem from a basis in purely time-reversible Newtonian mechanics? How can modern methods of chaotic instability analysis be applied to shockwave structure? We review all of these questions in the present Chapter. This general area looks particularly fascinating and fruitful for future research. Let us begin our investigation with a more detailed description of shock generation and analysis.

6.2 Equation of State Information from Shockwaves

Suppose a cold zero-energy zero-pressure fluid (or solid) impacts a fixed rigid wall at such a high particle speed u_p (typically 30km/sec in Ragan's Los Alamos experiments) that it is compressed to *three times its original density*. What happens? In the *laboratory frame* (of the stationary wall) no work is done. Thus the final internal energy, $E_{\text{hot}} = Nme_{\text{hot}}$, is equal to the initial kinetic energy:

$$E_{\text{hot}} = Nme_{\text{hot}} = Nm(u_p^2/2) \ .$$

Consider the threefold compression from V_{cold} to $V_{\text{hot}} = (V_{\text{cold}}/3)$, in the alternative frame of the cold moving matter. In that frame the pressure

P_hot at the oncoming wall does work $2(PV)_\text{hot}$ equal to the energy change. Thus a single measurement of the velocity change is enough to give us both the internal energy change and the pressure required for the threefold compression. For copper this is about 6TPa, or 60 Megabars, about 20 times the pressure at the center of the earth.

Now focus on the thin transition zone linking the cold matter to the hot. Energy conservation can be analyzed in a third coordinate frame, the frame of the shockwave itself, where the cold fluid or solid, passing through a stationary shockwave, is transformed to a hot high-pressure material. Consider a volume of the cold fluid, V_cold, through which a shockwave runs at the speed u_s. The time required for the cold material to pass through the wave, (V_cold/u_s), is (for a steady wave) evidently exactly the same as the time required for it to emerge hot, at the reduced velocity $u_s - u_p$:

$$(V_\text{cold}/u_s) = (V_\text{hot}/(u_s - u_p)) \ .$$

Here u_p is the "particle velocity", equal to the speed of cold matter approaching the wall or to the decelerated velocity reduction induced by the shockwave.

During this transit time the leading edge of the cold volume does work on the hot fluid to the right, $[P_\text{hot} V_\text{hot}/(u_s - u_p)]$. At the same time the work done by the trailing cold fluid on our sample is $[P_\text{cold} V_\text{cold}/u_s]$. The difference between the two is equal to the energy change in the process, internal plus kinetic:

$$[E_\text{hot} + (u_s - u_p)^2/2] - [E_\text{cold} + (u_s^2/2)] = (PV)_\text{cold} - (PV)_\text{hot} \ .$$

Introducing the enthalpy, $H \equiv E + PV$, we see that the shock process can be thought of as a steady conversion of part of the cold matter's kinetic energy into enthalpy:

$$H_\text{hot} - H_\text{cold} = [(u_s)^2 - (u_s - u_p)^2]/2 \ .$$

The process is *adiabatic* (because no heat flows in or out of the fluid) but *irreversible*, because the process is sudden rather than quasistatic. Evidently there *is* a *microscopic* heat flux *within* the shockwave itself so that two irreversible processes coexist there, the viscous dissipation of the velocity difference, and the conductive dissipation of the temperature difference. Remote measurements of temperature are not so easy as velocity measurements, but can be accomplished, in transparent materials, by analyzing the thermal radiation emitted by the shockfront.

Consider again this stationary shockwave problem in the coordinate system fixed on the wave. Cold matter approaches from the left, at a speed

u_s, the "shock velocity"; hot fluid departs toward the right, at a reduced speed $(u_s - u_p)$. The velocity change, $u_s \to (u_s - u_p)$, is the same as that in the stagnation experiment where cold matter is decelerated $u_p \to 0$ in its collision with a fixed wall. In the shock-centered coordinate system the mass flux from the left, $\rho_{\text{cold}} u_s$ is necessarily equal to the mass flux to the right, $\rho_{\text{hot}}(u_s - u_p)$:

$$\rho_{\text{cold}} u_s = \rho(x) u(x) = \rho_{\text{hot}}(u_s - u_p) \ .$$

This general relation, which applies anywhere *in* the shock, is a consequence of the fact that the flow into any Eulerian zone dx (motionless in the shock-fixed frame) is equal to the outflow, as there are no sources or sinks of mass within dx:

$$(\partial \rho / \partial t) = -(\partial [\rho u]/\partial x) \equiv 0 \ .$$

There is a similar relation for momentum conservation. The flow of x momentum in the x direction has two parts anywhere in the steady wave, the convective flow ρu^2 and the comoving flow P_{xx}. Evidently conservation of momentum guarantees that the momentum flux from the left is equal to that *in* the wave, in any fixed element dx, which is also equal to that within the hot fluid at the right:

$$P_{\text{cold}} + \rho_{\text{cold}} u_s^2 = P_{xx}(x) + \rho(x) u(x)^2 = P_{\text{hot}} + \rho_{\text{hot}}(u_s - u_p)^2 \ .$$

The energy flux is a little more complicated. During its passage through a fixed dx the energy changes in response to the pressure difference across the fixed element, $(\partial P_{xx}/\partial x)dx$ so that the constant energy flux is more accurately called an enthalpy flux:

$$\rho u [e(x) + (P_{xx}(x)/\rho(x)) + (u(x)^2/2)] + Q_x(x) \ .$$

Here Q_x is the heat flux vector in the propagation direction, the additional comoving, or *conductive*, contribution to the energy flow. Within the shockwave the internal energy per unit mass $e = e(x)$ is not necessarily equal to its thermodynamic value, though there is no special way of formulating or indicating this difference in present-day continuum mechanics.

6.3 Shockwave Conditions for Molecular Dynamics

We begin our exploration of shockwave structure by choosing a simple two-dimensional system modeling the twofold compression of a zero-pressure zero-temperature initial "cold" state. In the shock-fixed frame we insert cold particles at the left boundary with the speed $u_s = 1.930$ (for the triangular lattice structure shown at the left below). At the opposite right boundary the shocked hot fluid exits, at $u_s - u_p = 0.965 = (u_s/2)$. Near that righthand boundary, to prevent the exiting fluid's accelerating into a vacuum, we provide, every few timesteps, each particle with a new Maxwell-Boltzmann velocity, characteristic of the exiting hot fluid. Apart from these essential boundary details, required to simulate a steady state, the molecular dynamics program uses the usual equations of motion, the usual list of neighbors, and the usual Runge-Kutta integrator. For a force law we use a short-ranged repulsive pair potential with two vanishing derivatives at its cutoff length of unity,

$$\phi(r) = (10/\pi)(1-r)^3 \longrightarrow \int_0^1 2\pi\phi(r)r\,dr \equiv 1 \ .$$

With this choice the entering density (for a regular triangular lattice of particles with unit mass and an interparticle spacing of unity) is $\sqrt{4/3}$.

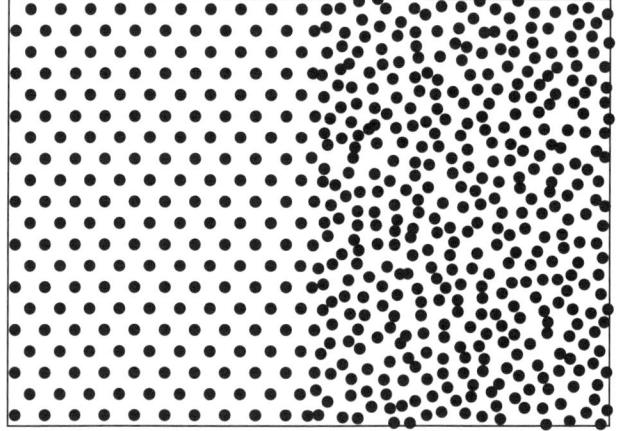

Figure 6.1. Portion of a molecular dynamics shockwave simulation. The incoming triangular lattice, with $u_s = 1.930$ and zero temperature, is converted to an outgoing hot fluid, with $u_s - u_p = 0.965$ and a temperature of 0.12. Periodic boundary conditions are applied on the horizontal boundaries at $\pm 10\sqrt{3/4}$. The horizontal offset between the odd and even rows of particles is $(1/2)$ and the nearest neighbor lattice spacing is unity.

Trial and error show that a shockwave speed of $u_s = 1.930$, twice the particle speed of $u_p = 0.965$, gives twofold compression of the triangular lattice, corresponding to a relatively strong shock in two dimensions. If the input stream has a speed of 1.94, the shockwave moves to the right, at speed 0.01. Likewise if the input stream is slowed to 1.92, the shockwave moves to the left, at speed 0.01.

Figure 6.1 shows a typical snapshot from such a flow. As time passes the shock front remains accurately fixed with the incoming and outgoing speeds of 1.930 and 0.965. Any precise analysis of the stationary spatial shock profile requires averaging. If the system is wide enough, time averaging can be avoided and spatial averaging in x is enough. The straightforward approach, using rectangular bins, can be substantially improved by using *smooth-particle weight functions* to define a grid of local twice-differentiable variables, either as functions of x or as functions of x and y.

For these local averages we use Lucy's weight function,

$$w(r < h) = \{(5/4h)_{D=1} \text{ or } (5/\pi h^2)_{D=2}\}[1 + 3z][1 - z]^3 \; ; \; z \equiv r/h \;.$$

This weight function is the simplest polynomial that provides two continuous derivatives everywhere and a cutoff at $r = h$. The two normalization constants, $(5/4h)$ and $(5/\pi h^2)$ correspond respectively to one-dimensional and to two-dimensional averages. The one-dimensional density of unit-mass particles at x in a system of height H is

$$\rho(x) \equiv (1/H) \sum_k w_{D=1}(|x_k - x|); \; \{|x_k - x| < h\} \;.$$

The two-dimensional density is likewise

$$\rho(x, y) = \rho(r) \equiv \sum_k w_{D=2}(|r_k - r|); \; \{|r_k - r| < h\} \;,$$

where the sum includes all particles within a radius h of the location r.

Smooth-particle averaging of particle properties is straightforward. The stream velocity at x, for instance, is the ratio of two local particle sums:

$$\rho(x)u(x) \equiv \sum_k v_k w(|x_k - x|) \longleftrightarrow$$

$$u(x) \equiv \sum_k v_k w(|x_k - x|) / \sum_k w(|x_k - x|) \;.$$

Smooth-particle averaging is a powerful investigative tool. Let us apply it to the question of shockwave stability. Unless these structures are stable, we have nothing to analyze!

6.4 Shockwave Stability

To judge the stability of planar shockwaves we must first *locate* the wave. The time-dependence of the density on the gridpoints $\{r_g\}$, follows from the usual Lucy-Monaghan smooth-particle recipe,

$$\{\rho(r_g)\} = \{m \sum_k w(|r_k - r_g|)\} \ .$$

Do "planar" shockwaves remain truly planar? That is, does the "shock width" tend to grow with system size, as do the spatial fluctuations of a particle in a two-dimensional solid, $\langle(\delta r)^2\rangle \propto \ln N$? We can find out by studying shock propagation in systems of increasing height, choosing an initial condition that should promote structural *instability* if it is present. For simplicity we consider the two-dimensional problem for which the shockwave is nominally a straight line. We avoid surface effects by choosing *periodic* boundaries parallel to the propagation direction, x.

A straightforward approach to stability begins by separating the cold and hot fluids with a *sinusoidal* shockwave boundary rather than a straight line. We designate six views of the system collectively as "Figure 6.2". The first of them shows the initial geometry near the center of a system $(x, y) = (0,0)$ of height $80\sqrt{3/4}$ with a shockwave amplitude amp = 12.0d00. To the left of the sine wave cold material moves right with velocity u_s. To the right an unstable phase (with a rectangular unit cell), compressed twofold, but uniaxially rather than homogeneously, and with an energy similar to the equilibrated shocked fluid, equilibrates in one or two atomic vibration times. In all six views particles with densities less than the mean,

$$\rho_{av} = [\rho_{\text{cold}} + \rho_{\text{hot}}]/2 = 1.5\rho_{\text{cold}} = 0.75\rho_{\text{hot}} \ ,$$

are shown as *filled* circles. *Open* circles show particles with densities greater than the mean. The sinewave initial condition is followed quickly by four snapshots, at times that a shockwave would take to move 2, 4, 12, and 18 particle diameters. The last snapshot corresponds to the shock traversal time for 120 diameters, the size of the picture frame. The intermediate snapshots show intricately contorted phase boundaries separating regions of wildly differing densities and temperatures. This whole history of the equilibration develops on a *sonic*, as opposed to diffusive, timescale. We describe both the one- and the two-dimensional spatial scales of the shockwave profile that results in the following Figures 6.3, 6.4, 6.5, and 6.6.

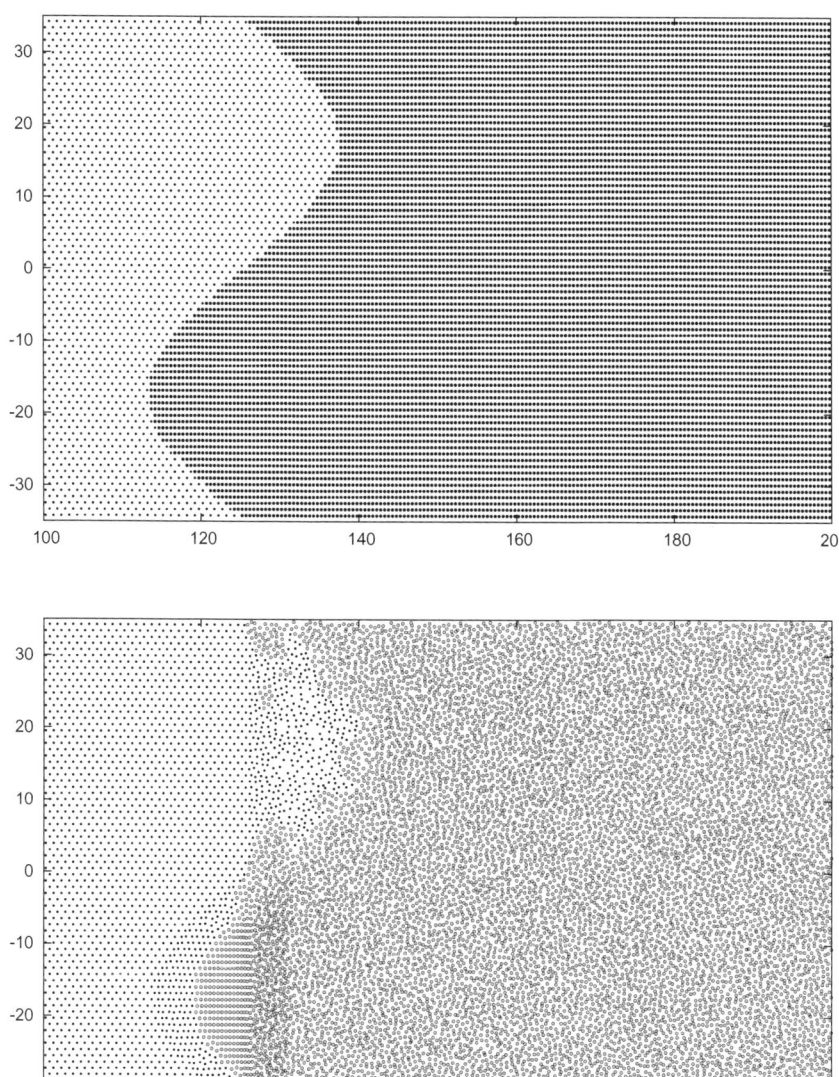

Figure 6.2. Initial crystalline state and fluid configuration at time 2.0. The boundaries at the top and bottom are *periodic*. The system height is $80\sqrt{3/4}$. Initially the cold stress-free material (at the left) has a triangular lattice, the "hot" (at the right) has an unstable rectangular lattice.

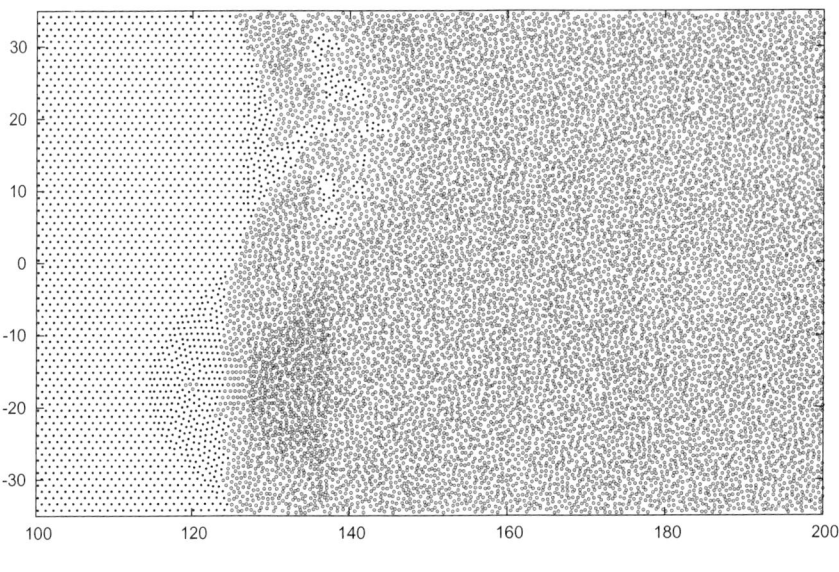

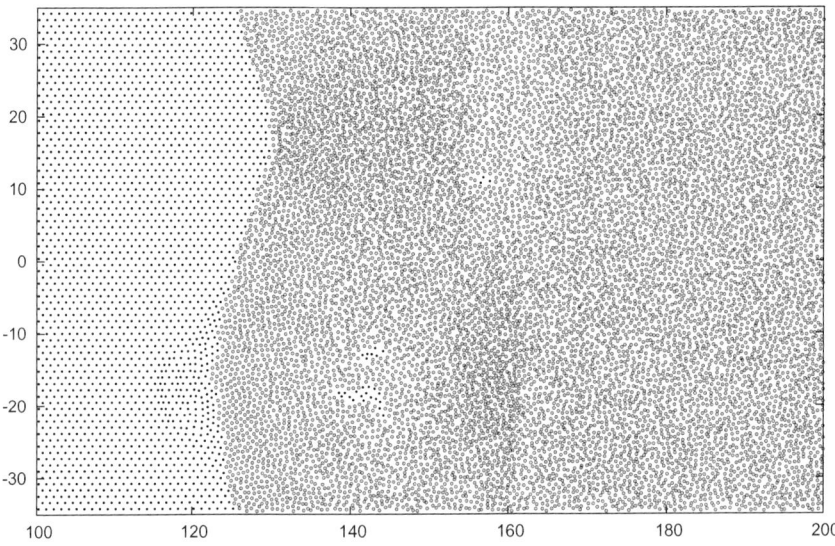

Figure 6.2. (*continued*). Fluid configurations at times 4.0 and 12.0. The average rightward velocity, $u_s = 1.930$, of the cold material (left) is twice as great as the hot fluid velocity, $u_s - u_p = u_p = 0.965$, (right).

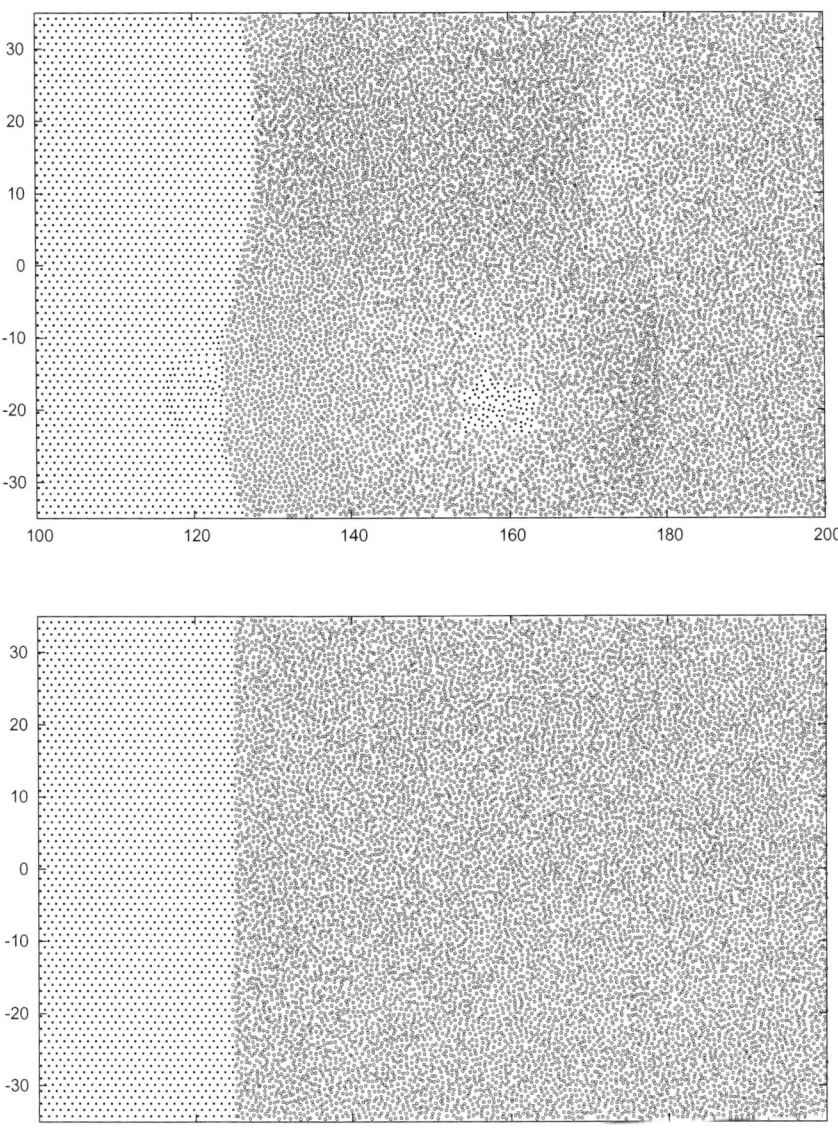

Figure 6.2. (*concluded*). Fluid configurations at times 18.0 and 120.0. The figure shows the evolution of a system during the decay of a sinusoidal shockwave toward planarity. The shapshots distinguish the particles with below and above average density. Notice the low-density tensile waves that appear in the third, fourth, and fifth snapshots.

To the right of the sine wave hot material moves rightward with the slower velocity $(u_s - u_p)$. If we choose an initial "hot" state with a particle-to-particle horizontal spacing, dx = 0.5d00, half that in the "cold" state, where dx = 1.0d00, then a suitable initial condition can be generated from the following Fortran fragment:

```
do iy = 1,ny
ytry = (iy - 0.5d00 - (ny/2.0d00))*dsqrt(0.75d00)
dx = 1.0d00
do ix = 1,lx+lx
if(2*(iy/2).ne.iy) then
calculation if iy is odd
if(ix.eq.1) xtry = +0.25d00*dx
if(xtry.gt.0.5d00*lx + amp*dsin(2*ytry*pi/ny)) dx = 0.5d00
if(ix.gt.1) xtry = x(index) + dx
if(xtry.le.elx + 3.0d00) then
index = index + 1
x(index) = xtry
y(index) = ytry
endif
calculation if iy is odd
endif

if(2*(iy/2).eq.iy) then
calculation if iy is even
if(ix.eq.1) xtry = -0.25d00*dx
if(xtry.gt.0.5d00*elx + amp*dsin(2*ytry*pi/ny)) dx = 0.5d00
if(ix.gt.1) xtry = x(index) + dx
if(xtry.le.elx + 3.0d00) then
index = index + 1
x(index) = xtry
y(index) = ytry
endif
calculation if iy is even
endif
enddo
enddo
```

Local equilibration happens very rapidly so that the decay of the sinewave can be determined accurately. Leading sections of the wave tend to slow while lagging sections of the wave tend to catch up. Although one might well expect an exponential decay of the wave amplitude with time, the decay here is underdamped. But in any case the sinewave profile simulations suggest that one-dimensional shockwaves are indeed stable.

A good choice of the Lucy function smoothing length is essential and requires good judgment. Too small a choice gives a density profile with unrealistic "wiggles" while too large an h loses the local character of the wave. Figure 6.3 shows the time development of a nominally one-dimensional shockwave profile using $h = 2, 3$, and 4 for a system height of 80 rows.

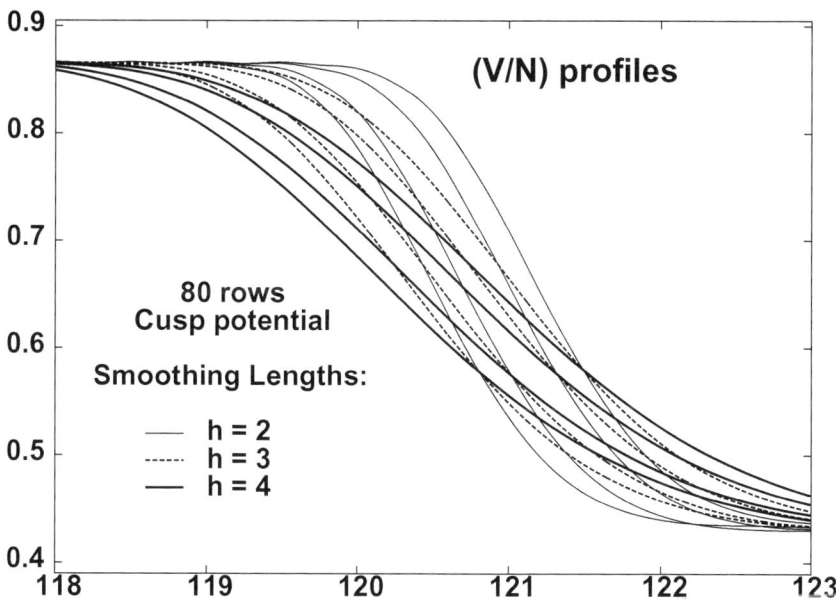

Figure 6.3. Snapshots of the density profile using Lucy function ranges of 2, 3, and 4. From left to right, the sets of three curves (for $h = 2, 3, 4$) correspond to times of 50,000Δt, 100,000Δt, 87,500Δt, and 25,000Δt.

The planar stability of the shockfront in time is well-established by the data shown in Figures 6.3 and 6.4, taken over time intervals corresponding to shockfront travel of *thousands* of particle diameters. There is no tendency for the shockfront, either in one dimension (Figure 6.3) or in two dimensions (Figure 6.4) to change shape.

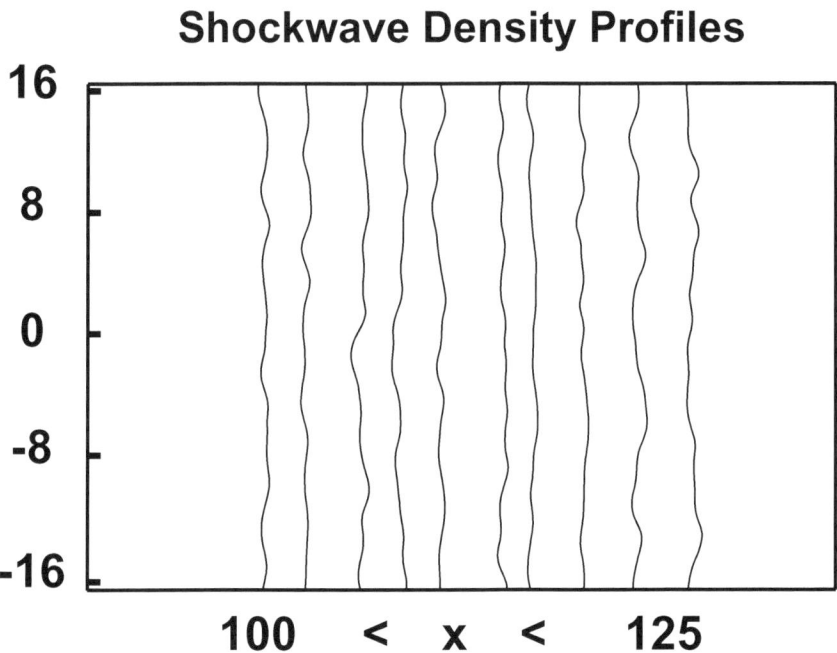

Figure 6.4. Snapshots of the left-moving two-dimensional density profile with an input velocity of 1.920 using Lucy's weight function with a range $h = 3$. The time interval between successive profiles is 234.2 and the periodic system height is $40\sqrt{3/4}$.

The stability of the shockfront to changes in system height is explored in Figures 6.5 and 6.6. In Figure 6.5 the time-dependence of the maximum slope $(\partial \rho/\partial x)_{\max}$, calculated using the one-dimensional Lucy density profile with $h = 3$, is displayed for system heights ranging from 10 to 160 rows of particles. The average values of this slope agree within two percent and show no tendency to change with system width. The evidence from these one-dimensional density profiles confirms the stability of one-dimensional fluid shockwaves in a two-dimensional fluid system.

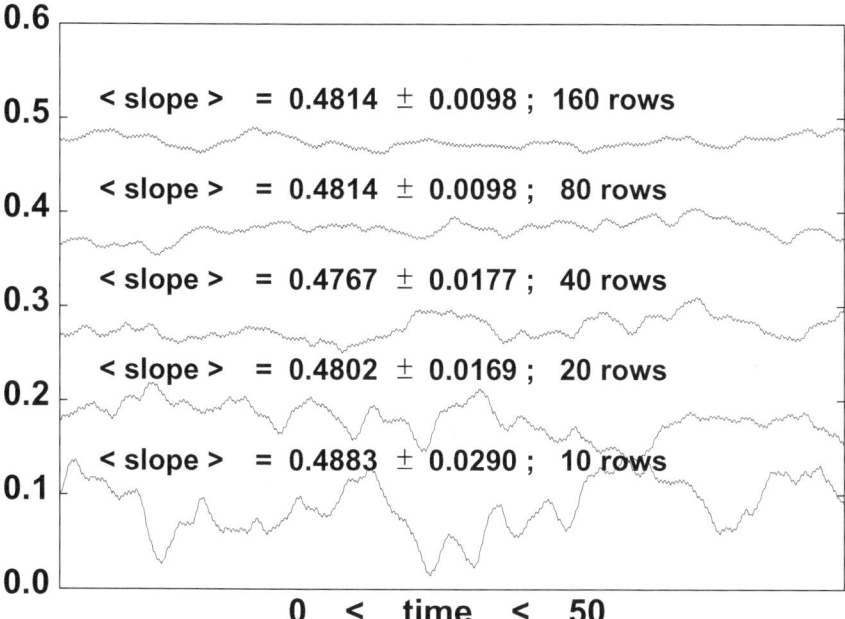

Figure 6.5. The curves show the time development of the maximum slope $(\partial \rho/\partial x)_{max}$ using one-dimensional Lucy averaging with $h = 3$. The rms deviations from the averages are indicated. For clarity the curves for 80-row, 40-row, 20-row, and 10-row systems have been displaced downward by 0.1, 0.2, 0.3, and 0.4 respectively.

Figure 6.6 explores the transverse fluctuation of the two-dimensional density contour corresponding to the mean density,

$$\rho_{1/2} = [\sqrt{4/3} + 2\sqrt{4/3}]/2 = \sqrt{3} \ .$$

The location $x_{1/2}$ of this mean density varies with y. Figure 6.6 displays the rms fluctuations for five system heights, averaged over y:

$$\sqrt{\langle x_{1/2}^2 \rangle - \langle x_{1/2} \rangle^2} \ ,$$

corresponding to a density-based fluctuation in the shockfront location. The fluctuations are quite small, an order of magnitude less than the shockwidth determined from the slope $(\partial \rho/\partial x)$ in Figure 6.5. Overall, the evidence for the stability of one-dimensional fluid shockwaves to time, system height, and transverse fluctuations is excellent, so that a mathematical treatment based on a one-dimensional profile is justified.

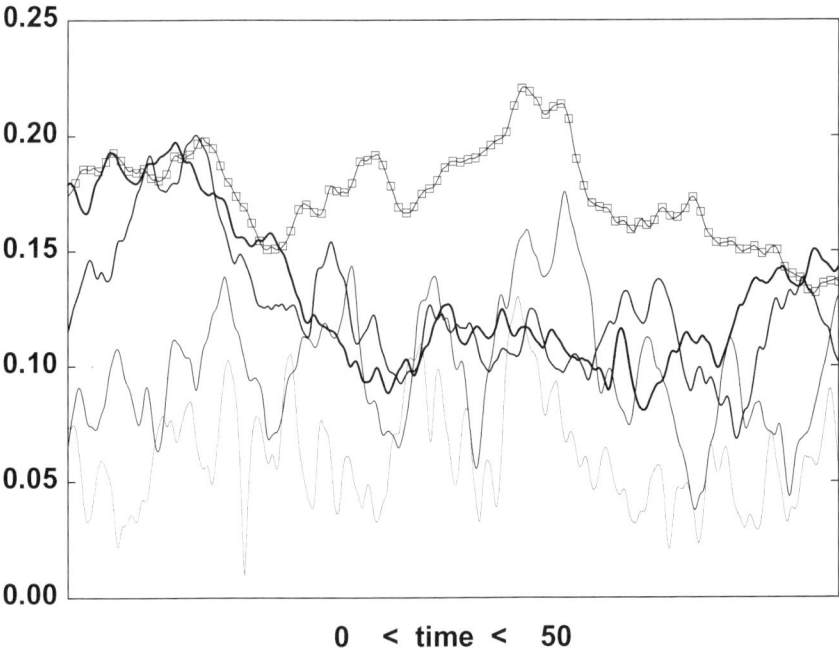

Figure 6.6. The five curves here correspond to the five system heights (y direction) from 10 rows (bottom) to 160 rows (top). Each of them shows the time-development of the rms fluctuation, averaged over y, of the mean-density shock location x, where $\rho(x,y) = (3/2)\sqrt{4/3} = \sqrt{3}$. The fluctuation in that density's contour value of x, $[\langle x^2 \rangle - \langle x \rangle^2]^{1/2}$, is shown here.

6.5 Thermodynamic Variables

Microscopic relations for velocity, energy, and pressure are straightforward once the question of *Temperature* is resolved. In (equilibrium) statistical mechanics there are many possible definitions. Among the kinetic definitions, based on moments of the velocity distribution function, it is evident that the *second* moment is the best choice,

$$kT_{xx} \equiv m\langle (v_x - \langle v_x \rangle)^2 \rangle; \; kT_{yy} \equiv m\langle (v_y - \langle v_y \rangle)^2 \rangle \;.$$

In a planar shockwave simulation where x is the propagation direction, one would probably set $\langle v_x, v_y \rangle \equiv (u, 0)$, with $u(x)$ the "stream velocity".

The average values $\langle v_x \rangle$ and $\langle v_x^2 \rangle$ can then be calculated in one of two ways, using either the one- or the two-dimensional weighting function, giving either $T_{xx}(x)$ or $T_{xx}(x,y)$.

"Configurational temperatures" created a flurry of excitement when Rugh rediscovered Landau and Lifshitz' identity:

$$kT \int \ldots \int \prod(dpdq) \nabla^2 \mathcal{H} e^{-\mathcal{H}/kT} = \int \ldots \int \prod(dpdq)(\nabla \mathcal{H})^2 e^{-\mathcal{H}/kT} \; .$$

Integrating the lefthand integral by parts gives the identity for any coordinate q, wholly independent of p. Rearrangement gives a definition of that coordinate's configurational temperature:

$$kT_C \equiv \langle (\nabla \mathcal{H})^2 \rangle / \langle \nabla^2 \mathcal{H} \rangle \; .$$

Though a velocity-independent temperature has some initial appeal, there are at least three downsides. (i) There is no configurational analog of the kinetic ideal-gas thermometer of Section 2.8. (ii) Steady rigid rotation gives centrifugal forces which would suggest, falsely, a radial temperature gradient. (iii) In addition, the configurational temperature can be, and often is, singular. For a typical two-body interaction the denominator,

$$\nabla^2 \mathcal{H} = \nabla^2 \phi = \phi'' + (D-1)(\phi'/r) \; \text{[in D dimensions]} \; ,$$

can vanish. Thus the local configurational temperature can *diverge*. For these reasons we restrict our nonequilibrium temperature definition to the traditional kinetic temperature. At a gridpoint r_g evaluating the kinetic definition is straightforward:

$$kT(r_g)/m \equiv \frac{\sum_k (v_k^2) w(|r_k - r_g|)}{\sum_k w(|r_k - r_g|)} - \left[\frac{\sum_k (v_k) w(|r_k - r_g|)}{\sum_k w(|r_k - r_g|)} \right]^2 \; .$$

Definitions *at* particles seem more problematic. An argument can certainly be made for omitting the "self" terms in computing particle temperatures.

6.6 Shockwave Profiles from Continuum Mechanics

Boundary conditions and a stable algorithm are the keys to most simulations. For continuum shockwaves, in shock-fixed coordinates, the boundary flows necessarily satisfy conservation of mass, momentum, and energy:

$$\{ \rho u, \; P_{xx} + \rho u^2, \; \rho u[e + (P_{xx}/\rho) + (u^2/2)] + Q_x \} \; \text{all constant} \; .$$

Specifying the initial state, including the velocity u, is sufficient information to find the final state. In this Section we sketch two different numerical techniques for finding continuum shockwave profiles. We then apply them to a variety of constitutive models.

For a "Newtonian fluid", viscous but without conductivity, the heat flux vanishes, $Q_x \equiv 0$. In this simplest case there is only one spatial derivative in the conservation laws, $P_{xx} \to (du/dx)$. Without conductivity (dT/dx) is absent. Once $u(x)$ is known the other hydrodynamic variables, and the derivative (du/dx) required to step forward or backward in x, can all be obtained by direct substitution into the conservation laws. The sequence of steps is as follows:

$$u(x) \to \rho(x) \to P_{xx} \to e(x) \ ;$$

$$\{\rho(x), e(x)\} \to P_{\text{eq}}(x) \ ;$$

$$\{P_{xx}(x), P_{\text{eq}}(x)\} \to (du/dx) \ .$$

The density $\rho(x)$ comes from the known mass flux, ρu. P_{xx} then follows from the known momentum flux, $P_{xx} + \rho u^2$. And so on. Notice that $e(x)$ and $\rho(x)$ must be sufficient information to give the equilibrium pressure $P_{\text{eq}}(x)$. This assumption can be *checked* by carrying out molecular dynamics studies, where all the variables can be defined and measured separately. We illustrate two solution techniques for this problem in the following Subsection. Much the simpler of the two is to integrate the single differential equation for (du/dx) *assuming* that the shockwave is steady. A more general approach solves the continuity equation, the equation of motion, and the energy equation simultaneously. An Eulerian grid is used and a relatively steady solution emerges from the time-dependent calculation.

For a Navier-Stokes-Fourier fluid, with both viscosity and heat conductivity, *two* equations need to be solved, one for (du/dx), and another for (dT/dx). With $u(x)$ and $T(x)$ known the sequence of steps is similar:

$$u(x) \to \rho(x) \to P_{xx}(x) \ ;$$

$$\{\rho(x), T(x)\} \to \{P_{\text{eq}}(x), e(x)\} \to \{(du/dx), (dT/dx)\} \ .$$

In order to solve the equations it is again necessary to assume that the energy has its equilibrium value, $e(x) \equiv e_{\text{eq}}(x)$.

Molecular dynamics, for dense fluids, and the Boltzmann equation, for a dilute gas, both show that the kinetic temperature is a tensor, with $T_{xx} \neq T_{yy}$. We will demonstrate that models can be developed in which the two temperatures obey separate kinetic equations. The corresponding set of differential equations is then expanded to three, for $\{(du/dx), (dT_{xx}/dx), (dT_{yy}/dx)\}$. It is again a necessary assumption for the models developed here that density and the two temperatures are enough to give the internal energy and the equilibrium pressure.

In addition to the tensor nature of temperature, molecular dynamics reveals (as is also clear from Green and Kubo's linear response theory) that stress and heat flux *lag behind* the strain rate and temperature gradient(s) assumed to "cause" them. Models capable of describing these delays incorporate phenomenological times familiar from Maxwell's model for gas-phase stress relaxation:

$$\sigma + \tau_\sigma \dot\sigma = \eta \dot\epsilon \;.$$

Here the stress σ follows the strain rate $\dot\epsilon$ in time. It is important to notice that the time derivative here is the *comoving* derivative, following the material motion, and *not* a fixed-in-space Eulerian derivative. The difference is clear in the shock problem, where the Eulerian time derivatives are all zero for steady shocks.

Until now, the shock structure problems we have considered can all be solved by hot-to-cold integration along the shock profile. The relaxation problems here require a different numerical technique for stability, with density and velocity computed on separate grids. As a result our most sophisticated model requires the solution of six differential equations, one each for $\{\rho, u, T_{xx}, T_{yy}, \sigma, Q_x\}$. In what follows we illustrate solutions for all of the models just described. The basic underlying equation of state used here for some simple example problems resembles van der Waals' equation, but with a purely-repulsive "cold curve", $P_{\text{cold}} \equiv (\rho^2/2)$ and a thermal energy composed of two parts, $e = kT_{xx} + kT_{yy}$. We will also compare a more sophisticated model directly to molecular dynamics shockwave simulations in what follows.

6.6.1 *Shockwave Profile with Shear Viscosity*

Shockwaves are dominated by viscosity. Heat conduction plays a minor role. To see this consider a shockwave in a material *without conductivity*, described by the following equation of state and constitutive relations:

$$P = \rho e; \; e = (\rho/2) + 2T; \; \eta = 1; \; \eta_V = 0; \; \kappa = 0 \;.$$

Choosing a cold fluid state ($\rho = 1; T = 0; u = 2$) it is an easy exercise to find the corresponding hot state for twofold compression: ($\rho = 2; T =$

($1/8$); $u = 1$) and the constant fluxes of mass, momentum, and energy:
$$\rho u = 2;\ P_{xx} + \rho u^2 = (9/2);\ \rho u[e + (P_{xx}/\rho) + (u^2/2)] + Q_x = 6\ ,$$
where the heat flux Q_x is zero for this problem. In two dimensions the Newtonian pressure tensor has the form:
$$P = P_{eq} - \eta[\nabla u + \nabla u^t] - \lambda(\nabla \cdot u)I\ ,$$
Where I is the unit tensor. As we have chosen zero bulk viscosity,
$$\eta_V = \eta + \lambda = 0 \longrightarrow \lambda = -\eta\ ,$$
We see that
$$P_{xx} = P_{eq} - \eta(du/dx);\ P_{yy} = P_{eq} + \eta(du/dx)\ .$$
A short computer program solving this problem by integrating a single ordinary differential equation, (du/dx), on the interval $[-10 < x < +10]$, is the following:

```
implicit double precision(a-h,o-z)
dimension yy(1),yyp(1)
you = 1.0000001d000
yy(1) = you
dx = -0.1d00
x = 10.0d00
do 30 it = 1,200
call rk(dx,yy,yyp)
x = x + dx
you = yy(1)
rho = 2.0d00/you
Pxx = 4.5d00 - 2.0d00*you
egy = 3.0d00 - (Pxx/rho) - 2.0d00/rho**2
Pee = rho*egy
dudx = Pee - Pxx
Pyy = Pee + dudx
Tee = (egy - 0.5d00*rho)/2.0d00
V = 1.0d00/rho
30 write(16,6) x,rho,you,Pxx,Pyy,egy,Pee,dudx,Tee,V
6 format(10f16.8)
stop
end
```

Note that the integration begins near the endpoint "hot" state and proceeds toward the "cold" state. As usual, the Runge-Kutta routine calls a righthand side routine which evaluates the current value of the derivative (du/dx) by using the known values of the mass, momentum, and energy fluxes $\{2, (9/2), 6\}$:

```
subroutine rhs(dx,yy,yyp)
implicit double precision (a-h,o-z)
dimension yy(1),yyp(1)
you = yy(1)
rho = 2.0d00/you
Pxx = 4.5d00 - 2.0d00*you
egy = 3.0d00 - (Pxx/rho) - 2.0d00/rho**2
Peq = rho*egy
yyp(1) = Peq - Pxx
return
end
```

Figure 6.7 on page 220 shows the mechanical and thermal variables computed in this way, with the coordinate system centered at the x coordinate corresponding to the mean speed, $u = (u_{\text{cold}} + u_{\text{hot}})/2 = (3/2)$. All of the hydrodynamic variables (with the exception of entropy, not shown here) vary smoothly through the shockwave. In classical mechanics the entropy diverges (to minus infinity) as $T \to 0$ so that the entropy production of this shockwave is infinite! It is easy to verify that doubling the viscosity doubles the distance scale so that the shock width is indeed proportional to viscosity. An interesting feature of the profile is that the transverse component of the pressure tensor has a minimum within the shock, for which that component even lies below the "cold curve" pressure, $P_{\text{cold}} - (\rho^2/2)$.

The hydrodynamic equations for a shockwave with unit conductivity and zero viscosity have no solution. Evidently it is the *viscosity* which is essential to stability in shockwaves. Even in the absence of conductivity the fluid eventually reaches the "right" temperature, 0.125 in Figure 6.7, due to the fact that the final density and pressure, $[\rho - 2, P = (5/2)]$, which follow from the conservation of mass and momentum, necessarily fix the *temperature* of the hot state, because that temperature must lie on the equilibrium equation of state surface.

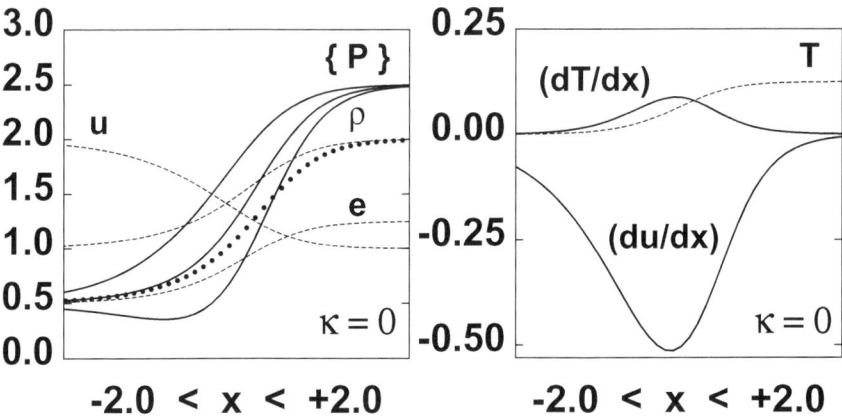

Figure 6.7. Mechanical and thermal variables for a shockwave without conductivity. The abscissa used here is a reduced distance scale, (x/η). The cold curve pressure (filled circles) typically exceeds P_{yy}. $P_{xx} > P_{eq} > P_{yy}$ are shown. This problem requires solving either a single ordinary differential equation for $u(x)$ or two sets of equations on a staggered grid, one for the density and the other for the remaining variables. These two methods agree.

6.6.2 Shockwave Profile with Viscosity and Conductivity

As a further example shown in Figure 6.8, let us consider the case in which the conductivity is also chosen equal to unity. The cold and hot states are exactly the same as before. It is only necessary to compute *two* derivatives, (du/dx) and (dT/dx) and to solve *two* differential equations assuming that the two variables yy(1) and yy(2), u and T, are known. The righthand side subroutine calculates the needed derivatives in seven lines:

```
you = yy(1)
Tee = yy(2)
rho = 2.0d00/you
```

```
Pxx = 4.5d00 - 2.0d00*you
egy = 0.5d00*rho + 2.0d00*Tee
yyp(1) = rho*egy - Pxx
yyp(2) = 2*(2*egy - (yyp(1)/rho) + 0.5d00*you**2) - 6.0d00
```

It is convenient to check the calculation by computing the three fluxes at each integration step. No doubt the reader has noted that, just as before, we start the numerical integration in the neighborhood of the "hot" state. Starting at the "cold" state is unsuccessful. The calculation soon "blows up". This instability was emphasized by Gilbarg and Paolucci in the early days of Navier-Stokes shockwave simulation. This instability only applies to the special *ordinary* differential equations describing a steady flow. The time-dependent *partial* differential equations (continuity, motion, and energy), can be solved numerically for the time-dependent evolving structure of a shock. With a reasonable initial condition, and the shockwave boundary conditions, the time-dependent approach replicates the physical process of shockwave formation.

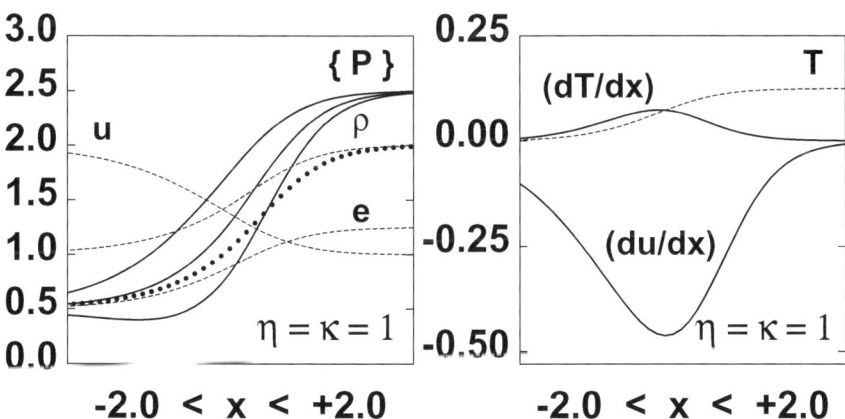

Figure 6.8. Thermal profiles for the van der Waals model with unit viscosity and unit conductivity. Unit conductivity increases the dissipation slightly, reducing the magnitudes of both the velocity and the temperature gradients relative to the zero-conductivity case of Figure 6.7.

6.6.3 Shockwave Profiles with Tensor Temperatures

Because molecular dynamics simulations show that *temperature is a tensor*, with $T_{xx} > T_{yy}$ in the shockwave, the question arises as to how this anisotropicity should be modeled. From the physical standpoint it seems clear that longitudinal-to-transverse collisional scattering has a tendency first to convert the kinetic energy into longitudinal temperature, and that only later does the transverse temperature respond. For low density gases the Boltzmann equation gives a good description of the disparity between T_{xx} and T_{yy}. For denser fluids something new is required. To describe the evolution of tensor temperatures the simplest path forward is to formulate separate kinetic equations for T_{xx} and T_{yy}.

Any kinetic equations are necessarily subject to the restrictions of the continuity, motion, and energy equations. Evidently the energy equation,

$$\rho \dot{e} = -P : \nabla u - \nabla \cdot Q = -P_{xx}(du/dx) - (dQ_x/dx) ,$$

has to be separated into potential and kinetic parts. A further requirement is that the two post-shock temperatures must eventually approach one another, $T_{xx} \longleftrightarrow T_{yy}$.

A simple model which achieves this goal begins by dividing the energy into three parts: a temperature-independent "cold curve" and longitudinal and transverse "thermal" parts. For the van der Waals' example with constant heat capacities:

$$e(\rho, T) = e_{\text{cold}}(\rho) + kT_{xx} + kT_{yy} .$$

$$e_{\text{cold}} = (\rho/2); \; P_{\text{cold}} = (\rho^2/2) .$$

For simplicity we allocate fixed portions of the work and heat to the x and y directions. The longitudinal portions are described by the parameters $(0 < \alpha, \beta < 1)$. We also include thermal relaxation, with a constant relaxation time τ_r.

$$\rho \dot{e}_{\text{cold}} = -P_{\text{cold}}(du/dx) ;$$

$$\rho \dot{T}_{xx} = -\alpha(P_{xx} - P_{\text{cold}})(du/dx) - \beta(dQ_x/dx) + \rho(T_{yy} - T_{xx})/\tau_r ;$$

$$\rho \dot{T}_{yy} = -(1-\alpha)(P_{xx} - P_{\text{cold}})(du/dx) - (1-\beta)(dQ_x/dx) + \rho(T_{xx} - T_{yy})/\tau_r .$$

It is also necessary to generalize Fourier's law to apply to the two-temperature case:

$$Q_x \equiv -\kappa_{xx}(dT_{xx}/dx) - \kappa_{xy}(dT_{yy}/dx) .$$

Consider, for illustration, the simplest case, in which all of the work and all of the heat contribute to the longitudinal part of the internal energy ($\alpha = \beta = 1$). The straightforward application of these ideas suggests solving three equations for $\{u, T_{xx}, T_{yy}\}$ with the usual Runge-Kutta algorithm:

$$(du/dx) = P_{\text{eq}} + \rho u^2 - (9/2) \ ;$$

$$(dT_{yy}/dx) = (\dot{T}_{yy}/u) = (T_{xx} - T_{yy})/(u\tau_r) \ ;$$

$$(dT_{xx}/dx) = (\dot{T}_{xx}/u) = \rho u[e + (P_{xx}/\rho) + (u^2/2)] - 6 + (T_{yy} - T_{xx})/(u\tau_r) \ .$$

Here $\eta = \kappa = 1$. Even with thermal relaxation and the asymmetric work and heat distributions, the mass, momentum, and energy flux conditions are exactly the same as in the simpler two cases:

$$\rho u = 2; \ P_{xx} + \rho u^2 = (9/2) \ ;$$

$$\rho u[e + (P_{xx}/\rho) + (u^2/2)] + Q_x = 6 \ .$$

The solution technique for these models is necessarily more complex than that for the Navier-Stokes-Fourier models. The thermal relaxation equation is unstable, when integrated backward in time.

When thermal relaxation is included we solve the full set of hydrodynamic equations, rather than the special steady-profile ordinary differential equations. The simplest way to do this is to introduce a staggered grid, with velocity and energy evaluated at the grid points, and density evaluated in the "cells" bounded by those gridpoints. We have used this same finite-difference technique for solving the Rayleigh-Bénard problem, for which the steady state assumption holds only at relatively small Rayleigh numbers. Before implementing the staggered-grid approach to shockwaves, let us also discuss and include a time lag between the hydrodynamic fluxes and forces, as suggested by Maxwell and Cattaneo.

6.6.4 Flow Algorithm with Maxwell-Cattaneo Time Delays

The usual hydrodynamic models give *instantaneous* responses to nonequilibrium gradients. By contrast, molecular dynamics reveals *time lags* between hydrodynamic causes and their effects. Strain rate typically precedes stress, and the temperature rises in advance of the heat flux. Both intuition and kinetic theory are consistent with these observed delays, which are on the order of a collision time.

Maxwell estimated the relaxation time τ required for air to respond to an instantaneous shear stress. His estimate of 200 picoseconds for air, is of the order of a mean collision time, $\tau \simeq (\eta/P)$, where η is the shear viscosity and P is the pressure. Such effects are essential to shockwave modeling, where the shock phenomenon itself takes place in a few collision times. We will take up that generalization here, by introducing time lags between strain rate and stress and between temperature gradients and the heat flux.

In a shockwave, the shear stress $(P_{yy} - P_{xx})/2$ lags behind the strain rate (du/dx) and the heat flux Q_x lags behind (dT_{xx}/dx) and (dT_{yy}/dx). These delays can be formulated as "Maxwell-Cattaneo equations":

$$\sigma + \tau_\sigma \dot\sigma = \eta \dot\epsilon; \; Q + \tau_Q \dot Q = -\kappa \nabla T \; ,$$

keeping in mind that two temperatures are required for a realistic description of shockwave structure.

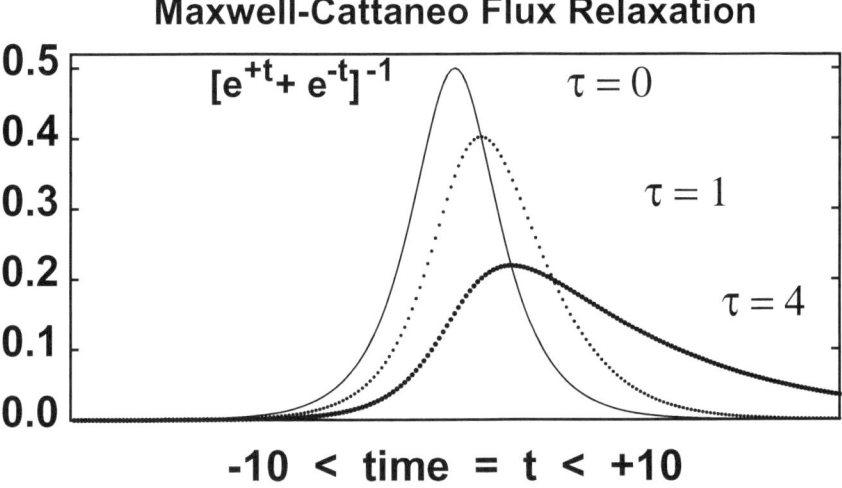

Figure 6.9. Stress response to a localized strain rate near $t = 0$ according to Maxwell's relaxation model with relaxation times $\tau = 0$, 1, and 4 and unit viscosity.

Figure 6.9 illustrates the response of stress to strain rate for three relaxation times, all for a strain rate taking place near time $t = 0$:

$$\dot\epsilon \equiv \frac{1}{e^{-t} + e^{+t}} \; .$$

For a relaxation time and a viscosity of unity, the responding stress can be expressed analytically:
$$\sigma + \dot\sigma = \dot\epsilon \longrightarrow \sigma(t) = e^{-t} \ln\sqrt{1 + e^{+2t}} \ .$$

An algorithm incorporating (i) tensor temperature, (ii) time delays, and (iii) thermal relaxation can be organized as follows. First, choose initial conditions for the density within the computational cells based on the Landau-Lifshitz solution, where the parameter wide is an estimate for the shock width:

```
do i = 1,nc
xc(i) = (i - 0.5d00*(nc+1))*dx
top = 1.0*dexp(-xc(i)/wide) + 2.0*dexp(+xc(i)/wide)
rc(i) = top/(dexp(-xc(i)/wide) + dexp(+xc(i)/wide))
enddo
```

Next, set the initial values of the velocity, energy, and temperatures at the computational nodes. For this we use an approximation to the energy based on neglecting the heat flux contribution to the energy flux:
$$\rho u[e + (P_{xx}/\rho) + (u^2/2)] + Q_x = 6; \ P_{xx} + \rho u^2 = (9/2) \longrightarrow$$
$$e \simeq 3 - (P_{xx}/\rho) - (u^2/2) = 3 - (9/2\rho) + (2/\rho^2) \ .$$

```
do i = 1,nn
xn(i) = (i - 0.5d00*(nn+1))*dx
top = 1.0*dexp(-xn(i)/wide) + 2.0*dexp(+xn(i)/wide)
rn(i) = top/(dexp(-xn(i)/wide) + dexp(+xn(i)/wide))
un(i) = 2.0d00/rn(i)
en(i) = 3.0d00 - (4.5d00/rn(i)) + (2.0d00/rn(i)**2)
txxn(i) = (en(i) - 0.5d00*rn(i))/2.0
tyyn(i) = (en(i) - 0.5d00*rn(i))/2.0
enddo
```

At the interior nodes we approximate both the heat flux (a sum of the T_{xx} and T_{yy} contributions) and the stress by the values they would have in the absence of relaxation:

```
xpart = -condx*(txxn(i+1) - txxn(i-1))/(dx+dx)
ypart = -condy*(tyyn(i+1) - tyyn(i-1))/(dx+dx)
qn(i) = xpart + ypart
sn(i) = visc*(uc(i) - uc(i-1))/dx
```

The complete description of a hydrodynamic state in this shockwave problem requires six variables, $\{\rho, u, T_{xx}, T_{yy}, Q_x, \sigma\}$. We choose to integrate the density in Nc cells and the other five variables at the Nn = Nc + 1 nodes. For a 2000-cell problem there are then $2000 + 5 \times 2001 = 12005$ ordinary differential equations reflecting the continuity, motion, and energy equations, as well as the division of the energy into three parts and the allocation of the energy changes between the x and y thermal parts of the energy.

All of this computational work needs to be done within the righthand side subroutine. A first step is to find the cell values of $\{u, T_{xx}, T_{yy}, Q_x, \sigma\}$ and the nodal values of density by averaging. The cell values and all but the first and last nodal values of density follow from interpolation. Given the density in the first and second cells one can use linear extrapolation to get the first nodal value:

```
rn(1) = (3.0d00*rc(1) - rc(2))/2.0d00.
```

With the state variables known the next step is to find the gradients of the density, temperatures, stress, and heat flux:

```
do i = 1,nc
drc(i) = (rn(i+1) - rn(i))/dx
duc(i) = (un(i+1) - un(i))/dx
dtxxc(i) = (txxn(i+1) - txxn(i))/dx
dtyyc(i) = (tyyn(i+1) - tyyn(i))/dx
dsc(i) = (sn(i+1) - sn(i))/dx
dqc(i) = (qn(i+1) - qn(i))/dx
enddo
```

Exactly similar equations provide the nodal values of the gradients. Next come the values of energy and the longitudinal pressure P_{xx}. We illustrate for the cell values:

```
do i = 2,nc-1
ec(i) = 0.5d00*rc(i) + txxc(i) + tyyc(i)
pc(i) = rc(i)*ec(i) - sc(i)
enddo
```

Next come the corresponding pressure and heat flux gradients:

```
do i = 2,nc-1
dpc(i) = (pn(i+1) - pn(i))/dx
dqc(i) = (qn(i+1) - qn(i))/dx
enddo
```

Now the time derivatives can be formulated:

```
do i = 2,nc-1
drcdt(i) = (rn(i)*un(i) - rn(i+1)*un(i+1))/dx
enddo
calculate drho/dt at the cell centers
calculate du/dt at the nodes
do i = 2,nn-1
dundt(i) = -un(i)*dun(i) - dpn(i)/rn(i)
do i = 2,nn-1
pthermal = pn(i) - 0.5d00*rn(i)*rn(i)
dtxxndt(i) = -un(i)*dtxxn(i)
&   - 1.0d00*(dun(i)/rn(i))*splitwx*pthermal
&   - 1.0d00*(dqn(i)/rn(i))*splitqx
&   + (tyyn(i) - txxn(i))/tauq
dtyyndt(i) = -un(i)*dtyyn(i)
&   - 1.0d00*(dun(i)/rn(i))*splitwy*pthermal
&   - 1.0d00*(dqn(i)/rn(i))*splitqy
&   + (txxn(i) - tyyn(i))/tauq
enddo
calculate de/dt at the nodes and don't include the potential!
calculate ds/dt at the nodes
```

```
do i = 2,nn-1
dsndt(i) = -un(i)*dsn(i) + (visc*dun(i) - sn(i))/taus
enddo
calculate ds/dt at the nodes
calculate dq/dt at the nodes
do i = 2,nn-1
dqndt(i) = -un(i)*dqn(i)
& -condx*(txxn(i+1) - txxn(i-1))/(2.0d00*taur*dx)
& -condy*(tyyn(i+1) - tyyn(i-1))/(2.0d00*taur*dx)
& -qn(i)/taur
enddo
calculate dq/dt at the nodes
```

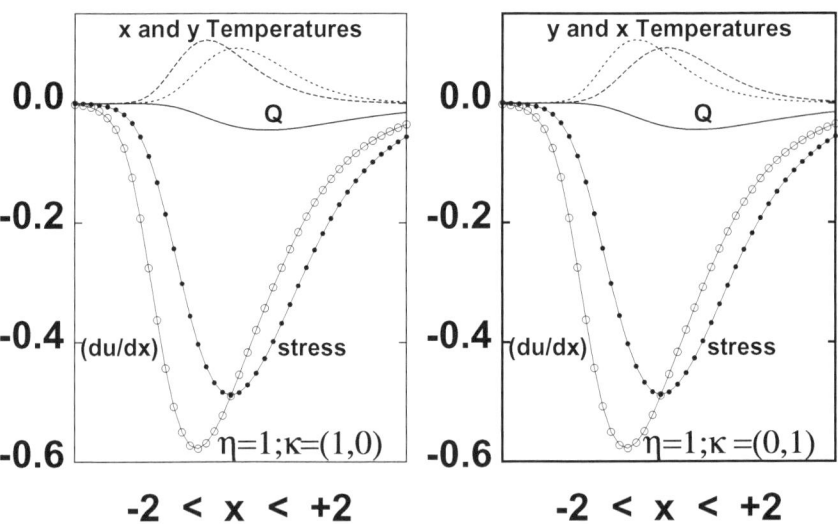

Figure 6.10. Profiles obtained with energy allocation and thermal relaxation. On the left the heat flux Q responds to (dT_{xx}/dx) and on the right to (dT_{yy}/dx). On the left the work and heat contribute to T_{xx}, which leads T_{yy} while on the right these relationships are all reversed. Notice that the stress and strain rate are independent of the thermal reversal. In both cases the strain rate (du/dx) leads the stress, $(\sigma_{xx} - \sigma_{yy})/2$. The relaxation times in the two examples shown here are $4\tau_\sigma = 4\tau_\kappa = \tau_T = 1$.

With these time derivatives complete we have all the information needed for the righthand side subroutine. Four such iterations allow the Runge-Kutta integrator to advance all the variables to the next timestep.

The flexibility of this algorithm is limited to "reasonable" values of the relaxation times. In retrospect this makes good sense. It is implausible that materials' responses can react to information arbitrarily far in the past. In fact the allowed memory for the delay-time equations of Maxwell and Cattaneo is short. It is possible to show that the viscous relaxation time τ_σ must be less than $(\eta/3G)$, where G is the high-frequency shear modulus.

The limitations on the partitioning of work and heat between the longitudinal and transverse temperatures are not at all stringent. Figure 6.10 shows typical profiles for the two extreme situations, all of the heat and work contributing to T_{xx} on the left, and all to T_{yy} on the right. Let us turn next to a comparison of this flexible model with profiles from molecular dynamics simulations.

6.7 Comparing Model Profiles with Molecular Dynamics

The molecular dynamics shockwave with twofold compression (see again Figure 6.1 on page 204) of the zero-pressure triangular lattice results in the following mass, momentum, and energy fluxes:

$$(\rho u)_{\text{cold}} = \rho u = (\rho u)_{\text{hot}} = 1.930 \times \sqrt{4/3} = 2.229 \ ;$$

$$P_{xx} + \rho u^2 = \sqrt{4/3}(1.930)^2 = 4.301 \ ;$$

$$\rho u[e + (P_{xx}/\rho) + (u^2/2)] + Q_x = \sqrt{4/3}(1.930)^3/2 = 4.151 \ .$$

Each particle in the wave can be assigned a density, velocity, pressure tensor, energy, and heat flux vector using Lucy's function for the smooth-particle averages, with a range $h = 3$. Figure 6.11 shows the mechanical and thermal variables which result. Although it is conventional to use the coordinate x as the independent variable for profiles, time could just as well be used, with the relationship

$$t = \int_0^x [dx'/u(x')] \ .$$

Such profiles would display the time-dependent hydrodynamic properties seen by a Lagrangian observer moving through the shockwave at the flow velocity $u(x(t))$.

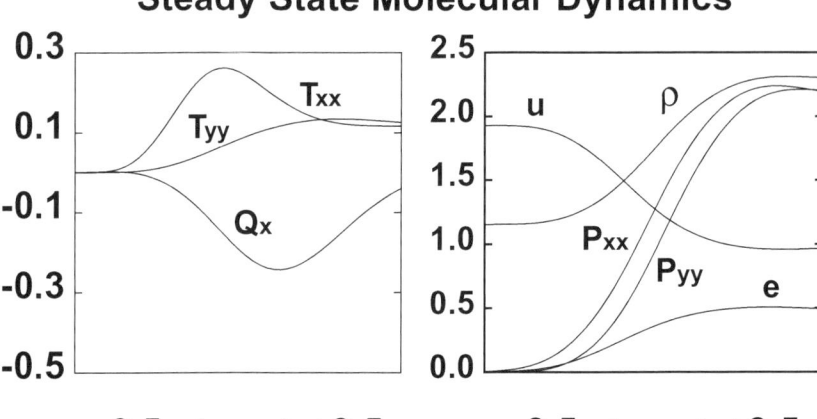

Figure 6.11. Molecular dynamics profiles of mechanical and thermal variables generated with Lucy's one-dimensional weight function

$$w(r < h = 3) = (5/12)[1 + r][1 - (r/3)]^3 .$$

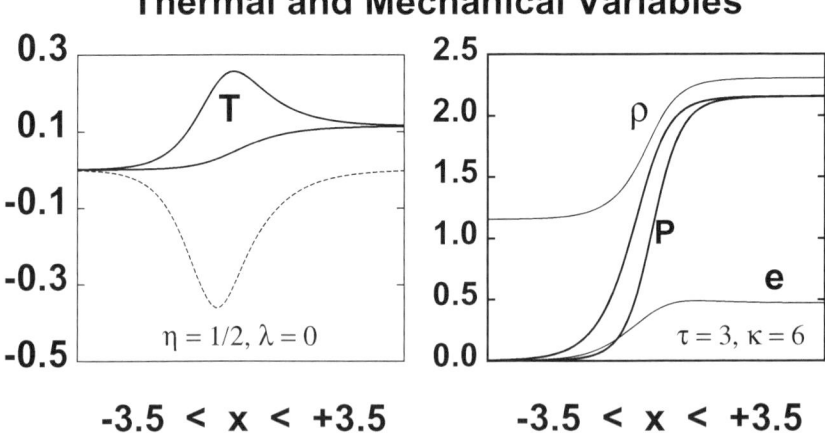

Figure 6.12. Mechanical and thermal variables for a one-dimensional shockwave using the triangular-lattice "cold" curve, an allocation of the work and heat to T_{xx}, and a tensor thermal conductivity with a thermal relaxation time of 3, as explained in the text.

To demonstrate the flexibility of our two-temperature time-delay constitutive model we show in Figure 6.12 *continuum* profiles calculated with constant parameters throughout the shockwave. The equation of state is based on Grüneisen's model and includes a "cold curve" energy and pressure appropriate to a triangular lattice. Let us express the energy and pressure for the cold curve in terms of the nearest-neighbor separation r:

$$r^2 \equiv (\sqrt{4/3}/\rho) \longrightarrow$$

$$e_{\text{cold}} = 3\phi = (30/\pi)(1-r)^3; \; P_{\text{cold}}v = -(3/2)r\phi' = (45/\pi)r(1-r)^2 \;.$$

The cold-curve energy and pressure are contributions of a single particle, Particle j, interacting with six neighbors (triangular-lattice, with only half of each ij interaction allocated to j), to the potential energy, Φ, and to the virial-theorem expression for the pressure-volume product, $PV = NPv$:

$$(\Phi/N)_{\text{cold}} = (1/2)\sum_k \phi_{jk}; \; (PV/N)_{\text{cold}} = (1/4)\sum_k (F \cdot r)_{jk} \;.$$

Routine molecular dynamics simulations then provide the thermal corrections to these estimates of energy and pressure. The temperature dependence of the corrections can be roughly approximated according to Grüneisen's linear model of thermal effects:

$$e = e_{\text{cold}} + 2T; \; P_{\text{eq}} = P_{\text{cold}} + 0.3(e - e_{\text{cold}})/v \;.$$

The transport properties can be expressed in terms of Newton's linear viscosity model,

$$P = [P_{\text{eq}} - \lambda \nabla \cdot u]I - \eta[\nabla u + \nabla u^t] \;,$$

where I is the unit tensor, and where we avoid a dip in P_{yy} within the shock by taking the *bulk* viscosity, $\eta_v = \lambda + \eta$, equal to the shear viscosity. The observed heat flow suggests that the heat flux responds primarily to the gradient of the transverse temperature T_{yy}. We use a generalization of Fourier's law appropriate to tensor temperature:

$$Q_x = -(3/4)[(dT_{xx}/dx) + 7(dT_{yy}/dx)] \;.$$

Finally, the rates at which work is done, $-P : \nabla v$, and heat is absorbed, $-\nabla \cdot Q$, need to be apportioned between the longitudinal and transverse directions. The observed disparity between T_{xx} and T_{yy} suggests that most of the energy flow be allocated to increasing T_{xx}:

$$2\rho \dot{T}_{xx} = -(P_{xx} - P_{\text{cold}})(du/dx) - (dQ_x/dx) + 2\rho(T_{yy} - T_{xx})/3 \;;$$

$$2\rho \dot{T}_{yy} = 2\rho(T_{xx} - T_{yy})/3 \;.$$

The combination of molecular dynamics with the generalizations of the simplest possible linear Navier-Stokes-Fourier model is a powerful tool, introducing flexibility into continuum models for nonequilibrium systems.

6.8 Lyapunov Instability in Strong Shockwaves

The irreversible nature of shockwaves is particularly dramatic for our strong shockwave example problems where the entropy increase is *infinite* due to the vanishing phase volume ($\otimes$) of the initial state. Of course bit-reversible integration algorithms make it possible to reverse the irreversible process exactly! In fact, just using sixteen-digit double-precision arithmetic makes it possible to reverse the shock process "for all practical purposes". Because the transition is so rapid we can be confident that the computed details are accurate.

Figures 6.13 and 6.14 show snapshots from two different Runge-Kutta integrations, one of them proceeeding forward only, the other reversed midway through the simulation. In the forward case, Figure 6.13, a 40×40 system of 1600 particles is followed forward in time from an initial condition with half the particles moving to the right at the particle velocity necessary for twofold compression. The other half, moving to the left, are shown as open circles, with the first frame of the figure at the point of maximum compression where the trailing edges of the two colliding blocks have just come to rest. The time is $(10/u_p) = 11.2$. Between 11.2 and 22.4 the system undergoes a free expansion from its initial hot state, filling the (periodic) container again at a time of 22.4. The following motion is shown in two more snapshots. In a few sound traversal times the initial kinetic energy per unit mass, is converted to internal energy through viscous dissipation; $(u_p^2/2) \longrightarrow e$.

Figure 6.14 shows what happens (with a good integrator) if the time is reversed (simply by changing the sign of all the momenta) at a time midway between the second and third frames, $t = 28$. The maximum-compression state, at $t = 11.2$, is accurately reproduced at $t = 44.8$. This accurate reversibility occurs despite the undoubted loss of information to Lyapunov instability during the shock-compression, free expansion, and reflected-shock processes. For visual purposes Runge-Kutta integration is sufficient. For even better reversibility, accurate within machine roundoff, a bit-reversible algorithm can be used.

Times 33.6 and 44.8, without reversal

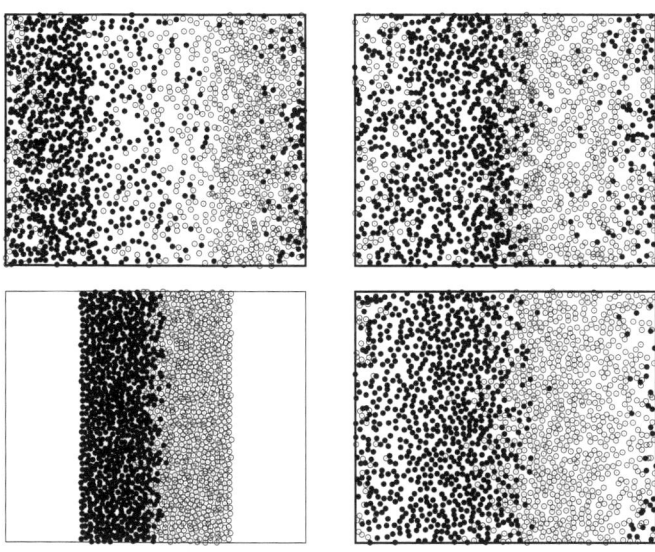

Times 11.2 and 22.4, speeds +/-0.875

Figure 6.13. Four snapshots from a twofold compression simulation with 1600 particles. The initial condition had the rightmost 800 particles moving left and the leftmost 800 moving right at sufficient speed to generate twofold compression (the bottom left illustration). The figure shows three more stages, equally spaced in time, of the expansion of the hot 1600 particles to fill the periodic container homogeneously.

Why isn't the effect of chaos (Lyapunov instability) noticeable here? A quantitative understanding can result from a perturbation-theory analysis of the separation of nearby phase-space trajectories as a function of the Runge-Kutta timestep Δt. Start out with two similar systems, differing in their phase-space coordinates by barely-significant differences of the order of machine roundoff. It is then straightforward to propagate two neighboring trajectories forward and to measure the time dependence of the phase-space separation between the two trajectories,

$$\Delta q(t) \equiv \sqrt{\sum_{k=1}^{N} \delta q_k^2}; \; \delta q^2 = \delta x^2 + \delta y^2 \; ; \; \Delta p(t) \equiv \sqrt{\sum_{k=1}^{N} \delta p_k^2}; \; \delta p^2 = \delta p_x^2 + \delta p_y^2 \; .$$

Times 33.6 and 44.8, reversed at 28.0

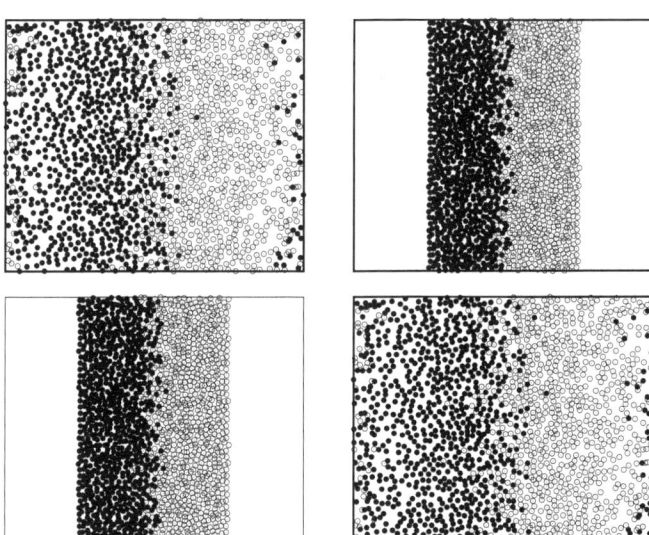

Times 11.2 and 22.4, speeds +/-0.875

Figure 6.14. Snapshots showing the progress of two shockwaves causing twofold compression of a square lattice ($\rho = 1 \to \rho = 2$) with 1600 particles and a Runge-Kutta timestep $\Delta t = 0.002$. At maximum compression the particle momenta are changed in sign so that the dynamics proceeds, to a good approximation, backwards, as can be seen comparing the top two snapshots to their predecessors at the bottom of the figure.

A typical Lyapunov unstable motion is eventually characterized by the exponential growth of both these separations. Figure 6.15 shows, for six different timesteps, a remarkably linear growth of the logarithms corresponding to the Lyapunov growth rates:

$$\Delta q(t) \simeq \Delta p(t) \simeq e^{\lambda_1 t}; \; \lambda_1 \simeq 0.7 \; .$$

Following the reversal at $t = 10$, the twin trajectories are reversed nearly perfectly, with errors too small to see. The point of diminishing return corresponds to a timestep of order 0.0002, with a formal single-step Runge-Kutta error of the order of $(0.0002)^5/5! \simeq 3 \times 10^{-21}$, a few orders of magnitude less than a typical double-precision error of 10^{-17}.

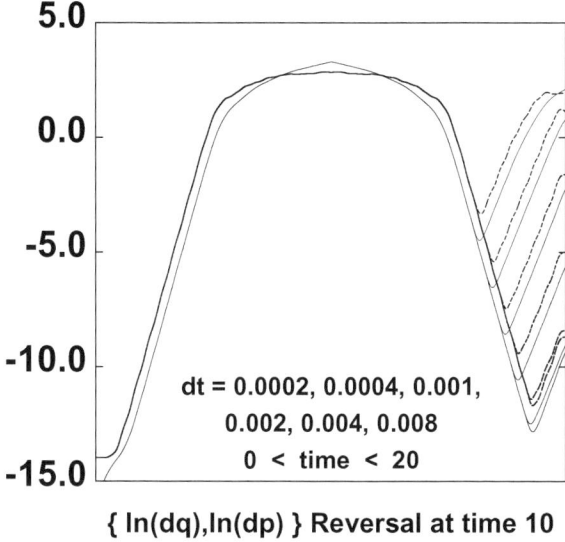

Figure 6.15. Logarithms of the separations of two nearby trajectories. The solid lines show configuration space, the dashed lines momentum space. The signs of the velocities change at the halfway point, $t = 10$, so that the trajectories (approximately) reverse, with the agreement improving with decreasing Δt until $\Delta t \simeq 0.0002$. In all these simulations the distance between the two trajectories is unconstrained.

This shock problem is composed of three types of particles: cold particles and hot particles, as well as the transitional particles located within the wave itself. It is tempting to look for the main locus of the Lyapunov instability. A straightforward approach examines individual particle's contributions to the largest Lyapunov exponent. These sensitivity studies can be carried out in configuration space, momentum space, or the full phase space. We found that there are only small differences among these three approaches. All three show that the sensitive particles, going forward in time, are quite distinct from those in the reversed motion.

This lack of symmetry is an "Arrow of Time", a dissimilarity between the forward and reversed stabilities in the shockwave problem, despite the time-reversibility of the Newtonian-Hamiltonian dynamics underlying the simulations. We reported on these results in Saint Petersburg in 2010 in "Three Lectures: NEMD, SPAM, and Shockwaves". Figures 20 and 21 in the Proceedings (our contribution is available from the arχiv) show the lack of symmetry quite clearly. To make this case free of the distracting long time irreversiblity introduced by Runge-Kutta integration, we undertook similar shockwave simulations based on Levesque and Verlet's bit-reversible scheme.

The bit-reversible approach makes it possible to integrate arbitrarily far into the past or future, reversibly, and to determine local Lyapunov exponents precisely for periodic cycles combining the forward and reversed dynamics. By computing the history of a satellite trajectory, constrained to a fixed separation from the bit-reversible trajectory, one can cycle the dynamics until the local Lyapunov exponents converge to machine accuracy.

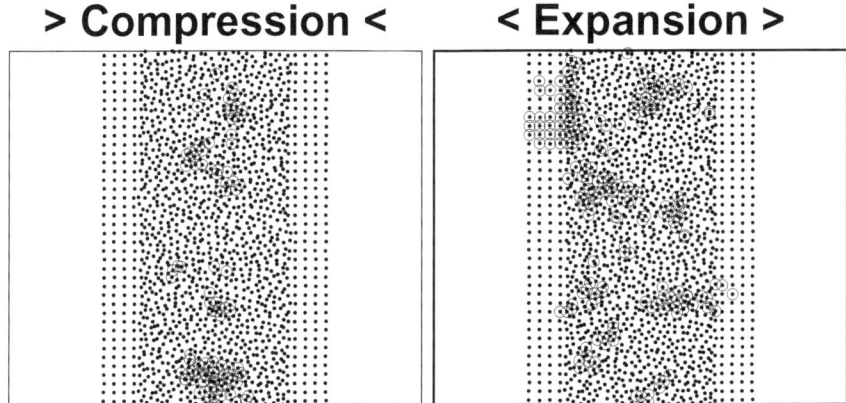

Figure 6.16. Particle positions at time 10.0 after 1000 timesteps with $\Delta t = 0.01$. Initially the 1600 particles move toward the center, at speed $u_p = 0.875$. For this speed the initial square-lattice arrangement of particles with pair potential $\phi = [1 - r^2]^4$ results in twofold compression at a time of $20/1.75 \simeq 12$. The particles making above-average contributions to λ_{forward} (left) and $\lambda_{\text{backward}}$ (right) are indicated by the larger open circles. The satellite trajectories forward and backward converged to machine accuracy after three compression/expansion cycles of 4000 timesteps each. The initial temperature is of order 0.00001.

Figure 6.16 illustrates the difference between the reference-to-satellite offset vectors going forward and backward in time and for precisely the same phase-space $\{q, p\}$ points. The momenta, required for the Runge-Kutta integration of the satellite trajectories forward from t to $t+\Delta t$ and backward from t to $t - \Delta t$, can be defined just as accurately as the Levesque-Verlet coordinates by choosing the linear combination of velocities that eliminates corrections of order Δt^2:

$$\frac{p(t)}{m} \equiv \frac{4}{3}\left[\frac{[q(t+\Delta t)-q(t-\Delta t)]}{2\Delta t}\right] - \frac{1}{3}\left[\frac{[q(t+2\Delta t)-q(t-2\Delta t)]}{4\Delta t}\right].$$

The figure shows the most sensitive particles for a configuration near the maximum compression. Forward in time 101 particles make an above-average contribution to λ_{forward}. Backward in time there are 212 such above-average particles. The reference trajectory for both of them is *exactly the same*.

This lack of symmetry between the forward and backward Lyapunov vectors is a highly interesting "Arrow of Time" reflecting the dependence of the present on the past rather than the future. The most important particles, from the standpoint of dissipation and information creation, are evidently those which have recently been shocked. In the reversed direction of time many more of the particles have been affected by the shock, including those in the "cold" fluid. Detailed study of such asymmetric Hamiltonian problems is bound to shed more light on the Second Law of Thermodynamics.

We emphasize here that the time-reversed shockwave states are unstable on the macroscopic level too. Because the shockwave creates entropy the time-reversed states cannot exist in nature. What happens instead follows from straightforward hydrodynamics: according to the Navier-Stokes-Fourier equations, a reversed shockwave spreads out and forms an adiabatic, nearly isentropic "rarefaction fan" linking the "initial hot" state to a set of expanded states with only a little more entropy.

An automotive analogy for the asymmetric time development of the shockwave problem may prove helpful to understanding these forward/backward differences. Imagine a speedy automobile traveling a highway with occasional sharp turns. A passenger is jostled on entering these turns and gradually recovers during the ensuing straightaways. The reference trajectory in Lyapunov analysis is analogous to the highway. The satellite trajectory, tethered at a fixed separation to the reference, is analogous to the passenger, and is always reacting to recent *past* changes in direction, regardless of time's direction.

6.9 Summary

Shockwaves can be a nuisance, complicating large-scale simulations by introducing singular regions with rapidly-changing hydrodynamic properties. When properly managed, and viewed closely, they provide an opportunity to characterize boundary-free nonlinear transport. Modeling shockwaves requires coming to grips with nonlocality in space and in time in the underlying constitutive relations. The limited stability of kinetic equations with delay times is likewise intriguing.

Shockwaves are an unusual and highly interesting phenomenon from the standpoint of time reversibility. The responsible dynamics is, in principle, reversible, and can even be made *exactly* so in practice by using one of the bit-reversible algorithms. Because the shockwave transition occurs so rapidly, at the smallest scale, double-precision arithmetic is enough to describe the process in a *nearly* reversible way, provided that the scale of the system is sufficiently small.

Consider again the prototypical demonstration, generating a pair of shockwaves using conservative mechanics to describe the highly-inelastic collision of two blocks of material. The stability of a molecular dynamics simulation of this process can be quantified by study of the growth of phase-space perturbations. Figure 6.15 shows the growth of a random offset between two otherwise identical trajectories. The instability of the motion can be described by observing those particles which contribute most to the largest Lyapunov exponent. The localization of the important particles in the vicinity of the shockfront shows a definite "Arrow of Time".

6.9.1 *Notes and References*

Our interest in shockwaves picked up with the publication of Klimenko and Dremin's molecular dynamics studies in 1978. Bill's first molecular dynamics efforts, mentioned in "Shockwaves in Condensed Media", presented by Russ Duff at the Paris conference *Behavior of Dense Media under High Dynamic Pressures* (1967) had been frustrated by the relatively frequent failure of the magnetic storage tapes at the "Rad Lab" in Livermore. By 1980 the hardware had improved and, with the stimulation of Brad Holian and Galen Straub, at Los Alamos, and the help of Bill Moran, at Livermore, we were able to get good solutions, using accurate equation of state and constitutive relations appropriate to a Lennard-Jones potential model of argon: "Shockwave Structure *via* Nonequilibrium Molecular Dynamics and Navier-Stokes Continuum Mechanics".

In comparing molecular dynamics with continuum mechanics it is essential to *define* a transformation method converting data at points into field variables. Lucy and Monaghan's idea of using *weight functions* is the best way of satisfying this need. Hardy later proposed a more complicated version of this idea, but the extra complexity of his approach has no advantages. See, for instance, Hardy's 1982 work, "Formulas for Determining Local Properties in Molecular Dynamics Simulations: Shockwaves".

In the course of updating the earlier versions of this book we decided to revisit the shockwave problem, concentrating on the tensor temperature that Mott-Smith had explained for gases ["The Solution of the Boltzmann Equation for a Shock Wave"] and that Oyeon Kum had observed in fluids ["Temperature Maxima in Stable Two-Dimensional Shock Waves"]. One thing led to another. Eventually we were able to solve continuum models which incorporated a tensor conductivity, thermal relaxation, and time-delayed responses of stress and heat flux to the driving forces, strain rate and temperature gradients. The evolution of these ideas over a period of two years can be traced through the corresponding arχiv contributions, two of which are joint works with Paco Uribe. Typically Paco solved the steady-state shockwave equations and we followed the alternative time-dependent path. Comparisons of the two approaches confirmed very good agreement.

Here are the arχiv references [Consult them for additional references and for additional examples of shockwave structure determinations.]:

0905.1913: "Tensor Temperature and Shockwave Stability in a Strong Two-Dimensional Shockwave"; also available from Physical Review E (2009).

0909.2882: "Shockwaves and Local Hydrodynamics; Failure of the Navier-Stokes Equations"; also available in the Festschrift celebrating Leopoldo García-Colín's 80th Birthday.

1001.1015: "Well-Posed Two-Temperature Constitutive Equations for Stable Dense Fluid Shockwaves using Molecular Dynamics and Generalizations of Navier-Stokes-Fourier Continuum Mechanics"; also available from Physical Review E (2010).

1005.1525: "Flexible Macroscopic Models for Dense-Fluid Shockwaves: Partitioning Heat and Work; Delaying Stress and Heat Flux; Two-Temperature Thermal Relaxation"; also available in the Proceedings of the 2010 conference at Saint Petersburg (once Leningrad), "Advanced Problems in Mechanics".

1008.4947: "Three Lectures: NEMD, SPAM, and Shockwaves"; also available in the Proceedings of the 2010 Granada Seminar *Nonequilibrium Statistical Physics Today*.

1102.2560: "Maxwell and Cattaneo's Time-Delay Ideas Applied to Shockwaves and the Rayleigh-Bénard Problem".

1112.5491: "Time's Arrow for Shockwaves; Bit-Reversible Lyapunov and 'Covariant' Vectors ; Symmetry Breaking".

Chapter 7
Macroscopic Computer Simulation

*An art can only be learned in the workshop
of those who are winning their bread by it.*

Samuel Butler

7.1 Introduction

For the most part, the microscopic details of macroscopic flows are an irrelevant distraction. By ignoring these atomistic details and focusing directly on macroscopic variables, one might expect to find a simpler implementation, free of distracting fluctuations. Reduced fluctuations should also lead to improved stability. But the opposite is true. Though the continuum approach is a simplification, the *partial differential equations* of continuum mechanics, with field variables depending upon space as well as time, incorporate a *continuum* of degrees of freedom. Microscopic motion equations are typically well-posed, with well-behaved solutions for as long a time as desired. Relatively simple algorithms are adequate. Macroscopic simulations can exhibit instabilities on a variety of length scales with a corresponding variety of frequencies and growth rates. The wide ranges of the length and time scales, from the free-path atomic-vibration scales of strong shockwaves up to the macroscopic scale, can cause resolution difficulties in addition to the instabilities. The demands of creating and tracking the progress of material surfaces and interfaces add to the complexity of algorithm development in the macroscopic case.

To simulate even the simplest of macroscopic flows we must solve the partial differential conservation equations for the space and time dependence of the density, velocity, and energy:

$$\dot{\rho} = -\rho \nabla \cdot u; \; \rho \dot{u} = \nabla \cdot \sigma = -\nabla \cdot P \; ;$$

$$\rho \dot{e} = \sigma : \nabla u - \nabla \cdot Q = -P : \nabla u - \nabla \cdot Q \; .$$

There are three prerequisites for solving these flow equations. First, we need to express the stress tensor σ (the pressure tensor is its negative, $P = -\sigma$) and the heat flux vector Q in terms of the time histories of the density, velocity, and energy fields. Then, we must supply initial conditions, along with boundary conditions.

It is typical also to insist that the macroscopic constitutive equations be "dissipative", obeying the Second Law of Thermodynamics, ensuring the decay of velocity and temperature gradients, so that boundary conditions driving the system away from equilibrium are usually necessary too. Macroscopic dissipation corresponds to the irreversible loss of information. That loss makes it impossible to reconstruct the past temperature and velocity gradients which have dissipated differences into internal energy, or "heat". Explicit dissipation—entropy production—ensures that the macroscopic equations are instantaneously irreversible. Entropy inexorably increases, without the need for any time averaging.

We must begin the solution process by reducing the continuum of degrees of freedom to a manageable number. There are several ways to convert the continuum equations into a discrete set: (i) expansion in orthogonal polynomials, such as Fourier's series; (ii) finite-difference approximations; (iii) finite-element approximations; and (iv) particle methods. Of these, particle methods are the simplest to implement. Particle methods closely resemble molecular dynamics, but with the interparticle "forces" directly incorporating the macroscopic constitutive relations for P and Q in terms of the velocity and temperature gradients.

There are in addition several useful approaches to simplified approximate forms of continuum mechanics. Relatively *slow* hydrodynamic flows can be treated as *incompressible*. The plastic (shear) flow of metals is also, to a good approximation, incompressible. Large-scale simulations of fluid motion—as in atmospheric turbulence studies—can often afford to ignore millimeter-scale dissipation caused by viscosity and heat conductivity, or can replace the dissipation with simple *ad hoc* models. Except for this last simplification, which makes numerical turbulence studies possible, such restrictions are specially useful in analytic studies, and are not specially relevant to computer simulations.

7.2 Continuity and Coordinate Systems

"Continuum" mechanics incorporates the underlying assumption that the "field variables" describing the continuum vary smoothly in space and time, so that the required space and time derivatives are well defined. With this assumption, the basic partial differential equations are simple to derive. The derivations proceed by considering the flows of the three conserved quantities: mass, momentum, and energy. These can be most simply expressed in either of two coordinate systems, the fixed "Eulerian" frame or the comoving "Lagrangian" frame. The steps are identical to those involved in demonstrating the compressible form of Liouville's Theorem. As an example, consider the "continuity equation" of Section 4.1. It describes density changes resulting from gradients in the mass flux—a vector (ρu) equal to the flow of mass, per unit area and time, at the location r:

$$(\partial \rho / \partial t)_r = -\nabla \cdot (\rho u) \longleftrightarrow \dot{\rho} = -\rho \nabla \cdot u \ .$$

The continuity equation, and its analogs for momentum and energy conservation, are completely independent of microscopic physics and should best be viewed as alternative *macroscopic* descriptions of experience. The underlying ideas are just two: (i) *differentiability* of material properties, with respect to space and time; (ii) *conservation* of mass, momentum, and energy.

To illustrate the source of the basic equations let us consider the flow of a material (solid, liquid, or gas) with a mass density $\rho(r,t)$ and a flow velocity $u(r,t)$. The Eulerian derivation is simplest. We choose our coordinates in the laboratory rest frame. Focus on a single square Eulerian sampling zone (often called a "bin", "cell", or "element") with area $\Delta x \Delta y$, centered on the location r and observed for a long enough time interval Δt, that any microscopic fluctuations can be ignored. This zone will experience four mass-flux contributions to its total mass $M = \rho \Delta x \Delta y$:

$$M(t + \Delta t) - M(t) = \Delta t (\partial (\rho \Delta x \Delta y)/\partial t)_r =$$

$$\Delta x[(\rho u_y)_{y-\Delta y/2} - (\rho u_y)_{y+\Delta y/2}]\Delta t + \Delta y[(\rho u_x)_{x-\Delta x/2} - (\rho u_x)_{x+\Delta x/2}]\Delta t \ .$$

Two assumptions are necessary to the derivation: (i) Δx and Δy are constant, and (ii) (ρu) is assumed to be "smooth", with well-defined space and time derivatives. With these assumptions, division by $\Delta t \Delta x \Delta y$, followed by a two-term series expansion of the difference terms, gives directly the "Eulerian form" of the partial differential continuity equation:

$$(\partial \rho/\partial t)_r = -\nabla \cdot (\rho u) \equiv -u \cdot \nabla \rho - \rho \nabla \cdot u \ ,$$

describing the change of density at a fixed location in space. It is equally interesting to ask for the change in density (or other field variables) while *following* the motion, just as the *comoving* time derivative in phase space can be evaluated in deriving Liouville's Theorems. The resulting "Lagrangian" time derivative gives the comoving time-dependence along a streamline, following the motion. For density, the Lagrangian form of the continuity equation is:

$$\dot{\rho} = (\partial \rho/\partial t)_r + u \cdot \nabla \rho = -\rho \nabla \cdot u \ ,$$

which is algebraically equivalent to the Eulerian form. In the older texts Lagrangian equations are often used to indicate and describe the motion of a material point as a function of its "original" coordinates at an arbitrary time origin. With the widespread knowledge that many flows involve chaos (and are therefore *exponentially* sensitive to small perturbations) this association of "new" with "old" coordinates assumes more than is reasonable, at least for fluids. At present, Lagrangian computer programs, designed to follow particular elements of material, are specially useful when large-scale deformations of several different materials prevail. The equations solved by such a program are the Lagrangian expressions for $\{\dot{\rho}, \dot{u}, \dot{e}\}$.

Eulerian computer programs, which are somewhat simpler to write unless it is necessary to keep track of moving interfaces, evaluate the partial time derivatives of these same variables in one or more fixed grids. This approach is natural for a problem with long-time circulation within fixed boundaries, such as Rayleigh-Bénard convection. A Lagrangian treatment of large-scale shears eventually leads to insuperable numerical problems. "Hybrid" methods advance the flow field using the Lagrangian equations and then interpolate the advanced field variables onto a fixed Eulerian grid.

In particular cases (incompressible flow is the best example) "spectral methods" representing the field variables by truncated Fourier series are cost effective. Still another approach to solving the continuum equations is a "particle" simulation method, in which individual representative particles have coordinates, velocities, and energies, which can be different to the field values of these variables at the same location. We outline this "SPAM" or "smooth-particle" algorithm in Section 7.7. Just as in Chapter 4 we then detail solutions of the simple Rayleigh-Bénard problem. At this Chapter's end we consider the merits and potential of all three numerical methods for simulating macroscopic flows, molecular dynamics, finite differences, and SPAM. For all three we consider the prototypical example problem, Rayleigh-Bénard convective heat-flow.

7.3 Macroscopic Flow Variables

A solution of the differential flow equations describing a continuum requires a complete description of the "state" of that continuum, including the distributions of mass, momentum, and energy. Predicting future flows of mass, momentum, and energy requires special constitutive laws for the comoving momentum and energy fluxes P and Q. P is the negative of the stress, σ. The mass flux ρu is the product of the local density and the fluid's velocity vector. The magnitude of the mass flux corresponds to the flow per unit area across an infinitesimal area normal to the fluid's flow velocity.

While the flow velocity $u(r,t)$ is most naturally measured relative to a fixed laboratory coordinate system, the momentum and energy fluxes (which are respectively tensors and vectors) are conventionally separated into two parts. The "comoving" momentum and energy fluxes—the fluxes relative to a Lagrangian coordinate system moving with the fluid velocity u—are the *pressure tensor P* and the *heat flux vector Q*, respectively. The additional "convective" momentum and energy fluxes comprise the "streaming" contributions, the tensor field ρuu for the momentum flux, and the vector field $\rho u[e + \frac{u^2}{2}]$ for the energy flux. In addition to these convective fluxes, energy changes due to the performance of work and the conduction of heat must be included too. These contributions to the energy flux are two vectors, $P \cdot u$ and the heat flux Q, respectively. Their divergences contribute to the rate-of-change of energy density.

Solutions of the continuum equation of motion,

$$\rho \dot{u} = \nabla \cdot \sigma = -\nabla \cdot P ,$$

and the continuum energy equation,

$$\rho \dot{e} = \sigma : \nabla u - \nabla \cdot Q = -P : \nabla u - \nabla \cdot Q ,$$

require *constitutive equations* for P and Q. P has both an equilibrium and a nonequilibrium part. The simplest equilibrium equation of state assumes a power-law dependence of pressure on density, $P \propto \rho^\gamma$. The most complex equations of state take the form of tables giving the dependence of the pressure on energy and density. The transport properties can be the simple Navier-Stokes and Fourier recipes, including Newtonian shear and bulk viscosities as well as Fourier's linear heat conductivity, or the much more elaborate constitutive relations describing anisotropic solids with tensor heat conductivities and stresses depending upon the orientation and past history of the material. The complexity of continuum constitutive relations is limited only by the imagination, causality, and the Second Law of Thermodynamics. Relations violating the Second Law result in prompt numerical instabilities because gradients tend to grow, exponentially fast, rather than to decay. We describe two traditional numerical methods for solving the continuum equations in the next two Sections. The smooth-particle method which follows them, in Section 7.6, is an excellent pedagogical bridge between the atomistic and continuum approaches.

7.4 Finite-Difference Methods

Although it seems most logical to formulate discretized solutions of the partial differential equations of continuum mechanics as sets of ordinary differential equations for the motion of nodes or the evolution of series-expansion coefficients, it is more usual to treat both time and space as discrete variables. The reason for this is pragmatic. The resulting simulations require less computer time and storage space. In the "doubly-discretized" case it is necessary to correlate the timestep Δt with the spatial increment Δx. In the limit that both Δt and Δx vanish, a proper correlation guarantees *stability* while maintaining *consistency* with the underlying partial differential equations. Typically Δt must be less than $(\Delta x/c)$, where c is the sound velocity, and also less than $(\Delta x^2/D)$, where D is a diffusion coefficient. Otherwise the solution technique is unstable or inaccurate. In applied continuum simulations it is usual to choose the *largest* possible timestep consistent with both the sonic and diffusive stability restrictions, because the goal is to reach a particular total time, known in advance.

Straightforward computational techniques can be based on Taylor's series expansions of the continuum equations. A network of sampling points, with a discrete spacing Δx, is carried forward through a series of discrete timesteps, with either a fixed or variable Δt. Some interesting more-elaborate variants can be developed, with the timestep itself varying in both space *and* time. Gradients throughout the network of points are approximated with low-order finite-difference formulas. Consider the evaluation of the temperature gradient required to drive a heat flow, according to Fourier's Law, $Q \equiv -\kappa \nabla T$. The required derivatives, (dT/dx), for instance, can be approximated by the centered "finite difference":

$$(dT/dx) \equiv \frac{[T_{+\Delta x/2} - T_{-\Delta x/2}]}{\Delta x} \longrightarrow \text{error} \propto \Delta x^2 T''' .$$

Improved accuracy can be achieved in either of two ways: by (i) reducing the mesh size or by (ii) including higher-order terms in the finite-difference expression:

$$(dT/dx) \equiv (9/8)\frac{[T_{+\Delta x/2} - T_{-\Delta x/2}]}{\Delta x} - (1/8)\frac{[T_{+3\Delta x/2} - T_{-3\Delta x/2}]}{3\Delta x}$$

$$\longrightarrow \text{error} \propto \Delta x^4 T''''' .$$

The extra effort required by higher-order methods is generally better spent increasing the resolution of the simpler second-order approximation. A higher-order expression can be used, from time to time, to estimate errors.

In Sections 4.1 and 7.2 we emphasized that the continuum equations can be written in terms of "Eulerian" derivatives, at fixed points in space, or in terms of the comoving "Lagrangian" derivatives, following the motion. There is a direct computational analog of this distinction. The sampling points, at which field variables like $\{\rho, u, e, ...\}$ are measured, can be fixed in space or can move with the material. The corresponding forms of the continuum equations then become evolution equations for the field variables at these sampling points. Both choices have advantages and disadvantages. The Eulerian approach makes it difficult to keep track of material interfaces and surfaces but avoids mesh-tangling. The Lagrangian approach takes care of material interfaces in a natural way—they become element boundaries—but can be defeated by large-scale deformations. Turbulent flows or large-scale plastic deformation eventually lead to the failure of Lagrangian approaches. Instead of tracking the motion accurately, the nodes tangle, and require "rezoning"—the construction of a new approximating grid—from time to time.

Occasionally, as in the study of steady shockwave propagation, a velocity different to either the laboratory frame or the material frame is advantageous. The continuum analog of the atomistic shockwave simulations of Section 5.7 and Chapter 6 can be based on a numerical solution of the continuum equations with the restrictions that the mass, momentum, and energy fluxes are all constant. For a shockwave propagating in the x direction, the corresponding flux components are:

$$\{\rho u_x,\ P_{xx} + \rho u_x^2,\ \rho u_x[e + (P_{xx}/\rho) + (u_x^2/2)] + Q_x\}\ ,$$

where u_x is the velocity component in the direction of shockwave motion. This one-dimensional problem is relatively elementary. It reduces to the solution of two coupled ordinary differential equations [for $\rho(x)$ and $T(x)$] when the continuity equation is used to eliminate the velocity u_x from the momentum-and-energy-conservation laws. In the more usual two- and-three-dimensional applications, the finite-difference and finite-element methods are simplest to employ if relatively low-order approximations to the space and time derivatives are used. Constructing and revising the meshes make up a large part of the programming work. Several solutions of the Rayleigh-Bénard problem using the finite-difference method can be found in Section 4.11.2 which begins on page 144.

7.5 Finite-Element Methods

Another approach can be based on the notion of a solution with small-scale "local" errors in both space and time, which is only correct in a spatially averaged sense. In standard finite-element terminology the cells and nodal points are referred to as elements and nodes. Rather than satisfying the continuity equation exactly, a "weak" (meaning spatially-averaged) solution ensures that the two (approximate) averages agree:

$$\langle (d\ln\rho/dt)\rangle_{\text{element}} \equiv \langle (\dot\rho/\rho)\rangle_{\text{element}} = \langle -\nabla\cdot u\rangle_{\text{element}}\ ,$$

or

$$\langle \dot\rho\rangle_{\text{element}} = \langle -\rho\nabla\cdot u\rangle_{\text{element}}\ ,$$

where the angular brackets indicate integration over the element volume and an average over a short time interval Δt. In some cases—elasticity theory and quantum mechanics are the best-known examples—such weak solutions satisfy a variational principle. The true energy of an elastic deformation must lie below the corresponding energy of any approximate deformation with the same boundary displacements. Likewise, the energy of the true ground-state quantum wave function must lie below the corresponding energy of *any* approximate wave function satisfying the same boundary conditions. More generally, as in either version of the continuity-equation example above, finite-element solutions are obtained by insisting that the differential equations' element averages be exactly correct. Again the equations chosen for solution can be either Eulerian, with the element fixed in space, or Lagrangian, corresponding to a comoving element following the flow.

In the Eulerian case the field variables are given throughout each finite element by appropriate interpolation or "shape" functions. In one dimension, for instance, density, velocity, and energy could be linearly interpolated between the two nodal points bounding an element. They could alternatively be represented by "cubic splines", guaranteeing the continuity of the first two derivatives at the element boundaries. In one space dimension the construction of a cubic spline is straightforward. Three adjacent nodal values give the second derivative at each node. Linear interpolation between these, followed by two space integrations matched to the two nearest nodal values, provides the cubic interpolating function. Use of this idea in solving the Rayleigh-Bénard problem requires very little additional computer time and reduces errors in the flow field's kinetic energy by about a factor of three. See Vic Castillo's 1999 PhD thesis (University of California at Davis/Livermore) for details and a host of examples.

In the Lagrangian case the field variables are again given throughout each element, but the values *at* the boundary nodes are used to advance the mesh forward in time. Just as in the Eulerian case, a shape function is required. A function based on one-dimensional cubic splines, guaranteeing continuous first and second derivatives, is desirable. The irregular geometry of the deforming Lagrangian zones requires more thought and creativity than does the Eulerian case. Fracture, sliding interfaces, and changing contacts between neighboring zones are the foci of much current research.

An advantage of both Eulerian and Lagragian finite-element techniques is that the elements need not be rectangular in shape. This provides the high degree of flexibility in modeling the irregular shapes needed to simulate complex manufacturing processes and engineering systems. A simple example is the Lagrangian modeling of the ball-plate impact simulation shown in Figure 7.1. The plate is filled in with rectangular elements and the ball is modeled with quadrilateral elements that begin with a rectangular zone at the origin.

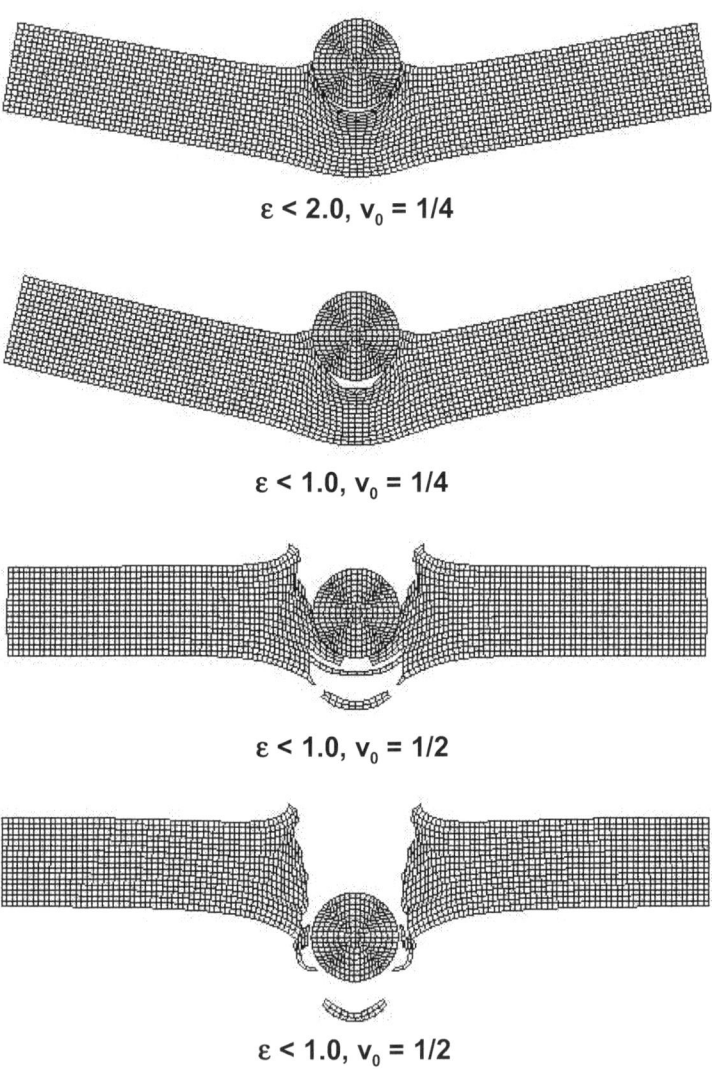

$\varepsilon < 2.0,\ v_0 = 1/4$

$\varepsilon < 1.0,\ v_0 = 1/4$

$\varepsilon < 1.0,\ v_0 = 1/2$

$\varepsilon < 1.0,\ v_0 = 1/2$

Figure 7.1. Snapshots taken from three finite-element simulations of the "Ball-Plate" problem with two different ball velocities v_0. The plastic strain at failure ϵ is greatest for the top simulation. The last two snapshots are taken from a simulation of complete penetration. For details see Section 9.10 in Bill's *Smooth Particle Applied Mechanics; the State of the Art*.

Both finite-element and finite-difference methods have tendencies toward instability unless special measures are employed. Shockwaves can be expected to develop due to the tendency of undamped waves to sharpen as time goes on. The sharpening is caused by the increase of signal speed with density, causing the higher-pressure part of the wave to overtake the slower components. Infinitely-sharp shock waves can be prevented by using an "artificial" (numerical) viscosity large enough to spread the shockwaves out over a few zones. This additional viscosity makes it possible to extend the length scale of simulation zones by eliminating the large-zone tendency toward turbulence. Unstable Lagrangian modes of deformation with picturesque and descriptive names ("butterfly", "hourglass", and so on) can likewise be prevented through the use of special artificial tensor viscosities. Figure 7.2 shows an example simulation computed both with and without hourglass control.

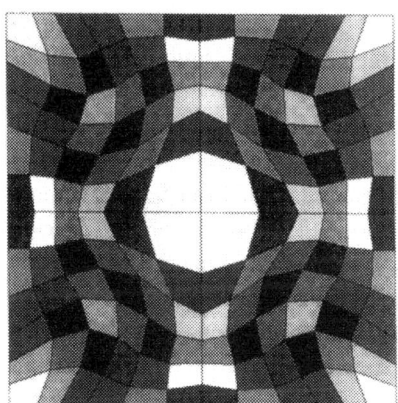 

Figure 7.2. Simulations with (left) and without (right) hourglass control. The initial condition was a periodic sinusoidal density profile with maximum density in the center of the mesh. All the zones have equal masses.

7.6 Smooth Particle Applied Mechanics [SPAM]

A regular spatial grid is not at all necessary for continuum simulations. Macroscopic continuum simulations can just as well be based on an "unstructured" moving spatial grid made up of "smooth particles". The smooth particles' equations of motion include averaged values of the macroscopic stress gradient evaluated at each particle's position.

It is appealing to solve continuum problems with a particle method. This is because the ordinary differential equations for particle motion are (i) relatively easy to solve, and (ii) like molecular dynamics, are free of some of the geometric instabilities that plague grid-based methods. Smooth-particle methods are based on two ideas: (i) continuum properties—density, velocity, energy, pressure, and heat flux, for instance—have first to be interpolated in space on a discrete but irregular particle grid ; (ii) the grid of *particles* then exchange energy and momentum according to the *continuum* constitutive relations, where local divergences of the pressure tensor and heat flux vector are evaluated at each particle's location.

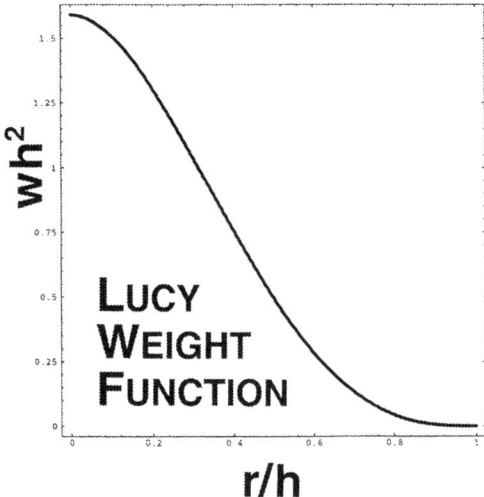

Figure 7.3. Lucy's weight function, normalized for two dimensions:
$$w(r < h) = \frac{5}{\pi h^2}[1 + 3(r/h)][1 - (r/h)]^3 .$$
In two dimensions the weight function normalization is $\int_0^h 2\pi r w(r) dr \equiv 1$. In typical simulations each particle interacts with a few dozen others.

In 1977, while both men were at Cambridge, Monaghan and Lucy independently discovered exactly the same practical scheme for interpolating continuum properties on a moving grid, and for exchanging momentum and energy among particles. The particle locations themselves define the moving grid, with spatially-averaged particle properties simultaneously representing the underlying continuum. The spatial influence of each moving particle is described by a normalized short-ranged *weight function* $w(r)$. Useful weight functions need to have at least two continuous derivatives so as to represent solutions of diffusive continuum equations, which incorporate two space derivatives. The simplest example of such a weight function is Lucy's quartic function, shown in Figure 7.3. This same function was used to generate the density profiles of Figure 3.3 on pages 102 and 103.

At any point in space, r, local values of the mass, momentum, and energy densities are calculated by summing up the contributions of nearby particles $\{j\}$. The local mass density $\rho(r) \equiv \rho_r$ for instance, is a superposition of contributions from every particle lying within the range h of the point r in question. Though r is typically the location of a grid point, exactly the same idea can be applied at the particle locations. The mass density *at* the location of Particle i (for simplicity we assume all particles have the same mass m) then follows from a special case of the definition:

$$\rho_r = \sum_j m w_{rj} \longrightarrow \rho_i = \sum_j m w(r_{ij}) = \sum_j m w_{ij} .$$

In such smooth-particle pair sums a typical smooth particle interacts with perhaps twenty of its neighbors. The very smooth character of w, with both ∇w and $\nabla \nabla w$ continuous, guarantees the resulting continuity of both first and second spatial derivatives, such as $\nabla \rho$, ∇u, $\nabla^2 T$, and $\nabla \cdot P$.

The smooth-particle continuum equation of motion, in the "Lagrangian" form following the motion, $\rho \ddot{r} = -\nabla \cdot P$, likewise illustrates the simplicity of the gradient operation using smooth particles. The divergence of the pressure tensor, evaluated *at* the location of Particle i, for instance, becomes a sum of *individual-particle* pressure tensors multiplied by weight-function gradients for all particles within range of Particle i:

$$\{\ddot{r}_i = -\sum_j m[(P/\rho^2)_i + (P/\rho^2)_j] \cdot \nabla_i w_{ij}\} .$$

This particular symmetrized form is a specially nice choice, because it guarantees the conservation of linear (but not angular) momentum.

For a fluid isentrope, the pressure in the inviscid smooth-particle equation of motion is hydrostatic, and can therefore be viewed as the volume derivative of a density-dependent specific energy $e(\rho)$:

$$P = -(\partial E/\partial V)_S = +\rho^2 (de/d\rho)_s \longrightarrow$$
$$\{\ddot{r}_i = \sum_j m[-(de/d\rho)_i - (de/d\rho)_i]\nabla_i w_{ij}\} .$$

It is interesting that this same set of smooth-particle *continuum* motion equations is *identical* to that introduced by Foiles, Baskes, and Daw to describe the dynamics of metal *atoms*. In their "embedded-atom" theory the energies of metal atoms depend upon the local electronic density.

The smooth-particle approach to solving problems in continuum mechanics has a 35-year history of applications to a wide range of flows, as well as to complex deformations in solid mechanics. For references to applications, see recent reviews of the method. There is a comprehensive description of the numerical aspects of the method in the Los Alamos report written by Crotzer, Dilts, Knapp, Morris, Swift, and Wingate. Monaghan has been particularly creative in modifying the basic smooth-particle algorithm so as to improve its accuracy and eliminate its instabilities.

Smooth particles exhibit interesting numerical features due to the discretization of space—artificial viscosity, artificial heat conductivity, artificial yield strength, and artificial surface tension. All of these artificial effects vanish for sufficiently small weight function ranges and sufficiently many particles. Two examples illustrate these ideas. Smooth-particle simulation of the shear of an ideal gas, with no transport coefficients whatever, gives rise to an *effective* shear viscosity coefficient and a corresponding turbulent Reynolds' stress. Likewise, smooth-particle simulations of liquid drops can produce oscillations driven by an artificial surface tension. The resulting drop oscillations are described perfectly by Rayleigh's theory. The artificial numerical transport properties can be understood by considering again the smooth-particle equations of motion:

$$\{\ddot{r}_i = -\sum_j m[(P/\rho^2)_i + (P/\rho^2)_j] \cdot \nabla_i w_{ij}\} \ .$$

Provided that the quotient (P/ρ^2) is slowly varying in space, the equations of motion,

$$\{\ddot{r}_i \propto -\sum_j \nabla_i w_{ij}\} \ ,$$

become those of ordinary molecular dynamics, so that the "artificial" viscosities and conductivities are simply Green-Kubo transport coefficients for a fluid with a pair potential proportional to the smooth-particle weight function $w(r)$. Let us turn to the details of the smooth particle applied mechanics "SPAM" algorithm.

7.7 A SPAM Algorithm for Rayleigh-Bénard Convection

Here we illustrate the smoothed-particle method—often abbreviated, as "sph" for "smooth-particle hydrodynamics", or "SPAM" for "Smooth Particle Applied Mechanics"—for solving the convective Rayleigh-Bénard problem. In this approach the particles themselves define a (moving and deforming Lagrangian) grid, upon which and within which the continuum properties are defined as weighted sums of particle properties. Although the instabilities encountered with finite-difference or finite-element methods are avoided by using smooth particles, the method requires creativity to adapt it to more-complex boundary conditions, especially those with free surfaces or material interfaces.

Let us begin with the details of the smoothed-particle approach to the Rayleigh-Bénard convective flow problem. The atomistic and finite-difference approaches to this problem were treated in Section 4.11. Molecular dynamics is straightforward, but particularly tedious, requiring as it does tens of thousands of particles in two dimensions and tens of *millions* of particles in three dimensions. The Eulerian finite-difference method illustrated in Section 4.11.2 is straightforward, with horizontal and vertical spatial derivatives replaced by equivalent differences. To ensure numerical stability it was necessary to define *two* different spatial grids, one for energy and velocity and another for density, stress, and heat flux. Although this complication is not explicitly included in molecular dynamics, it is certainly reminiscent of the leapfrog algorithm as well as the ideas that potential-energy contributions of particle pairs to the pressure and heat flux are thought of as shared quantitites localized in space *between* the two members of each pair.

7.7.1 *Initial Conditions*

To begin, the timestep, boundary conditions [including the gravitational field strength gee, and the transport coefficients, $(\eta, \kappa) \to$ (vis, cap) for a fluid with Newtonian shear viscosity and Fourier heat conduction] are specified. We consider a rectangular system with N = nx*ny particles. Then each of the smooth particles has its own location, velocity, and energy. These variables are all stored in the vector yy which is updated at each timestep by Runge-Kutta integration based on four evaluations of the derivative vector yyp:

$$yy(t) \longrightarrow yyp(t) \longrightarrow yy(t+dt) .$$

Here is the corresponding initialization for a 50×50 system of 2500 particles ($\text{nx} * \text{ny} = 50 \times 50$) with a Rayleigh Number of 10,000 and the origin at the center of the domain:

```
dt = 0.01d00
time = 0.0d00
Thot = 1.5d00
Tcold = 0.5d00
gee = (Thot - Tcold)/ny
vis = 0.5d00
cap = 0.5d00
index = 0
do j = 1,ny
do i = 1,nx
index = index + 1
x(index) = i - 0.5d00*(nx + 1)
y(index) = j - 0.5d00*(ny + 1)
vx(index) = 0.0d00
vy(index) = 0.0d00
avT = 0.5d00*(Thot + Tcold)
e(index) = avT - (Thot - Tcold)*(y(index)/ny)
yy(index) = x(index)
yy(index+N) = y(index)
yy(index+N+N) = vx(index)
yy(index+N+N+N) = vy(index)
yy(index+N+N+N+N) = e(index)
enddo
enddo
```

For a Navier-Stokes-Fourier fluid these data are enough to evaluate the righthand sides of the ordinary differential equations evolving $\{\dot{x}, \dot{y}, \dot{v}_x, \dot{v}_y, \dot{e}\}$. The three-step evaluation of the righthand sides begins with computing the particle densities, continues with an evaluation of the velocity and temperature gradients, and concludes with an evaluation of the resulting time derivatives. The reader should note that we typically use $\{v\}$ for the particle velocities and u for the hydrodynamic "stream velocity", which is a weighted average of the particle velocities.

7.7.2 SPAM Evaluation of the Particle Densities

The first step is the calculation of the smooth-particle density at each particle, calling the weight function w(r,h):

```
function w(r,h)
implicit double precision(a-h,o-z)
z = r/h
pi = 3.141592653589793d00
c = 5.0d00/(h*h*pi)
w = c*(1.0d00 + 3.0d00*z)*(1.0d00 - z)**3
return
end
```

The contribution of each particle to its *own* density is w(0.0d00,h). [For simplicity our particles all have unit mass.] Then, for each pair of particles separated by less than the range h there is an additional contribution to the density of both members of the pair:

```
if(rij.lt.h) then
rho(i) = rho(i) + w(rij,h)
rho(j) = rho(j) + w(rij,h)
endif
```

With periodic boundaries at (-nx/2.0d00) and (+nx/2.0d00) the nearest-image distance between Particles i and j includes the coding:

```
xij = x(i) - x(j)
if(xij.lt.-nx/2.0d00) then xij = xij + nx
if(xij.gt.+nx/2.0d00) then xij = xij - nx
```

Two additional kinds of pairs result near "mirror boundaries". Any particle at a distance $r < h/2$ from a mirror boundary has a contribution w(2.0d00*r,h), from its mirror image on the other side of the mirror, added to its density. Consider also the distance rij from Particle i to the mirror image of Particle j (this distance rij is exactly equal to the distance rji from Particle j to the mirror image of i). If rij < h then both rho(i) and rho(j) have contributions w(rij,h) added to their densities. With all the bulk, periodic, and mirror contributions added in, the densities are complete.

7.7.3 SPAM Evaluation of $\{\nabla u\}$ and $\{\nabla T\}$

The velocity and temperature gradients are calculated next. To illustrate for velocity, the algorithm is based on the useful identity:

$$\rho \nabla u \equiv \nabla(\rho u) - u \nabla \rho \ .$$

For temperature exactly the same steps are followed, but starting with the identity:

$$\rho \nabla T \equiv \nabla(\rho T) - T \nabla \rho \ .$$

Then, the smooth-particle definitions,

$$\rho(r) \equiv m \sum_j w_{rj}; \ \rho(r)u(r) \equiv m \sum_j w_{rj} v_j \ ,$$

can be differentiated with respect to the location of the field point r, with all the $\{r_j\}$ and $\{v_j\}$ held constant, so as to give the smooth-particle representations:

$$\nabla \rho = m \sum_j w'_{rj}[(r-r_j)/|r-r_j|]; \ \nabla(\rho u) = m \sum_j w'_{rj} v_j[(r-r_j)/|r-r_j|] \ .$$

Here the chain rule is used to evaluate the derivative of w_{rj} with respect to r. An important point to keep in mind is reinforced by our notation. The smooth-particle average stream velocity $u(r_j)$ at the location of Particle j *differs* from the *particle* property v_j. This difference can be used in formulating definitions of the local kinetic temperature.

Choosing the location $r \to r_i$ and representing the distance between Particles i and j by $|r_{ij}|$ gives smooth-particle representations of the gradients at the location of Particle i:

$$[\nabla \rho]_i = m \sum_j w'_{ij}[(r_{ij})/|r_{ij}|]; \ [\nabla(\rho u)]_i = m \sum_j w'_{ij} v_j[(r_{ij})/|r_{ij}|] \ .$$

Multiplying the first of these by v_i [but *using the approximation* $u(r_i) \simeq v_i$ on the lefthand side] and subtracting from the second gives a form for the density gradients at Particle i and at Particle j:

$$[\rho \nabla u]_i = m \sum_j (v_j - v_i) w'_{ij}[(r_i - r_j)/|r_{ij}|] \longleftrightarrow$$

$$[\rho \nabla u]_j = m \sum_i (v_i - v_j) w'_{ij}[(r_j - r_i)/|r_{ij}|] \ .$$

Finally, symmetrizing by replacing ρ_i and ρ_j with either $\rho_{ij} \equiv \sqrt{\rho_i \rho_j}$ or $\rho_{ij} \equiv (\rho_i + \rho_j)/2$ gives smooth-particle *definitions* of the velocity and temperature gradients *at* each particle:

$$(\nabla u)_i \equiv m \sum_j (v_j - v_i) w'_{ij}[(r_i - r_j)/(|(r_{ij}|\rho_{ij})];$$

$$(\nabla T)_i \equiv m \sum_j (T_j - T_i) w'_{ij}[(r_i - r_j)/(|r_{ij}|\rho_{ij})] \, .$$

Despite the crude replacement of v_i by $u(r_i)$ these gradient definitions are particularly "natural" and useful in that there is no contribution to the gradient (of u or of T) from particle pairs sharing the same value (of velocity or temperature). All four components of ∇u (nine in three dimensions) need to be calculated for use in the smooth-particle analogs of the equation of motion and the energy equation:

$$(\partial u/\partial t) = -u \cdot \nabla u - g - (1/\rho)(\nabla \cdot P) \, ;$$

$$(\partial e/\partial t) = -u \cdot \nabla e - (1/\rho)(\nabla u : P + \nabla \cdot Q) \, .$$

The *bulk* contributions to ∇u and ∇T are calculated as sums over all `npair` interacting pairs, as follows:

```
do ip = 1,npair
i = ipair(ip)
j = jpair(ip)
dx = x(i) - x(j)
if(dx.lt.-nx/2.0d00) dx = dx + nx
if(dx.gt.+nx/2.0d00) dx = dx - nx
dy = y(i) - y(j)
r = dsqrt(dx*dx + dy*dy)
if(r.lt.h) then
rhoij = (rho(i)+rho(j))/2.0d00
dvxdx(i) = dvxdx(i) - (vx(i) - vx(j))*dx*wpor(r,h)/rhoij
dvxdy(i) = dvxdy(i) - (vx(i) - vx(j))*dy*wpor(r,h)/rhoij
dvydx(i) = dvydx(i) - (vy(i) - vy(j))*dx*wpor(r,h)/rhoij
dvydy(i) = dvydy(i) - (vy(i) - vy(j))*dy*wpor(r,h)/rhoij
dvxdx(j) = dvxdx(j) - (vx(i) - vx(j))*dx*wpor(r,h)/rhoij
dvxdy(j) = dvxdy(j) - (vx(i) - vx(j))*dy*wpor(r,h)/rhoij
dvydx(j) = dvydx(j) - (vy(i) - vy(j))*dx*wpor(r,h)/rhoij
dvydy(j) = dvydy(j) - (vy(i) - vy(j))*dy*wpor(r,h)/rhoij
```

```
dedx(i) = dedx(i) - (e(i) - e(j))*dx*wpor(r,h)/rhoij
dedy(i) = dedy(i) - (e(i) - e(j))*dy*wpor(r,h)/rhoij
dedx(j) = dedx(j) - (e(i) - e(j))*dx*wpor(r,h)/rhoij
dedy(j) = dedy(j) - (e(i) - e(j))*dy*wpor(r,h)/rhoij
endif
enddo
```

Again, for simplicity we choose a particle mass of unity. For convenience the additional function `wpor(r,h)`,

$$(dw/dr)/r = -(60/h^4\pi)[1-(r/h)]^2 \;,$$

has been used here. Just as in the case of the density the contributions to all these gradients from each of the periodic and mirror boundaries need to be included. The mirror-image values of `v` and `T` correspond to bulk values in the periodic case and to the specified boundary values in the mirror-boundary case.

7.7.4 SPAM Evaluation of the Constitutive Relations

With the gradients calculated, the individual particle pressure tensors and heat flux vectors follow from Newtonian viscosity and Fourier heat conduction:

```
do i = 1,N
Pxx(i) = rho(i)*e(i) - vis*(dvxdx(i) - dvydy(i))
Pxy(i) = 0.0d00   - vis*(dvxdy(i) + dvydx(i))
Pyy(i) = rho(i)*e(i) - vis*(dvydy(i) - dvxdx(i))
Qx(i)  = -cap*dedx(i)
Qy(i)  = -cap*dedy(i)
enddo
```

We consider here the simplest case, an ideal gas with constant shear viscosity and heat conductivity and with vanishing bulk viscosity. For this choice the trace of the pressure tensor is twice the equilibrium value,

$$P_{xx} + P_{yy} = 2\rho e = 2\rho kT \;.$$

Though we have illustrated the SPAM algorithm for a Navier-Stokes-Fourier fluid, it is quite possible to generalize this approach to more complicated constitutive relations. Consider, as an example, a fluid with Maxwell-

Cattaneo relaxation of the heat flux:
$$Q + \tau \dot{Q} = -\kappa \nabla T .$$
The heat-flux components Q_x and Q_y are simply added to the five variables characterizing each particle and the number of equations is increased from five to seven per particle:
$$\{\dot{x}, \dot{y}, \dot{v}_x, \dot{v}_y, \dot{e}\} \longrightarrow \{\dot{x}, \dot{y}, \dot{v}_x, \dot{v}_y, \dot{e}, \dot{Q}_x, \dot{Q}_y\}$$

The equations for the evolution of momentum and energy can be written in a conservative form by using another pair of gradient identities from continuum mechanics:
$$(1/\rho)\nabla \cdot P = \nabla \cdot (P/\rho) + (P/\rho^2) \cdot \nabla \rho ;$$
$$(1/\rho)\nabla \cdot Q = \nabla \cdot (Q/\rho) + (Q/\rho^2) \cdot \nabla \rho .$$
In smooth-particle form these lead to particle expressions for the divergence of P and Q:
$$[\nabla \cdot P]_i \equiv m^2 \sum_j [(P/\rho^2)_i + (P/\rho^2)_j] \cdot w'_{ij}[(r_i - r_j)/|r_{ij}|] ;$$
$$[\nabla \cdot Q]_i \equiv m^2 \sum_j [(Q/\rho^2)_i + (Q/\rho^2)_j] \cdot w'_{ij}[(r_i - r_j)/|r_{ij}|] .$$
The continuum "work term" in the energy evolution equation,
$$-(P/\rho) : \nabla u \equiv$$
$$-[P_{xx}(\partial u_x/\partial x) + P_{xy}(\partial u_x/\partial y) + P_{yx}(\partial u_y/\partial x) + P_{yy}(\partial u_y/\partial y)]/\rho$$
is treated similarly. For stability against a diverging rotational acceleration P_{xy} and P_{yx} are necessarily equal. The "bulk" contributions to the velocity and energy evolution are calculated as follows:

```
do ip = 1,npair
i = ipair(ip)
j = jpair(ip)
dx = x(i) - x(j)
dy = y(i) - y(j)
if(dx.lt.-nx/2.0d00) dx = dx + nx
if(dx.gt.+nx/2.0d00) dx = dx - nx
r = dsqrt(dx*dx + dy*dy)
if(r.lt.h) then
```

```
xxx = ((Pxx(i)/rho(i)**2) + (Pxx(j)/rho(j)**2))*wpor(r,h)*dx
xyx = ((Pxy(i)/rho(i)**2) + (Pxy(j)/rho(j)**2))*wpor(r,h)*dx
xyy = ((Pxy(i)/rho(i)**2) + (Pxy(j)/rho(j)**2))*wpor(r,h)*dy
yyy = ((Pyy(i)/rho(i)**2) + (Pyy(j)/rho(j)**2))*wpor(r,h)*dy
vxdot(i) = vxdot(i) - (xxx + xyy)
vxdot(j) = vxdot(j) + (xxx + xyy)
vydot(i) = vydot(i) - (xyx + yyy)
vydot(j) = vydot(j) + (xyx + yyy)
edot(i) = edot(i) + (xxx + xyy)*(vx(i) - vx(j))/2.0d00
edot(i) = edot(i) + (xyy + yyy)*(vy(i) - vy(j))/2.0d00
edot(j) = edot(j) + (xxx + xyy)*(vx(i) - vx(j))/2.0d00
edot(j) = edot(j) + (xyy + yyy)*(vy(i) - vy(j))/2.0d00
xxQ = -((Qx(i)/rho(i)**2) + (Qx(j)/rho(j)**2))*dx*wpor(r,h)
yyQ = -((Qy(i)/rho(i)**2) + (Qy(j)/rho(j)**2))*dy*wpor(r,h)
edot(i) = edot(i) + xxQ + yyQ
edot(j) = edot(j) - xxQ - yyQ
endif
enddo
```

Just as with the density and the gradients, all of the similar contributions from the periodic and mirror interactions must be added to the righthand sides of the evolution equations for $\{\dot{v}, \dot{e}\}$.

The smooth-particle approach gives the field variables *everywhere*, not just at the particle locations. As a consequence, it can be used to interpolate the particle properties onto a regular grid. Using the same form for the weighting function but a different range, w(r,h) can be used to generate "arrow plots" of the velocity field, as well as contour plots of the local hydrodynamic variables. We describe next some simple applications and extensions of this method to the Rayleigh-Bénard problem.

7.8 Applications of SPAM to Rayleigh-Bénard Flows

We considered both molecular dynamics and finite-difference continuum mechanics in Chapter 4. With the smooth-particle algorithm detailed here we have in all three distinct methods for solving fluid convection problems. Molecular dynamics, the most "fundamental" of the three, is not particularly useful as a solution technique. Even with tens of thousand of particles, or millions in three dimensions, the instantaneous state suffers from large fluctuations. Fluctuations are absent in the usual finite-difference solutions of continuum mechanics and errors vary in a regular way with the grid spacing. Simple roll patterns can be reproduced with continuum simulations using just a few thousand degrees of freedom.

Smooth-particle mechanics has several advantages over the other two methods. The required number of degrees of freedom is small and the programming is relatively simple (apart from the boundary conditions), much like molecular dynamics. *Snapshots* of smooth-particle calculations are adequate for analysis so that a separate time average can be omitted. Fortunately smooth-particle and finite-difference methods both assume constitutive relations, rather than interparticle force laws, as a representation of the underlying fluid. Even with this similarity there is a wide variety of boundary conditions available for either method. Let us consider finite-difference continuum mechanics first.

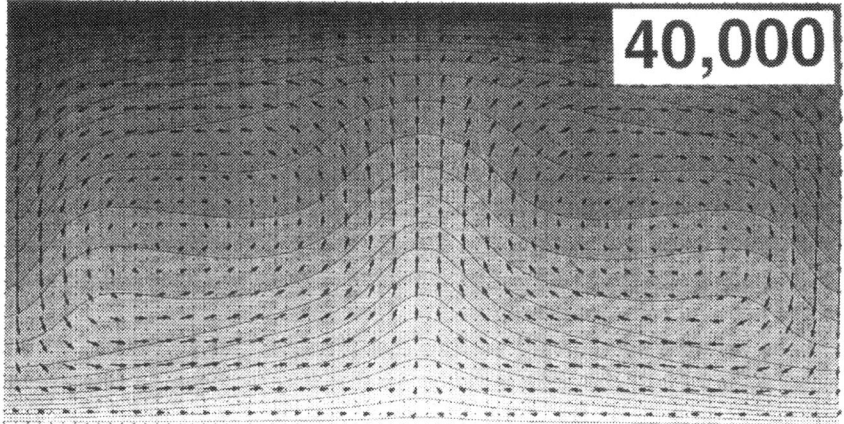

Figure 7.4. Instantaneous flow velocities and corresponding temperature contours in a stationary Rayleigh-Bénard flow. The Rayleigh Number is 40,000. The fluid is an ideal gas with constant transport coefficients.

Two kinds of boundaries are common. *Periodic* boundaries are simplest to implement. Rigid boundaries are implemented by specifying fixed nodal coordinates. Figures 7.4 through 7.6, taken from Vic Castillo's UCDavis PhD thesis, illustrate the complexity and details that result for a simple continuum simulation with periodic vertical boundaries and rigid thermal boundaries at the top and bottom. Those boundary conditions provide a progression from static conduction to steady convection, to periodically varying rolls, to time-varying, and ultimately chaotic, thermal plumes.

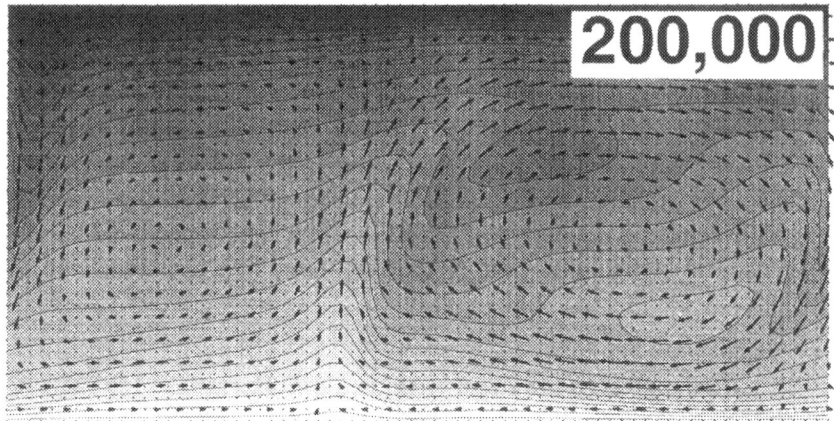

Figure 7.5. Instantaneous flow velocities and temperature contours in an oscillating Rayleigh-Bénard flow. The Rayleigh Number is 200,000. The fluid is an ideal gas with constant viscosity and heat conductivity.

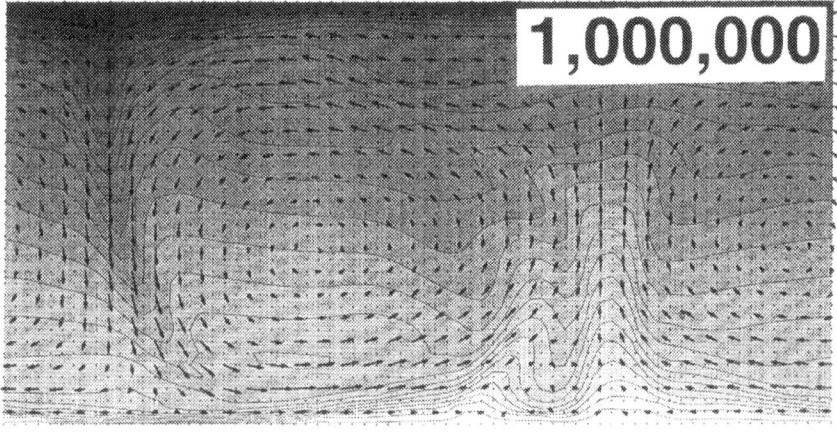

Figure 7.6. Flow velocities and temperature contours in a chaotic Rayleigh-Bénard flow with thermal plumes. The Rayleigh Number is 1,000,000. The fluid is an ideal gas with constant viscosity and heat conductivity.

If the vertical boundaries are stationary, with specified temperature and zero velocity, we saw in Chapter 4 that relatively complex solutions, periodic in time but with varying numbers of rolls, can result. Strict periodicity in time is ruled out for chaotic particle methods, including molecular dynamics and SPAM and the boundary conditions are more varied. Mirror boundaries, with image particles taking on either boundary values of u and T or complementary values, with, for example,

$$v_i + v_i^{\text{mirror}} \equiv 2u^{\text{boundary}}; \; T_i + T_i^{\text{mirror}} \equiv 2T^{\text{boundary}},$$

can be used. Figures 7.7 and 7.8 illustrate snapshots generated with this approach and taken from Oyeon Kum's UCDavis PhD thesis, with the image particles chosen with boundary values of both velocity and temperature.

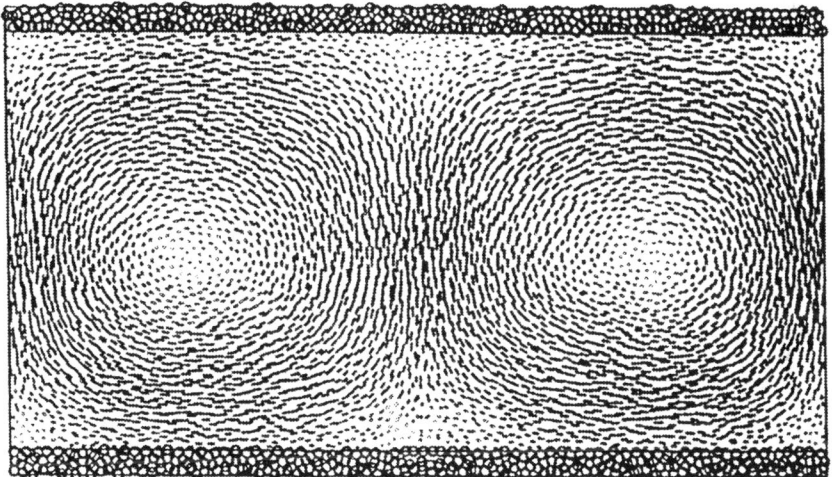

Figure 7.7. Reflected image particles, indicated by circles, follow the motion of corresponding bulk smooth particles. The 5000 bulk particles shown here simulate the Rayleigh-Bénard flow of an ideal gas with constant transport coefficients at a Rayleigh Number of 10,000. For details see the 1995 Physical Review E paper by Kum, Hoover, and Posch.

In the next Section we consider a particular SPAM problem, Rayleigh-Bénard flow with $\mathcal{R} = 10,000$, the ideal-gas constitutive relations, $PV = NkT = Ne$, augmented with Newtonian shear viscosity and Fourier heat conduction, and with a gravitational field strength, $|g| = [T_{\text{hot}} - T_{\text{cold}}]/n_y$

chosen to minimize the density gradient, $\rho(y) \simeq 1$. This problem has a periodic two-roll structure when vertical periodic boundaries are used.

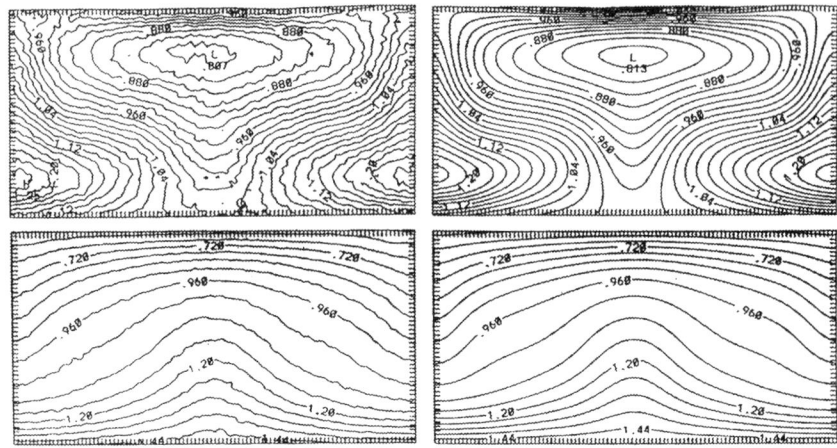

Figure 7.8. Contours of density (above) and temperature (below) for a stationary Rayleigh-Bénard flow with a Rayleigh Number of 2000. The fluid is an ideal gas with constant viscosity and heat conductivity. On the left are shown contours corresponding to an instantaneous snapshot from a smooth-particle simulation. Contours from a corresponding conventional Eulerian finite-difference simulation appear on the right.

7.8.1 SPAM with and without a Core Potential

A peculiarity of the smooth-particle method is a tendency of the particles toward clumping together, sometimes forming long-lived highly-overlapping dimer particles. The relatively weak nature of the effective repulsion between particles is responsible. Consider a situation in which the thermodynamic state varies slowly in space so that the pressure and density gradients are "small". Then the smooth-particle equations of motion approach those of molecular dynamics, with the pair potential proportional to the weighting function $w(r < h)$:

$$\dot{v}_i = -m \sum_j [(P/\rho^2)_i + (P/\rho^2)_j] \cdot \nabla w_{ij} \simeq -\langle 2mP/\rho^2 \rangle \sum_j \nabla w_{ij} .$$

Because the "force" from Lucy's weight function is a maximum at $r = (h/3)$ the sufficiently-energetic dynamics of a Lucy fluid favors small-r overlaps.

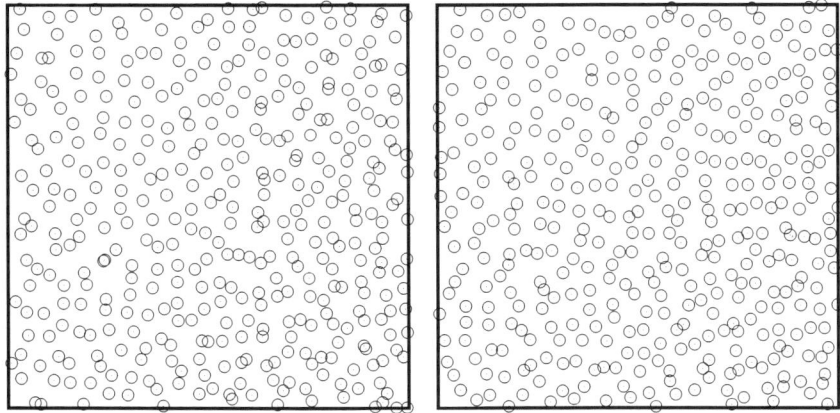

Figure 7.9. Particle configurations, both with and without a repulsive core potential, for a horizontally periodic Rayleigh-Bénard flow with Rayleigh Number $\mathcal{R} = 10,000$. The 20×20 region shown here is from the center of a 50×50 simulation. The circles represent the size of the core region.

A more regular arrangement of particles results if a "core" potential is used to discourage this clumping tendency:

$$\phi_{\text{core}}(r < 1/2) = 100[(1/2)^2 - r^2)]^4 \ .$$

Figure 7.9 shows typical particle configurations from a 20×20 square region centered at $y = 0$ containing about 400 particles. The overall number of core overlaps $\{r_{ij} < 1/2\}$ using the core potential is reduced by roughly a factor of three, giving a more regular interpolation grid. The equilibrated configurations used in Figure 7.9 were generated both with and without a core potential, using a collision diameter $\sigma = (1/2)$. The core should not be chosen so large that the flow is disturbed. For our example Rayleigh-Bénard problem, with unit number density, a collision diameter on the order of unity would cause the smooth-particle fluid to freeze.

Figure 7.10 shows that the kinetic energy is essentially unchanged by the presence of a core potential. The righthand panel of the figure shows the number of core overlaps as a function of time both with, and without, the use of the core potential in the equations of motion.

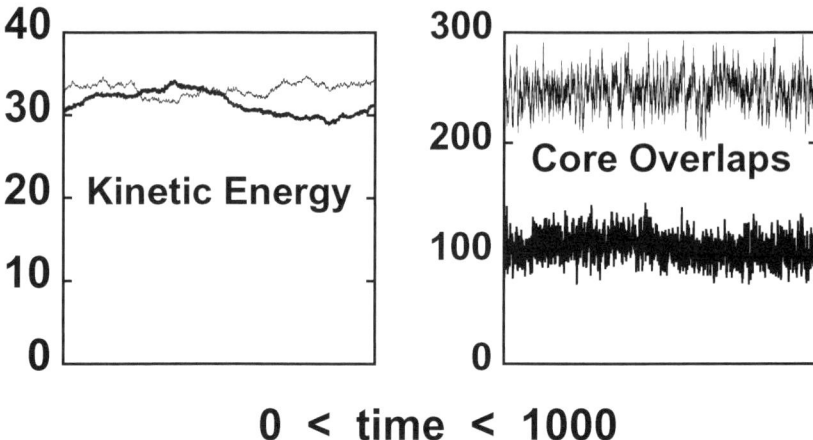

Figure 7.10. Time dependence of the kinetic energy and number of particle pairs with $r_{ij} < (1/2)$ both with and without a short-ranged repulsive core potential. The vertical boundaries are periodic and the Rayleigh number is 10,000.

7.8.2 SPAM and Kinetic-Energy Fluctuations

The smooth-particle fluid exhibits field-variable fluctuations reminiscent of the thermal fluctuations of kinetic theory. While the kinetic-theory fluctuations are necessarily of order kT, SPAM fluctuations can be reduced by increasing the smoothing length h. When the smoothing length h is sufficiently large, the thermal fluctuations can be ignored. On the other hand, if the smoothing length is small the SPAM fluctuations provide an effective kinetic temperature,

$$\rho \left[\langle v^2 \rangle_h - \langle v \rangle_h^2 \right] = \rho k T(r, h)/m \ .$$

By extrapolating the relatively well-behaved large-h flow field toward $h = 0$ we can extract the purely-hydrodynamic behavior from the smooth-particle flow data. To illustrate this idea, we consider the same 50×50 problem at a Rayleigh number of 10,000. With periodic vertical boundaries the kinetic energy calculated in the usual hydrodynamic way, $\int \rho v^2 dr/2$, varies smoothly with h. The dependence on the smoothing length is almost quadratic. Figure 7.11 shows flow fields, for the *same* particle data, for both $h = 3$ and $h = 12$. The main difference is the suppression of fluctuations for the larger weighting function range.

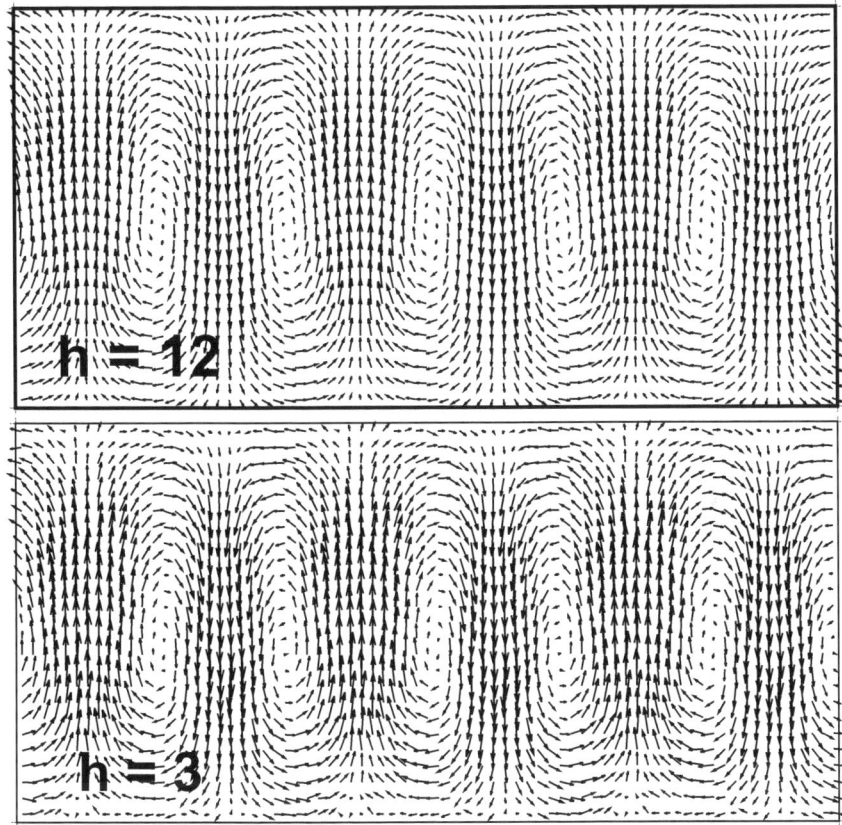

N = 2500, three periods

Figure 7.11. Comparison of smooth-particle velocity arrows computed with $h = 3$ and $h = 12$. These flow fields are snapshots, without any time averaging. The Rayleigh number is 10,000. Images of the central two roll system are plotted to the left and right to make these 150×50 images.

Figure 7.12, on the next page, shows the h-dependence of the usual continuum kinetic energy, $\int dx \int dy \rho u^2(x,y)/2$, at four different times. The larger-h values, which vary nearly linearly with h give (as h approaches zero) the hydrodynamic kinetic energy. The additional small-h energy is the thermal fluctuation energy of the SPAM fluid.

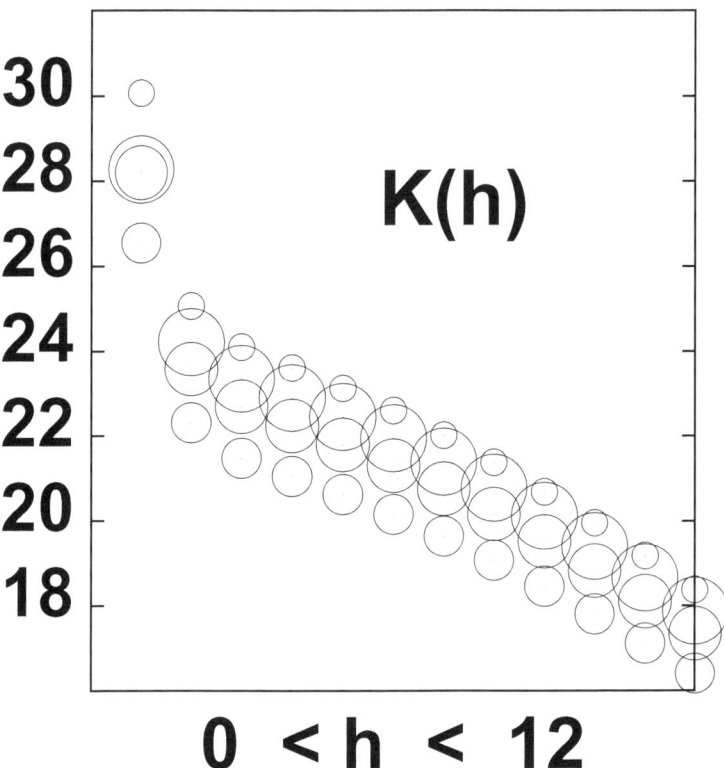

Figure 7.12. Dependence of the smooth-particle kinetic energy on the range of the weight function h. At each of four times the kinetic energy has been calculated for twelve values of h, ranging from 1 to 12. The four circle sizes correspond to data calculated at these four different times. The side boundaries are periodic and the Rayleigh number is $\mathcal{R} = 10,000$.

7.9 Summary

Simulating macroscopic flows presents additional difficulties and instabilities not present in atomistic simulations. Spatial discretization introduces special possibilities for numerical difficulties in Eulerian simulations which follow interfaces and in Lagrangian simulations which seek to follow the flow. These difficulties are usually overcome by introducing *ad hoc* viscosities to slow or eliminate the instabilities. Rescaling Lagrangian simulations, by introducing a new grid from time to time, is a relatively expensive remedy for a highly-distorted, or even tangled, Lagrangian mesh. The errors due to discretization can introduce paradoxical transport and constitutive properties leading to artificial dissipation and structure.

Because macroscopic simulations involve more complex properties (stress and heat flux, rather than just coordinates and velocities) and also require interpolation to find fluxes and derivatives, the numerical work is increased by roughly an order of magnitude. Smooth-particle methods provide an interesting compromise, allowing continuum constitutive relations to dictate the motion of an interpenetrating set of representative particles. The relative simplicity of smooth-particle algorithms recommends them for student use. Stability under compression is improved while the method is prone to instability under tension (where the weight function acts like an attractive force). Accuracy is somewhat hindered by the enhanced fluctuations intrinsic to particle methods as well as by the difficulties in locating interfaces and preventing interpenetration. This approach, dating back to Lucy and Monaghan's 1977 work, suggests many extensions of continuum mechanics.

The very smooth twice-differentiable behavior of smooth-particle sums opens the way to defining local variables and to optimizing their definition through the range of the weighting function h. Analyses of far-from-equilibrium atomistic systems—shockwaves are the best example—can suggest new forms of the constitutive equations taking the different time symmetries of the macroscopic and microscopic viewpoints into account.

7.9.1 *Notes and References*

Monaghan has been much more prolific than Lucy, with his publications readily available on his website. Lucy's discovery of smooth-particle techniques is recorded in his "Numerical Approach to the Testing of the Fission Hypothesis". Over the past ten years interest in smooth-particle methods has spread. We have introduced the method to graduate students successfully and recorded some of our experience with the method in *Smooth Particle Applied Mechanics; the State of the Art*. There are many hydrodynamic problems for which comparisons of SPAM, molecular dynamics, and finite-difference techniques will prove fruitful.

SPAM presents a means for introducing students to computational methods in continuum mechanics and contains many relatively unexplored areas suited to thesis work. In particular the thermal fluctuations in SPAM, due to the differences between the field variable u and the individual particles' $\{v\}$, provide a link between molecular dynamics, where the thermal fluctuations come at much greater computational cost, and standard continuum mechanics, where the fluctuations are absent.

A virtually unexplored area is the dependence of the constitutive relations on the range (or possibly even the form) of the smooth-particle weight function. Because the weight function provides one or more adjustable parameters there is a strong possibility that the nonequilibrium properties can be simplified by a good choice of weight function. The ability of smooth particles to describe Rayleigh-Bénard flows with snapshots, as opposed to the time exposures required with molecular dynamics, argues for choosing smooth particles as the bridge linking continuum and particle techniques.

Chapter 8
Chaos, Lyapunov Instability, Fractals

The brain is wider than the sky,
for put them side by side.
The one the other will include
with ease, and you beside.

Emily Dickinson

8.1 Introduction

The mathematical ideas necessary to a quantitative description of chaos languished during the years that relativity and quantum mechanics attracted the attention of most physicists. Poincaré, Lyapunov, and Krylov had already discussed and detailed the exponential growth of small perturbations. Cantor had described and displayed fractal sets exhibiting fractional dimensionality. These two elements—exponential growth and fractals—are the common features of time-reversible dissipative flows. With time symmetry present the equations of motion guarantee that any system exhibiting phase-space shrinkage (to its attractor) must also have a similar, but concealed, region of unstable phase-space growth (*from* its repeller). The complete structure can be thought of as a "simple" flow, from the repeller source to the attractor sink. The simplicity is actually more than

a little illusory because both the attractor and repeller sets are typically ergodic, dense everywhere. As this flow passes from the expanding region near the repeller to the shrinking region near the attractor, the sum of local Lyapunov exponents shifts from positive to negative:
$$\sum \lambda(t) \equiv (\dot{\otimes}/\otimes) < 0 \ .$$
When it became possible to use computers to generate and visualize solutions of coupled nonlinear equations, widespread realization dawned that unpredictability, based on sensitivity to small perturbations, was characteristic of such problems. Lorenz' Attractor, published in 1963 and discussed in Section 4.11.1, soon became a widely-known example of unpredictability. Lorenz documented the sensitive dependence on the initial conditions displayed by his simple model for Rayleigh-Bénard convection. This model is the set of three coupled quadratic equations:
$$\{\dot{x} = -\sigma(x-y); \ \dot{y} = Rx - y - xz; \ \dot{z} = xy - bz\} \ .$$
He concluded that nature could hardly be less complex than this "simple" system, making the prediction of weather problematic. He showed that successive maxima of his variable z (proportional to the vertical heat transfer rate) have nearly the same form as iterates of the irreversible "tent map",
$$z'(0 < z < 1) = 1 - |1 - 2z| \ ,$$
which, like the Baker Maps, creates and destroys information by shifting less-significant bits to more-significant positions, from which they are ultimately discarded.

Weather has no monopoly on the instabilities we call "chaos" or "Lyapunov instability". *Whenever* pervasive small-scale microscopic divergences are localized, on a larger macroscopic scale, chaos is the result. The growth of very small chaotic perturbations is exponential—Lyapunov unstable. The localization of the chaotic motion imposed by geometric or energetic constraints, then leads to the complex structures familiar from nature, on the one hand, or computer displays, on the other. We avoid here a common alternative phrase for chaos, "sensitive dependence on initial conditions", because the initial conditions in computer simulations are never precisely sharp. They are limited by the number of digits carried. Further, the numerical methods which advance the time continually inject additional "information" or "noise" or "errors" as the calculation proceeds. We are certainly free to imagine the hypothetical existence of an idealized "exact solution" of the motion equations. But such a solution has no real existence, even in principle—any feasible approximate numerical "solution" would be continually perturbed away from it. Joseph Ford repeatedly emphasized, as had Maxwell a century earlier, that there is no way, even in principle, to construct a (precise) solution of the classical motion equations in the presence of deterministic chaos. Such solutions are intrinsically unknown and unpredictable. For any fixed level of uncertainty, the number of digits kept, in the initial conditions and in the computational algorithm, both increase in proportion to the total time for which an approximate numerical "solution" is required.

For applications to reproducible real-world problems the lack of a "true" infinitely-precise solution is unimportant. A good approximation suffices. The computational situation resembles that facing experimentalists, for whom it is commonplace that the irrelevant experimental details are neither interesting nor reproducible. The successful experimentalist, like his computational counterpart, selects those features common to a wide body of potential experiments for compilation, analysis, and understanding.

From a more theoretical outlook the instabilities which amplify small perturbations are extremely interesting. Our current understanding is based on dynamical and topological ideas combined with the computers and the algorithms necessary to try them out. Simple topological analyses of Lyapunov instability follow the distortions of a small comoving "ball" or hypersphere of neighboring solutions. For short times the moving fine-grained ball becomes a rotating hyperellipsoid, with well-defined principal axes. The relative rates of growth and decay parallel to these rotating axes define the local "Lyapunov Spectrum", which is intimately related to the macroscopic dissipation through the instantaneous sum rule:

$$(\dot{\otimes}/\otimes) \equiv \sum \lambda_\mathrm{L} .$$

As usual $\otimes$ and $\dot{\otimes}$ denote a small comoving volume element, and its time derivative following the motion, in the phase space. In order that the local exponents be well-defined, the comoving corotating vectors defining them—see Section 8.4—need to have been followed forward in time from an initial trajectory point sufficiently far in the past. Though the local exponents $\{\lambda_\mathrm{L}\}$ do depend upon the chosen coordinate system, their long-time averages, the Lyapunov Spectrum $\{\lambda \equiv \langle \lambda_\mathrm{L} \rangle\}$ do not. It is typical of the local exponents that their time dependence is regular while their spatial variation is singular. This singular nature reflects the bifurcations, both future and past, that frustrate predictability in chaotic systems.

In "dissipative" systems the summed spectrum of the (time-averaged) Lyapunov exponents is negative, signaling an overall decreasing phase volume ($\dot{\otimes} < 0$). Obviously, the long-time limit of that comoving fine-grained volume must vanish *despite* the exponential divergence of small perturbations! Lorenz' system of three differential equations is among the simplest of the continuous systems that show this behavior. For a wide range of parameter values Lorenz' system produces chaos, with exponential growth of perturbations in a confined region of (x, y, z) space. His system, which includes the decays:

$$\{\dot{x} \propto -x;\ \dot{y} \propto -y;\ \dot{z} \propto -z\}\ ,$$

is obviously not time-reversible. A somewhat simpler set of three differential equations, which *is* time-reversible, and which also shows chaos for some initial conditions, but not for others, is the "Nosé-Hoover oscillator":

$$\{\dot{q} = p/m;\ \dot{p} = -\kappa q - \zeta p\ ;\ \dot{\zeta} = [(p^2/mkT) - 1\ ]/\tau^2\}\ .$$

The control variable, or "friction coefficient" ζ maintains the long-time average temperature, $T = \langle p^2/mk \rangle$. The time-reversed trajectory is constructed by the transformation:

$$(+q, +p, +\zeta) \longrightarrow (+q, -p, -\zeta)\ ,$$

coupled with a reversal of the time ordering of the points. The chaos of the Nosé-Hoover oscillator is not "confined" in the usual sense because the stationary density distribution,

$$f(q, p, \zeta) \propto e^{-\kappa q^2/2kT} e^{-p^2/2mkT} e^{-\tau^2 \zeta^2/2}\ ,$$

extends over *all* values of the three dependent variables. This regular distribution gives way to a complex multifractal, or a limit cycle, if the temperature T varies with q.

For systems with a physical interpretation, such as Lorenz' dissipative model for Rayleigh-Bénard convection and the Nosé-Hoover oscillator with a variable temperature, strange attractors have a physical significance. They indicate first of all the rarity of the represented chaotic states. In most cases the rate of collapse of their phase volume corresponds also to the external dissipation rate $(\dot{S}/k)$ and to the information loss rate. Both features of these strange attractors are common to large classes of physical problems. External dissipation and an exponentially fast collapse of the extension in phase $\otimes$ both correspond to macroscopic irreversible behavior, as described by the Second Law of Thermodynamics.

In this Chapter we describe the tools necessary to take the measure of chaos: the Lyapunov spectrum and the fractal dimensionalities which characterize nonequilibrium phase-space structures. Let us begin by describing the structure of the stage upon which this drama unfolds.

8.2 Continuum Mathematics

The *discontinuous* nature of computation might seem unsuited to the exploration of fine-grained fractals generated by *continuous* equations. But a little probing reveals this apparent problem to be an insubstantial straw man. There is no compelling reason, other than the pursuit of simplicity, to distinguish between the analytic and numerical descriptions of physical systems. The *analytic* work is simplest for the purely-hypothetical infinitely precise continuous descriptions of coordinates, velocities, and time. Some of the blind alleys constructed with this description by mathematicians searching for Cantor's paradise were illuminated and mapped by Bridgman in 1934.

Numerical work requires a *discretized* description of all these variables. Computation is simplest for a fixed (finite) word length corresponding to about sixteen decimal digits. From an operational standpoint the fully continuous case can consistently be regarded as a limiting case of the natural computational representation of space and time by rational numbers. The movable scalable finite grid of computer numbers contains equal fractions of binary 0's and 1's, and has exactly the same *fractal dimensionalities* (see Section 8.5), both box-counting and information, as does the underlying hypothetical continuum. Wherever one looks, and however finely, there are sixteen orders of magnitude with continuum properties before the level of neglect of 10^{-16} is reached. Any admixture of 0's and 1's *other* than equal fractions corresponds to a set with a family resemblance to Cantor's, with a singular distribution and a fractional information dimension, relative to the continuum.

Numerical solutions of differential equations naturally occupy the nodes of a computational lattice, while the probability density, or measure, generated by accumulating a long trajectory is often more naturally viewed as a continuous function. Probability density can be thought of as "coarse-grained", averaged over (infinitesimal) regions "between" adjacent nodes of the computational lattice. It is interesting that coarse-graining was a familiar idea to Boltzmann and Gibbs half a century before the construction of the first electronic computers. It is just as *natural* to measure and analyze the fractal distributions generated by computers with arrays of boxes as it was *natural* to discover and apply integral calculus.

8.3 Chaos

Systems with a "chaotic" time dependence display a disorderly lack of regularity and predictability, due to the pervasive global exponential growth of small perturbations and the absence of periodicity. Isolated local exponential behavior is not enough for chaos. The inverted rigid pendulum, for instance, is perfectly predictable and periodic. Its motion is not "chaotic". Two coupled rigid pendula, one supporting the other, *are* enough for chaos. Two coupled pendula behave unpredictably whenever they lie within the chaotic regions of their four-dimensional phase space. The corresponding system trajectory develops in a three-dimensional constant-energy subspace. If the trajectory were ever exactly to intersect itself the motion would in theory repeat. But the inevitable uncertainty in any trajectory, either physical or numerical, is amplified so as to avoid this repetition. Evidently the distinction between a trajectory which "eventually" intersects itself and one which "never" does is at best an ill-posed problem for the philosophers, as it lies beyond computation.

Maxwell, Boltzmann, and Poincaré had a crude knowledge of what is now called "chaos". The knowledge was of little use until it could be implemented on the fast computers which became available at the close of the Second World War. Lorenz' well-known butterfly-shaped attractor captured the interest and imagination of computational scientists. Consider again Lorenz' set of three differential equations in the variables (x, y, z):

$$\{\dot{x} = -\sigma(x - y);\ \dot{y} = Rx - y - xz;\ \dot{z} = xy - bz\}\ .$$

He discovered that perturbations $\{(\delta x, \delta y, \delta z)\}$ typically grew exponentially in the time, exhibiting "Lyapunov instability". Nevertheless, despite the exponential growth the motion is confined to a fractional-dimensional region. An example, for $\{\sigma = 10,\ R = 28,\ b = 8/3\}$, appears in Figure 4.5. As a result of Lorenz' work a new kind of mathematical object, a "strange attractor", became familiar to physicists. The name emphasizes the *simultaneous* coexistence of exponential divergence with a fine-grained contracting volume which has a vanishing limit. For *interesting* strange attractors the traditional perturbation expansions about *fixed* points, classified as elliptical, parabolic, or hyperbolic, are of little use. Furthermore, many sets of motion equations, like those of the Nosé-Hoover oscillator, for which we make the simplest choice, unity, for the force constant, the mass, the relaxation time, and the temperature:

$$\{\dot{q} = p;\ \dot{p} = -q - \zeta p;\ \dot{\zeta} = p^2 - 1\}\ ,$$

have no *fixed* points at all.

Purely Hamiltonian chaos has attracted considerable attention to a variety of physics problems over a wide range of space scales. Small-scale particle accelerators and large-scale astrophysical systems can both exhibit complex structures incorporating stable as well as unstable regions in their solution spaces. Although fascinating, such intricate structures are not useful for validating or understanding statistical mechanics, where it is fervently desired that any interesting *macroscopic* results not vary with the microscopic initial conditions. Though general proofs establishing the ergodicity property understandably elude the efforts of mathematicians, by now there are many small nonequilibrium systems for which computation has demonstrated ergodicity. This is a *typical* feature of long-time-averaged nonequilibrium steady states. And provided that any reasonable initial condition generates the *same* strange attractor, the time-reversibility of the equations of motion implies ergodicity, because any trajectory point lying in the past corresponds also to a possible trajectory "image point"—a point with all the velocities reversed—which will necessarily be encountered in the future. Because exhaustive sampling is the only way to check ergodicity, systems chosen for investigation must have only a few degrees of freedom. For the simplest of systems, some theoretical analysis is possible—Sinai is credited with proving the quasiergodicity of hard-disk and hard-sphere systems at equilibrium. In either the conservative or the dissipative case, chaos can be quantified through the spectrum of Lyapunov exponents, to which we turn next.

8.4 The Spectrum of Lyapunov Exponents

Let us consider stationary ergodic flows. Exponential instabilities for these flows are described by the Lyapunov exponents $\{\lambda\}$, as indicated in Figure 5.2 on page 175. These "Lyapunov exponents" are long-time averages of "local" exponents $\{\lambda_L\}$, giving the orthogonal growth and decay rates of corotating basis vectors, one for each independent direction in the embedding space. Conventionally the exponent with the largest long-time averaged value is λ_1, that with the second-largest average is λ_2, and so on. One of the exponents, which corresponds to the growth rate along the direction of motion in the phase space, has a time-averaged value of zero. Because the *ordering* of the local exponents changes (the local exponents likewise typically change sign) the numbering convention, with $\lambda_i \geq \lambda_{i+1}$

requires *global* information. For short times, the local exponents depend upon the initial conditions. The Lorenz model has three such exponents. Though their sum is fixed,

$$(\dot{\otimes}/\otimes) = \sum \lambda_{\rm L} = -\sigma - 1 - b \; ,$$

the three individual local exponents fluctuate wildly, but smoothly, with time, and show all possible combinations of the signs of λ_1 and λ_2, but with λ_3 consistently negative. The smooth time dependence contrasts with their highly-singular spatial dependence. See Figure 8.1.

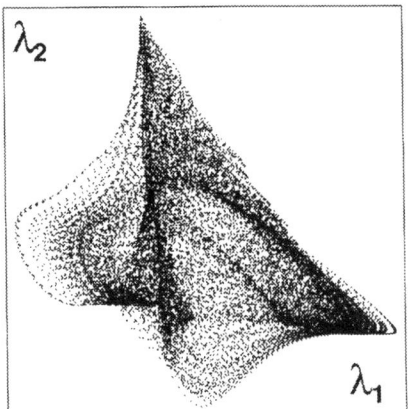 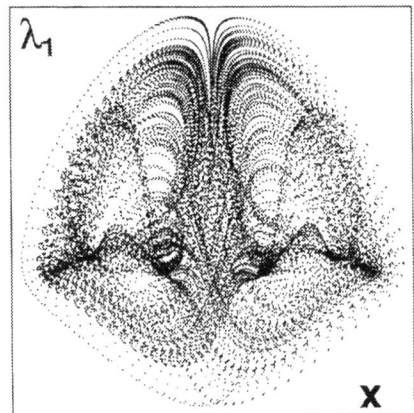

Figure 8.1. Distribution of the local Lorenz attractor Lyapunov exponents. The attractor parameters are $\{\sigma = 10, \; R = 28, \; b = 8/3\}$. The left view shows the correlation $\{(-10 < \lambda_1 < +10), \; (-6 < \lambda_2 < +8)\}$ linking the local exponents, λ_1 and λ_2. The right view shows the variation of the local value of the largest exponent $(-10 < \lambda_1 < +10)$ with Lorenz' variable $(-20 < x < +20)$. The sum of all the local exponents is a constant of the motion, $-(\sigma + 1 + b) = -(41/3)$.

Even for conservative dynamical systems the individual local values depend upon the chosen coordinate system, while the long-time averages do not. Thus the local exponents do not really characterize the underlying physical system uniquely. To avoid clutter we will often use the same notation $\{\lambda\}$ for both the local and the time-averaged Lyapunov exponents, trusting to the reader to distinguish the two concepts by context.

Let us consider the calculation of the complete Lyapunov spectrum for a general dynamical system, with motion equations $\dot{x} = \mathcal{F}(x)$. The time-dependent variables x define a vector locating the system in its embedding space. Begin with a "reference trajectory", a solution of the equations starting from a definite point, but sufficiently long ago that the identity of this initial point is unimportant. In the neighborhood of this solution of the motion equations, the equations of motion can be linearized, so that the vector separation, from the reference trajectory to any nearby "satellite trajectory" solution, is the infinitesimal vector $\delta(t)$, which has a motion equation of its own, determined by the first derivative of the reference system's motion equations:

$$\dot{\delta} \equiv D \cdot \delta; \; D \equiv (\partial \mathcal{F}/\partial x) \;.$$

For clarity, let us consider a simple example of this general approach. For the Nosé-Hoover oscillator of the last section, governed by the equations,

$$\mathcal{F} = \{\dot{q} = p; \; \dot{p} = -q - \zeta p; \; \dot{\zeta} = p^2 - 1\} \;,$$

the *relative* motion of a satellite trajectory, with "offset vector",

$$\delta \equiv (\delta q, \delta p, \delta \zeta) \equiv (q, p, \zeta)_{\text{satellite}} - (q, p, \zeta)_{\text{reference}} \;,$$

follows the *linearized* motion equations:

$$\{\dot{\delta q} = \delta p; \; \dot{\delta p} = -\delta q - (\zeta \delta p + p \delta \zeta); \; \dot{\delta \zeta} = 2p\delta p\} \;.$$

Notice that these motion equations are time-reversible provided that the original flow was. Thus the reversed evolution of a corotating and comoving hyperellipsoid, following the flow forward in time, generates an image of the repeller, with the signs of all the local Lyapunov exponents reversed $\{+\lambda\} \longrightarrow \{-\lambda\}$. Because the reversed trajectory is actually unstable, these hypothetical precisely-reversible repeller exponents $\{-\lambda\}$ are not observable.

For our Nosé-Hoover oscillator example the "dynamical matrix" $D = [(\partial \mathcal{F}/\partial x)]$ governing the evolution of the offset vector δ is:

$$D = \begin{bmatrix} 0 & 1 & 0 \\ -1 & -\zeta & -p \\ 0 & 2p & 0 \end{bmatrix} \;.$$

The maximum Lyapunov exponent is the (time-averaged) rate of growth of the unconstrained infinitesimal vector δ and describes the rate of divergence of a nearby satellite trajectory away from the reference trajectory. Spotswood Stoddard and Joseph Ford used this idea to compute the largest Lyapunov exponent for a dense Lennard-Jones fluid in 1973. The second-largest Lyapunov exponent, λ_2, is defined in the same way, but for an offset vector δ_2 which is required to be *orthogonal* to δ_1. This idea generalizes to the computation of the entire spectrum. The exponents have an interesting topological interpretation: the sum of the first n time-averaged exponents gives the global growth rate of an $(n+1)$-point n-dimensional object in the embedding space. We discuss the consequences of this observation in the following Section.

In following the offset vectors $\{\delta\}$ linking the reference trajectory to the satellites (one satellite trajectory for each exponent) forward in time it is absolutely necessary to rescale them, either continuously, or at fixed time intervals. Otherwise their size will rapidly grow or shrink beyond the possible limits of computational precision. This rescaling operation is analogous to the velocity rescaling used in isokinetic thermostats. The orthogonality of the vectors also needs to be explicitly maintained.

The evolution of the local Lyapunov exponents can then be followed by solving a coupled set of equations, with the orthogonality and rescaling restrictions included, using a separate offset vector δ for each Lyapunov exponent. If the original dynamical system has N independent variables, then a system of $N(N+1)$ equations needs to be solved for the N variables defining the reference trajectory and the N^2 offset-vector components of the N satellite trajectories. In the Nosé-Hoover oscillator example, the complete spectrum of three exponents requires the solution of twelve ordinary differential equations, three for the reference trajectory and nine for the three orthogonal satellite trajectories. If N is not too large an elegant Lagrange-multiplier approach can be used to impose both the N length constraints $\{\delta_i^2 \equiv 1\}$ and the $N(N-1)/2$ orthogonality conditions $\{\delta_i \cdot \delta_j \equiv 0\}$. The general idea is clear enough from the three-exponent example:

$$\dot{\delta}_1 = D \cdot \delta_1 - \lambda_{11}\delta_1 \ ;$$
$$\dot{\delta}_2 = D \cdot \delta_2 - \lambda_{21}\delta_1 - \lambda_{22}\delta_2 \ ;$$
$$\dot{\delta}_3 = D \cdot \delta_3 - \lambda_{31}\delta_1 - \lambda_{32}\delta_2 - \lambda_{33}\delta_3 \ .$$

For convenience, let the constant infinitesimal scalar lengths of all the offset vectors be ϵ. Then the *diagonal* Lagrange multipliers, $\{\lambda_{ii}\}$, which exactly compensate for the stretching and shrinking of the unconstrained vectors, can be calculated from the N constant-length conditions:

$$\{(d/dt)\delta_i^2 = 0 = 2\delta_i \cdot \dot{\delta}_i \longrightarrow \lambda_{ii} = (\delta_i/\epsilon) \cdot D \cdot (\delta_i/\epsilon)\} \ .$$

The off-diagonal Lagrange multipliers are chosen to keep the offset vectors $\{\delta_1, \delta_2, \delta_3, \dots\}$ orthogonal. These off-diagonal multipliers follow from the $N(N-1)/2$ conditions:

$$\{(d/dt)\delta_i \cdot \delta_j = 0 = \dot{\delta}_i \cdot \delta_j + \delta_i \cdot \dot{\delta}_j \longrightarrow$$

$$\lambda_{ij} = (\delta_j/\epsilon) \cdot D \cdot (\delta_i/\epsilon) + (\delta_i/\epsilon) \cdot D \cdot (\delta_j/\epsilon)\} \ .$$

The Lyapunov exponents are themselves the long-time averages of the diagonal Lagrange Multipliers:

$$\lambda_1 = \langle \lambda_{11} \rangle; \ \lambda_2 = \langle \lambda_{22} \rangle; \ \lambda_3 = \langle \lambda_{33} \rangle \ .$$

The spectrum of exponents can, in some exceptional circumstances, exhibit interesting symmetry properties. In Hamiltonian mechanics, with steady boundary conditions, the reversibility of the equations of motion, together with the lack of any dissipation or explicit time dependence, indicates that any expanding direction is converted to a contracting direction in the time-reversed flow. This reversed flow and the forward flow are both equally stable. Thus Hamiltonian Lyapunov spectra are made up of *pairs* of exponents $\{+\lambda, -\lambda\}$. If all the momenta are similarly thermostated, with the same friction coefficient, the result can be to shift each pair of exponents by the same amount, $-\zeta$, a rule called "conjugate pairing" by its discoverers, Denis Evans and Gary Morriss.

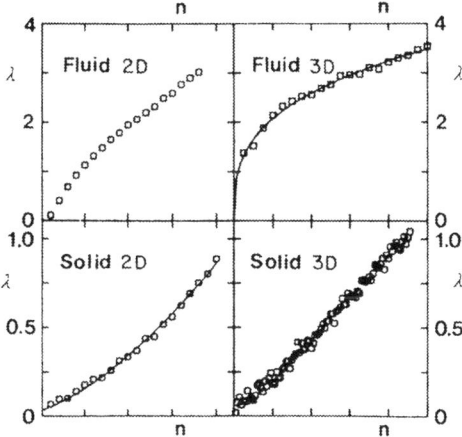

Figure 8.2. Typical many-body Lyapunov spectra for both two-dimensional and three-dimensional fluids and solids. The positive half of the spectrum is shown here.

Figure 8.2 illustrates typical equilibrium spectra for fluids and solids in two and three space dimensions. The *fluid* spectra look much like the Debye vibrational spectra of *solid*-state physics. Although the dynamical matrix D is common to both approaches (Lyapunov spectra and vibrational frequencies) the exact connection between them is just now beginning to be understood. Posch and Dellago have established that the individual *mode* components $\{\delta q, \delta p\}$ oscillate with a common phase in the long-wavelength Lyapunov eigenvectors. This is quite different to long-wavelength sound waves, for which δq precedes δp by a phase shift of $(\pi/2)$.

From the theoretical standpoint of kinetic theory, hard disks and spheres are somewhat simpler to treat than are soft particles. Recent results from simulations and kinetic theory, carried out by Dellago, Posch, and their coworkers, agree well.

8.5 Fractal Dimensions

The two- or three-dimensional macroscopic mass, momentum, and energy densities from continuum mechanics, as well as the many-dimensional microscopic probability densities from Gibbs' equilibrium statistical mechanics, are typically smooth and differentiable, with the same dimensionalities as the embedding dimensions $\{D_E\}$ of the spaces in which they are embedded. Away from equilibrium *probability densities* behave in a different way—they are typically multifractals, with *fractional* dimensionality. These reduced fractal dimensions are most easily defined by generalizing the usual notion of one-, two-, and three-dimensional objects to include objects whose dimensionality is not an integer. Suppose that a sufficiently compact geometric "object" with dimensionality D covers an arbitrarily large number $\#$ of sufficiently-small cells (or "boxes"), of infinitesimal width ϵ in an embedding space with dimensionality D_E greater than D. The number covered is of order ϵ^{-D} for small enough cells. For example, a straight *line* piercing a cube composed of 10^6 small cubes, intersects a number of these smaller cubes of order 100. A *plane* piercing the same large cube intersects on the order of 10^4 of the small cubes. Thus the ("box-counting") dimension of ordinary one-, two-, and three-dimensional objects can be found by taking the limiting small-box ratio:

$$\ln(\#)/\ln(1/\epsilon) \longrightarrow \ln(\epsilon^{-D})/\ln(\epsilon^{-1}) \equiv D_{\mathrm{BC}},$$

where $\#$ is the number of infinitesimal cubes intersecting the embedded object. This definition can be generalized to nonintegral cases; the resulting box-counting dimension is sometimes termed the "Hausdorff dimension". Here we avoid that latter terminology, so as to avoid confusion with two more interesting and useful ideas, the "information dimension" and the

"Kaplan-Yorke dimension" defined below. For the attractors representing typical time-reversible nonequilibrium systems the information dimension is sensitive to dissipation while the box-counting dimension is not.

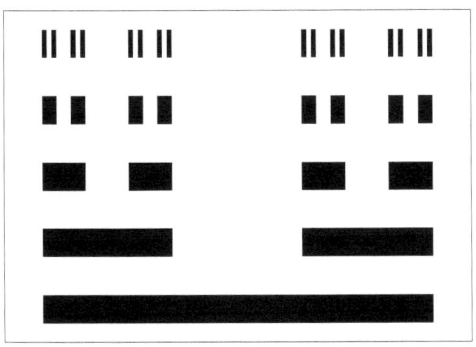

Five Stage Cantor Set

Figure 8.3. Development of the 0.63093-dimensional Cantor's Set by the repeated removal of the middle-third segments of the unit interval.

Fractals are distributions of points in which the density of nearby points varies as a power law in the vicinity of each point. If the power law itself varies from point to point the distribution is termed "multifractal". Most textbook examples of fractals contain holes, while those representing physical systems on digital computers appear instead to represent continuous, and often *ergodic*, multifractals. The simplest fractal, *with holes*, is probably Cantor's set. This set can be constructed recursively, as is indicated in Figure 8.3. It is made up of all those base-3 numbers which contain only 0's and 2's in their base-3 expansion. To construct a coarse-grained approximation to this set, with N ternary digits, divide the unit interval,

$$0.000000 \cdots = 0 < x < 1 = 0.222222 \ldots \text{[base 3]},$$

into $3N$ equal bins (or "boxes") of width 3^{-N}. Then discard any number with a "1" in its first N digits. Evidently the entire interval from $(1/3)$ to $(2/3)$ is discarded, and so not included in Cantor's set, for all such numbers begin with the digit 1: 0.1 ... . Likewise all the numbers from $(1/9)$ to $(2/9)$ and $(7/9)$ to $(8/9)$ are excluded, for they have a 1 in the second position: 0.01 ... and 0.21 ... . For *any* set of $3N$ similar boxes, the number of occupied boxes is only $2N$, so that the box-counting dimension of this Cantor set is:

$$D_{\text{BC}} = \ln(\#)/\ln(1/\epsilon) = (N \ln 2)/(N \ln 3) = 0.63093 .$$

The set is "self similar". Enlarged versions of each of the 2^N intervals isomorphic to $0 < x < 3^{-N}$, when scaled up by factors of 3^N, appear identical to the Cantor's set occupying the entire unit interval $0 < x < 1$.

Cantor's Set is a *mathematical object*, not a physical one. The *density* of points in the Cantor set has no special physical significance. By way of contrast, time series of phase-space points, generated by solving the equations of motion for many-body systems, represent equally-likely dynamical states, and can be used to define a variety of fractal dimensions. Though the data represent samples from a hypothetical one-dimensional trajectory in phase space, sufficient points can be generated to characterize the many-dimensional coarse-grained probability density in phase space.

The most useful of the many fractal dimensions characterizing that density is the "information dimension" D_I, which gives the limiting (small-box) dependence of the box probability "prob" on the box size ϵ. If the probabilities vary as the Dth power of the box size, $\{\text{prob} \simeq (\epsilon^D)\}$, then the averaged variation defines the information dimension as follows:

$$D_I = \langle \ln(\text{prob}) \rangle / \ln(\epsilon) \equiv \sum \text{prob} \ln \text{prob} / \ln(\epsilon) \longrightarrow \langle \ln(\epsilon^D) \rangle / \ln(\epsilon) = D_I .$$

The box probability "prob" is the product of the box volume, $\propto \epsilon^{D_E}$, where D_E is the embedding dimension, and the probability density f. As a fringe benefit, this information dimension gives explicitly the dependence of the coarse-grained Gibbs' entropy on the box size:

$$[S_{\text{Gibbs}}(\epsilon) - S_{\text{ideal}}] = -k \langle \ln(f/f_{\text{ideal}}) \rangle_\epsilon ,$$

where S_{ideal} and f_{ideal} refer to the uniform probability density characteristic of an ideal gas. Other dimensions can be obtained by using formal "measures" which are *powers* of the actual probability, proportional to the number of pairs of points, or triples of points, ..., found in each box. Farmer, Ott, and Yorke suggested that the box-counting and information dimensions have special significance for physically interesting attractors. The box-counting dimension is necessarily unchanged from its equilibrium value in an ergodic system. The information dimension *does* change. It is uniquely important for its connection to dissipation, Liouville's Theorem, the Lyapunov spectrum, and Gibbs' entropy.

The "Kaplan-Yorke dimension" D_{KY} is the best estimate for the information dimension in many-dimensional systems with phase spaces too extensive for computing box probabilities. The Kaplan-Yorke dimension is based on a fundamental and natural idea—*a comoving hypervolume having the dimensionality of the attractor can have no long-time tendency to grow or to shrink.* Thus the number of terms at which the linearly-interpolated Lyapunov-exponent sum changes from positive to negative is the Kaplan-Yorke estimate for the information dimension $D_{\rm I} \simeq D_{\rm KY}$. This correspondence has been proved true for the Baker Maps though it is definitely wrong, by about ten percent, for one carefully-investigated four-dimensional oscillator system. This oscillator system is described in this Chapter, in Section 8.10.2.

Because the information dimension for a multifractal with a *varying* probability density,

$$\langle (\langle {\rm prob} \rangle + \delta {\rm prob}) \ln (\langle {\rm prob} \rangle + \delta {\rm prob}) \rangle \ ,$$

is minimal when the probability fluctuation δprob vanishes, $D_{\rm I}$ is necessarily less than the box-counting dimension $D_{\rm BC}$ for *multifractal* distributions.

Are the instantaneous Lyapunov exponents and the various fractal dimensions point functions? Yes, they are, for a specified coordinate system. Christoph Dellago and Bill studied this carefully for both continuous and impulsive motion equations. But Hamiltonian systems, for example, can be described by *any* convenient generalized coordinates $\{q\}$ and the corresponding generalized momenta $\{(\partial \mathcal{L}/\partial \dot{q}) \equiv p\}$. For these systems the sums of all the Lyapunov exponents vanish, and so long as the boundary conditions are stationary, the exponents come in pairs $\{(+\lambda, -\lambda)\}$.

But, even in this simplest case, the *local* exponents, like the local fractal dimensions, depend upon the choice of coordinates, and so are joint properties of the representation and the system. To take an extreme example, consider the one-dimensional harmonic oscillator, with the parameter-dependent Hamiltonian

$$2\mathcal{H}(s) \equiv (q/s)^2 + (sp)^2 \ .$$

Though both the Lyapunov exponents vanish for the oscillator, the *local* exponents vary with the parameter s and with the phase of the motion. In this case it is possible to show that the mean squared local exponents, averaged over the orbit, can be made arbitrarily large:

$$2\langle \lambda^2 \rangle = (s^{+1} - s^{-1})^2 \ .$$

Nonhamiltonian thermomechanical systems can likewise be described with a variety of physical coordinates. Typically the total time-rate-of-change of the extension in phase space has physical significance,

$$-(\dot{\otimes}/\otimes) = (\dot{S}_{\rm external}/k) \ ,$$

but the individual exponents need not. Thus the individual local Lyapunov exponents are not point functions unless a particular coordinate system is specified. With Nosé-Hoover reservoirs their sum does give the logarithmic rate of phase-volume loss, and the corresponding dissipation, but the local individual contributions to that loss depend upon the chosen coordinate system.

In cases which do not correspond to physical systems, there is no "natural" alternative set of equations. But new rotated coordinate combinations such as $\{(x+y)/\sqrt{2}, (x-y)/\sqrt{2}\}$ could be introduced, and would lead in general to new values for the local Lyapunov exponents. Lorenz' well-known butterfly-shaped attractor stems from the irreversible motion equations of continuum hydrodynamics. The irreversible nature of those equations is reflected in its localized attractor structure, which is very different to the ergodic repeller-attractor pairs found in reversible dynamical simulations.

8.6 A Simple Ergodic Fractal

Numerical solutions of the time-reversible dynamical equations describing nonequilibrium thermomechanical problems are *never* time-reversible. The many-to-one dissipative nature of nonequilibrium flows prevents exact computational reversibility. Nevertheless, the necessarily-approximate computer-generated solutions agree on two points: *ergodicity* and *multifractality* are generic characteristics of many dissipative nonequilibrium stationary states, quite independently of the details distinguishing the underlying computer algorithm from its fellows. *Ergodicity* implies that the computational time series approximating the attractor (eventually) covers *every* allowable point while *multifractality* suggests an extremely singular discontinuous structure. How can *both* properties be present simultaneously? This paradox is best understood through an example.

Let us consider sets made up of asymmetric one-dimensional random walks, with the probability of a step to the right arbitrarily chosen to be twice that of a step to the left. Each of the 2^N possible N-step walks can be represented as an N-digit binary number on the unit interval. If probabilities of $(2/3)$ are associated with the zeros (steps to the right) and $(1/3)$ with the ones (steps to the left) in each binary fraction B, the resulting walk probabilities $\{\text{prob}(B)\}$ are normalized in the usual way:

$$\text{prob}(B) \equiv (2/3)^{N_0}(1/3)^{N_1} \longrightarrow$$

$$\sum_B \text{prob}(B) = \prod_N [(2/3) + (1/3)] = [(2/3) + (1/3)]^N \equiv 1 \ .$$

Consider now the distribution of probabilities as a function defined on the unit interval. The distribution of walk probabilities is clearly enough a good model for an erqodic distribution, with density everywhere and a point-to-point spacing as small as desired, $\Delta B = \epsilon = 2^{-N}$. The probabilities themselves have a wildly-discontinuous fractal character. Figure 8.4 shows the *integrated* cumulative probability $C(B)$:

$$C(B) \equiv \int_0^B \text{prob}(B')dB' \ .$$

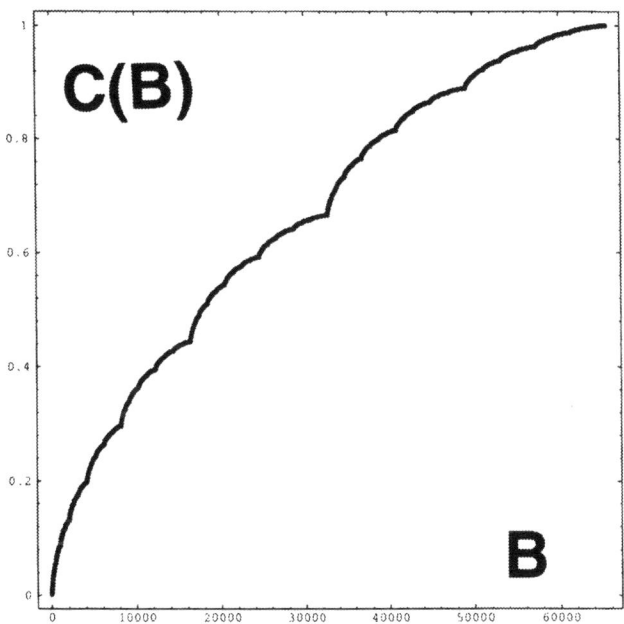

Figure 8.4. Cumulative probability $C(B)$ using 65,536 sampling bins. The underlying fractal distribution of random-walk probabilities is "ergodic", with density everywhere, and has an information dimension of 0.918296.

The fractal character of the underlying distribution $\text{prob}(B)$ can be established by calculating its information dimension D_I from the averaged value of the coarse-grained probabilities,

$$D_I = -\langle \ln \text{prob}/\ln \epsilon \rangle = [(2/3)\ln(2/3)+(1/3)\ln(1/3)\,]/\ln(1/2) = 0.918296 \ .$$

Thus the small-ϵ limit of this distribution not only has density everywhere but is also fractal.

8.7 Fractal Attractor-Repeller Pairs

That time-reversible motion equations can lead to irreversible behavior is paradoxical. But there are many examples which are above reproach— the shockwave problems in Chapter 6 come immediately to mind. The time-reversible dissipative Baker map, discussed in Chapter 2, generates a strange attractor with an equilibrium value for the box-counting dimension, which is exactly two, but with an information dimension of approximately 1.7337. The same multifractal results from any reasonable initial condition, as a time average, or, from any distribution, as a long-time limit. The resulting fractal attractor corresponds to dissipation, with the loss of one-third of a "bit" of "information" for every forward iteration of the map, $M(q, p) \to (q', p')$. Should we attempt to recover the past history at any particular point (q, p), by iterating the map backward in time rather than forward, starting with $M(q, -p)$, exactly the *same* attractor—not its mirror image—is the inevitable result. Nevertheless, the time-reversed set of points, $\{q, -p\}$ *is* the mirror-image of the attractor. This mirror image is an exactly similar geometric object, the *repeller*, satisfying exactly the same map, but with the reversed and unstable Lyapunov spectrum,

$$\{+\lambda_1 = +0.6365, +\lambda_2 = -0.8676\} \longleftrightarrow \{\lambda_R\} = \{-0.6365, +0.8676\} \ .$$

The underlying dissipative Baker map is time-reversible, just as is its Lyapunov spectrum. Evidently the "repeller" is *unstable*, just as the attractor is stable. Because the repeller cannot be observed, and has measure zero, the reversed states have no physical significance. In many-body phase space the similarly unobservable repeller states are those which would collectively violate the Second Law of Thermodynamics. The dissipative Baker Map illustrates another paradoxical characteristic of attractor-repeller pairs. Not only are the box-counting dimensions of the attractor and repeller equal. Both objects occupy *exactly the same* boxes. Only the box *weights* are different. The summed-up attractor weights approach unity while the summed-up repeller weights approach zero as the resolution length ϵ decreases to zero. Evidently the distinction between an "attractor box" and a "repeller box" has no operational significance!

In phase space, continuing collapse, to an attractor, is the path of least resistance for a time-reversible nonequilibrium steady state. Its reverse, sustained expansion, is obviously impossible in a confined space. This important symmetry breaking, caricatured in dissipative maps, is common to all nonequilibrium sets of time-reversible differential equations, such as those used in simulating the Rayleigh-Bénard flows with atoms in Section 4.9. Such a flow, when traced backward in time, would converge to an unstable entropy-consuming image of the flow forward in time. Because this artificial repeller is actually unstable (in the sense that nearby trajectories are repelled) relative to its parent attractor, it can *only* be generated formally, and artificially, by storing attractor states and playing them backwards. In simulations of systems incorporating heat reservoirs, the relative instability of the repeller revealed by its Lyapunov spectrum corresponds to the impossibility of systems' long-time averages violating the Second Law of Thermodynamics, as detailed in the following Section. The attractor states, generated going *forward* in time, are the *only* states which can be observed with a finite information supply.

Let us consider a simple time-reversible "attractor", which degenerates to a one-dimensional line in three-dimensional space. The (q, v, ζ) space for a falling thermostated particle illustrated in Section 5.4, is such an example. The equations of motion,

$$\{\dot{q} = v;\ \dot{v} = 1 - \zeta v;\ \dot{\zeta} = v^2 - 1\}\ ,$$

are time-reversible, with both the velocity v and the friction coefficient ζ changing sign in the reversed trajectory. The "attractor" for this motion is the line $(q, +1, +1)$. The "repeller" analog is the time-reversed line $(q, -1, -1)$. All possible flows in (q, v, ζ) space are portions of trajectories connecting the two one-dimensional structures.

Time reversibility requires reversing the system boundary velocities too. Thus the repeller corresponding to a steady many-body shear flow, with a *positive* strain rate $(dv_x/dy) = \dot{\epsilon} > 0$, for instance, differs from the instantaneous time-reversed configuration, which would have a *negative* strain rate incorporated in its boundaries. The attractor and repeller for any specific flow, necessarily occupy the *same* phase space with the *same* boundary conditions including the *same* strain rate $\dot{\epsilon}$. The phase-space repeller for a shear flow with positive strain rate corresponds to the instantaneous time reversal of a *mirror-image* steady shear flow with the *negative* strain rate $(dv_x/dy) = -\dot{\epsilon} < 0$. The phase-space flow then occurs *from* the repeller *to* the attractor with identical boundary conditions throughout.

8.8 A Global Second Law from Reversible Chaos

Begin with the language and ideas provided by chaos theory—in particular the concepts of fractal distributions and Lyapunov spectra. Then consider a conservative Newtonian system and drive that system away from equilibrium with a specific algorithmic representation of heat reservoirs—reservoirs controlled by Nosé-Hoover thermostat forces. The result is a simple, but compelling, *proof* of a Global *time-averaged* microscopic version of the macroscopic Second Law of Thermodynamics:

> **Long-time-averaged time-reversible nonequilibrium steady-state flows invariably generate external entropy and correspond to fine-grained contracting flows from a fractal repeller to its mirror-image strange attractor.**

The one-way nature of this global Second Law emerges naturally, for computer simulations of simple experiments involving mass flows, shear flows, or heat flows, despite the formal time reversibility of the Newtonian and Nosé-Hoover equations of motion. Two assumptions are required for the proof: (i) *existence* of any long-time-averaged macroscopic quantities of interest; (ii) *independence* of these averaged quantities to the detailed nature of the microscopic initial conditions.

Sources and sinks of energy are necessary to drive or to moderate any nonequilibrium flow. In order for heat transfer to occur at least one of these sources and sinks is necessarily a Nosé-Hoover heat reservoir. The two required nonequilibrium assumptions have corresponding analogs at equilibrium: (i) Observable macroscopic quantities exist (so that the system is stationary and *stable* from the thermodynamic point of view); (ii) Long-time averages converge to the phase-space averages used in Gibbs' statistical mechanics. The two analogous *nonequilibrium* assumptions are most easily motivated by the observation that both laboratory experiments and computer experiments typically (i) give *definite* results, and (ii) these results are *reproducible*, despite differences in the details of the initial conditions and the makeup of the thermostats providing nonequilibrium temperature control.

Although the proof sketched below applies to *any* thermomechanical Newtonian system, the thermal driving of its reservoirs must be provided by the differentiable Nosé-Hoover equations of motion. With Nosé-Hoover heat reservoirs it is simple to evaluate the phase-space collapse rate from the equations of motion. The calculation follows directly from the local

phase-space velocity divergence, $\nabla \cdot v$:
$$-(\dot{f}/f) \equiv (\dot{\otimes}/\otimes) \equiv (\nabla \cdot v) = \sum \lambda_L \ .$$

It should be emphasized that although the individual local Lyapunov exponents $\{\lambda_L\}$ *do* depend upon the chosen coordinate system, *their sum does not*. Only the complete sum is crucial to the argument which follows.

The Nosé-Hoover heat-reservoir forces $\{-\zeta p\}$ describe one or more reservoir regions with specified temperatures maintained by the friction coefficients $\{\zeta\}$. These coefficients—$\{\zeta_{cold}, \zeta_{hot}\}$ in the case of a simple two-reservoir heat-transfer simulation—are *independent* variables, so that the phase-space velocity divergence is a simple sum of friction coefficients:

$$\nabla \cdot v \equiv \sum_q (\partial \dot{q}/\partial q) + \sum_p (\partial \dot{p}/\partial p) + \sum_\zeta (\partial \dot{\zeta}/\partial \zeta) =$$

$$\sum_q (0) + \sum_p (-\zeta) + \sum_\zeta (0) \ .$$

Note that the one nonzero sum $\sum_p (-\zeta)$ includes only those degrees of freedom belonging to thermostated regions. Independence of averages to the initial conditions implies that exactly the same expression necessarily holds as a time average, giving the averaged comoving velocity divergence $\langle \nabla \cdot v \rangle$ in terms of the long-time average of the summed-up friction coefficients. In the "extended" (q, p, ζ) phase space Liouville's compressible Theorem applies. It relates the changing probability density, following the flow, to the velocity divergence and so to the local Lyapunov exponents $\{\lambda_L\}$ and the changing extension in phase $\otimes$:

$$(d\ln f/dt) \equiv -(\dot{\otimes}/\otimes) \equiv -(\nabla \cdot v) = +\sum_p \zeta = -\sum \lambda_L \longrightarrow$$

$$\langle d\ln f/dt \rangle \equiv -\langle \dot{\otimes}/\otimes \rangle \equiv -\langle \nabla \cdot v \rangle = +\left\langle \sum_p \zeta \right\rangle = -\left\langle \sum \lambda_L \right\rangle \ .$$

It is crucial to see that the long-time averages of each of the five terms in these equations *must* be positive, corresponding to *collapse* (of comoving phase-space hypervolume), rather than negative, which would correspond to divergence. To see this, consider the time development of a small compact hypervolume $\otimes$ in the extended phase space. For definiteness imagine the hypervolume to be a hypersphere initially containing a constant normalized density function,

$$f(q, p, \zeta, t = 0) \equiv f_0 \equiv (1/\otimes) \ ,$$

for all those points "inside" the hypervolume $\otimes$. This fine-grained probability density has an integral of unity and vanishes outside $\otimes$. It corresponds physically to a nonequilibrium ensemble of similarly-prepared systems.

Now apply Liouville's *compressible* Theorem. According to that theorem, the local expansion or contraction of the hypervolume and the local increase or decrease of the density $f(q,p,\zeta,t)$ responds to the local summed-up friction coefficients $\sum_p \zeta$. The small hypervolume, initially a hypersphere, will rapidly be transformed to a hyperellipsoid. The corresponding initial growth and decay rates are described by the local Lyapunov exponents. At somewhat longer times [of order $(q/\dot{q}) \simeq (p/\dot{p})$] the small ellipsoid bends and stretches. The local growth rate and density become nonuniform throughout the moving hypervolume. At much longer times—of order $(\ln L)/\lambda_1$—where L is *macroscopic*—the fibrillating hypervolume element traverses the space. At *very* long times we have already assumed that the distribution becomes *uniform*, in a sufficiently coarse-grained sense, making it possible to calculate convergent steady averages *independent* of the initial location of the hyperspherical extension in phase $\otimes$:

$$\langle \dot{\otimes}/\otimes \rangle_{\text{local}} \equiv \langle \dot{\otimes}/\otimes \rangle_{\text{global}} .$$

Throughout its time development the comoving "fine-grained" density responds to the summed-up local friction coefficients as is described by the summed-up local Lyapunov exponents. Consider now the time-averaged density change following a particular solution of the equations of motion for a time t. According to Liouville's Compressible Theorem, applied to a nonequilibrium Newtonian system interacting with one or more Nosé-Hoover heat reservoirs:

$$(d \ln f/dt) \equiv \sum_p \zeta \longrightarrow f(t) = f_0 e^{\int_0^t \sum \zeta dt'} = f_0 e^{t \langle \sum \zeta \rangle} .$$

Because it is assumed that the boundary conditions are fixed, so that a steady state results, there are only three possibilities for the long-time limit of such a comoving fine-grained density:

$$\langle \ln f_{\text{steady}}(t \to \infty) \rangle \longrightarrow \{-\infty, \ln f_0, +\infty\} ,$$

These three cases correspond to three possibilities for the time-averaged friction-coefficient sum:

$$\left\{ \left\langle \sum \zeta \right\rangle < 0 ; \left\langle \sum \zeta \right\rangle = 0 ; \left\langle \sum \zeta \right\rangle > 0 \right\} .$$

The *first* possibility, that $\langle \sum \zeta \rangle$ is negative, implies, through Liouville's Theorem, the *divergence* of the comoving volume element (which has to be independent of the hypersphere's initial location), which is inconsistent with a convergent stable solution. Thus the first "possibility" must be ruled out.

The *second* possibility, that $\langle \sum \zeta \rangle$ has a stable average value of 0, *must* correspond to equilibrium. This is obvious for the simplest case, a Newtonian system coupled to exactly two heat reservoirs, one at T_cold and one at T_hot. Then the zero-sum condition,
$$\langle \sum_p \zeta \rangle = \langle (\#\zeta)_\text{cold} \rangle + \langle (\#\zeta)_\text{hot} \rangle = 0 \;,$$
is inconsistent with the long-time-averaged energy-balance condition:
$$\langle (\#\zeta)_\text{cold} T_\text{cold} \rangle + \langle (\#\zeta)_\text{hot} T_\text{hot} \rangle \equiv 0 \;,$$
unless the two temperatures are equal, in which case the averaged heat transfers necessarily vanish. We conclude that a simple system with heat transfer between two heat reservoirs can have neither a negative nor a vanishing friction-coefficient sum. There seems to be no direct argument to show that the zero-sum condition must correspond to equilibrium when more than two reservoirs are present.

Only the *third* and final possibility, $\langle \sum \zeta \rangle > 0$, remains. This possibility is thereby necessarily established, away from equilibrium, and implies that *the Gibbs entropy approaches $-\infty$ in a nonequilibrium steady state*:
$$\langle (\dot{S}_\text{Gibbs}/k) \rangle \equiv -\langle (d \ln f / dt) \rangle = \langle (\dot{\otimes}/\otimes) \rangle = -\langle \sum \zeta \rangle < 0 \Longrightarrow$$
$$(S_\text{Gibbs}/k) \longrightarrow -\infty \;!$$
A similar result also holds if the boundary conditions vary periodically with time.

This divergence of the Gibbs entropy corresponds to the vanishing fraction of the phase space effectively occcupied by the attractor's "core",
$$\langle \otimes \rangle_\text{steady} \simeq \langle \otimes \rangle_\text{equilibrium}^{D_\text{I}/D_E} \;,$$
where D_E is the "equilibrium" or "embedding" dimension of the full phase space.

For two heat reservoirs the proof just sketched establishes that ζ_hot is *necessarily* negative, and ζ_cold is *necessarily* positive, corresponding to the usual flow of heat, from hot to cold. With more than two heat reservoirs, the various directions of the individual flows cannot generally be determined in advance, but it must *necessarily* still be true that the full friction-coefficient sum, over all thermostated degrees of freedom, can only have a positive long-time average, corresponding to the eventual collapse of phase-space density onto an attractor. Let us reïterate the conclusion, the global Second Law of Thermodynamics:

Long-time-averaged time-reversible nonequilibrium steady-state flows invariably generate external entropy and correspond to fine-grained contracting flows from a fractal repeller to its mirror-image strange attractor.

Although the two divergences, of the "fine-grained" phase-space density and the corresponding Gibbs' entropy, seemed odd in 1986, when the first simulations studying this effect were carried out, a variety of succeeding low-dimensional simulations confirmed a relatively-simple geometric interpretation. In every case, the limiting steady-state probability density $f(q, p, \zeta)$ collapses and diverges on a fractal "strange attractor",

$$\langle f \rangle \longrightarrow +\infty; \; \langle \otimes \rangle \longrightarrow 0; \; S_{\text{Gibbs}} = -k \langle \ln f_N \rangle \longrightarrow -\infty \; .$$

The Kaplan-Yorke estimate of the attractor's "information dimension" is also *necessarily* strictly less than that of the original equilibrium extended phase space.

The information dimension D_{I} reflects the way in which the coarse-grained entropy,

$$S_{\text{CG}}(\epsilon) = -k \langle \ln \text{prob}_{\text{CG}} \rangle_\epsilon \; ,$$

depends upon the box size ϵ and the corresponding coarse-grained probabilities $\{\text{prob}_{\text{CG}} \propto \epsilon^{D_{\text{I}}}\}$ (for small ϵ), as we saw in Section 8.5:

$$D_{\text{I}} \equiv \langle \ln \text{prob}_{\text{CG}} \rangle_\epsilon / \ln(\epsilon) \equiv \sum \text{prob}_{\text{CG}} \ln \text{prob}_{\text{CG}} / \ln(\epsilon) \; .$$

The fractal explanation of the probability density's divergence has additional physical significance. *It indicates the extreme rarity of nonequilibrium states*. These core states have *zero measure* relative to the continuously-distributed equilibrium states. This means that *any* fixed fraction of the total nonequilibrium measure (half, nine tenths, ninety nine percent, ...) can be found within a *vanishing* fraction of the total number of boxes in the limit that the box size approaches zero. The averaged rates of divergence of $[\ln f]$ and S_{Gibbs} correspond also to a (long-time-averaged) *negative* Lyapunov sum, so that any comoving hypervolume element eventually collapses, with a volume of order $e^{-\Delta S/k}$, where ΔS is the time-integrated net entropy increase of the external Nosé-Hoover heat reservoir(s). Suppose now, as is usual, that the equations of motion are time-reversible. Then, the time-reversed trajectory, with the momenta and friction coefficients changed in sign, has a *positive* time-averaged Lyapunov sum, corresponding to instability and a diverging hypervolume element.

Such hypothetical time-reversed trajectories cannot be observed, even though they are formally legitimate "solutions" of the equations of motion.

This symmetry breaking, with forward attractor trajectories collapsing in a stable way and reversed repeller trajectories unstable and unobservable, *is* the mechanical equivalent of the Second Law of Thermodynamics. Before considering detailed examples, let us briefly consider the relation between the fine-grained entropy change from Liouville's compressible theorem and the macroscopic coarse-grained entropy production.

8.9 Coarse-Grained and Fine-Grained Entropy

For a nearly homogeneous system described by Newtonian viscosity and Fourier's heat conduction, a phenomenological system-entropy production

$$(\dot S_{\text{prod}}) = [(\eta/T)\dot\epsilon^2 + \kappa(\nabla \ln T)^2]V \; ,$$

is often arbitrarily introduced in order to avoid any net entropy change *within* a stationary nonequilibrium state. Such an internal macroscopic entropy production should be chosen to offset the *decrease* with time of the *fine-grained* Gibbs' entropy discussed in the last Section:

$$(\dot S_{\text{prod}}/k) = +d\langle \ln f\rangle/dt = -\dot S_{\text{Gibbs}}/k \; .$$

The continual evolution of the fine-grained entropy in a "steady" state describes the penetration of information and phase-space structure to smaller and smaller scales. Because these scales have no physical significance below the limits of observation, and further have no computational significance below the level of precision carried, it is tempting to define a "coarse-grained" entropy based on an averaged probability density function. Because we assume the accurate convergence of macroscopic quantities at long times, the corresponding coarse-grained entropy $S_{\text{CG}}(\epsilon)$ must eventually converge too, for any fixed box size ϵ. This coarse-grained entropy has the substantial drawback that its box-dependent definition is quite arbitrary, and is even unbounded from below, away from equilibrium.

Such a coarse-grained entropy has sometimes been used to fill the perceived need for an entropy away from equilibrium. In his boxed notes at the Yale University library Gibbs repeatedly muses over such a nonequilibrium entropy. He even mentions two specific examples, steady flows with density or temperature gradients. Evidently Gibbs could find no convincing answer to this question. If there *were* a well-defined nonequilibrium entropy we could perhaps avoid the fractal divergence of the Gibbs entropy. Gibbs' fine-grained entropy inexorably *decreases* toward minus infinity in nonequilibrium steady states. To save the notion of a finite fixed steady-state entropy, it is necessary to somehow compensate for the actual decrease. In continuum mechanics, entropy *production* is traditionally introduced, arbitrarily, for this reason. It is evident, with the coarse-grained entropy fixed and constant, while Gibbs' fine-grained entropy decreases, that the phenomenological entropy production can be defined so as to exactly cancel the loss of Gibbs' entropy:

$$0 \equiv \langle \dot{S}_{\text{CG}} \rangle \equiv \langle \dot{S}_{\text{Gibbs}} + \dot{S}_{\text{prod}} \rangle \longrightarrow \langle \dot{S}_{\text{prod}} \rangle = -\langle \dot{S}_{\text{Gibbs}} \rangle \;.$$

In a series of very interesting papers, Tél, Vollmer, and Breymann have extended this idea to open systems, in which an additional entropy change, due to mass flows, is included. See their paper "Transient Chaos: the Origin of Transport in Driven Systems".

By connecting the fine-grained change of Gibbs' entropy with the phenomenological entropy production of irreversible thermodynamics, we see that the microscopic approach emphasizes the loss of extension in phase due to nonequilibrium constraints while the macroscopic approach ascribes the external entropy change of heat reservoirs to an internal entropy production. Let us clarify and illustrate the concepts of fractal dimension and Lyapunov spectra by developing and applying our algorithmic tools to a series of five illustrative examples.

We begin with a cautionary warmup exercise, the conducting Nosé-Hoover oscillator, and then analyze a better-behaved doubly-thermostated oscillator, determining its four Lyapunov exponents by three related, but different, algorithms. Next we compare polar and Cartesian descriptions of an eight-dimensional double-pendulum problem, exhibiting local Lyapunov spectra that depend upon the chosen coordinate system. We complete this set of examples with determinations of the information dimension for the two-dimensional Galton Board problem and its many-body color-conductivity analog, with over 100,000 phase-space dimensions. All of these problems are suggestive of good research projects for the future.

8.10 Oscillators, Lyapunov Algorithms, Fractal Dimensions

8.10.1 *A Thought-Provoking Oscillator Exercise*

We have chosen a doubly-thermostated oscillator for a detailed demonstration of Lyapunov-exponent algorithms. An interesting and thought-provoking warmup exercise for that four-dimensional oscillator is the "simpler" and innocent-appearing three-dimensional Nosé-Hoover oscillator with a coordinate-dependent temperature varying from 0.6 to 1.4:

$$\{\dot{q} = p;\ \dot{p} = -q - \zeta p;\ \dot{\zeta} = p^2 - T(q)\}\ ;$$

$$T(q) = 1 + \epsilon \tanh(q);\ \epsilon = 0.40\ .$$

The equilibrium version of this oscillator, with $\epsilon = 0$, a Hamiltonian problem with both periodic and chaotic solutions, was thoroughly investigated in 1985, but the nonequilibrium problem is not nearly so well known and evidently has some very interesting features. For large ϵ, with 0.5 a good choice, the three-dimensional oscillator problem has a simple limit cycle with $\epsilon = 0.5 \to \langle\zeta\rangle = 0.294$. For smaller values of ϵ, which is the maximum temperature gradient, fractal strange attractors can appear.

Two Million Steps, 0.001 and 0.01

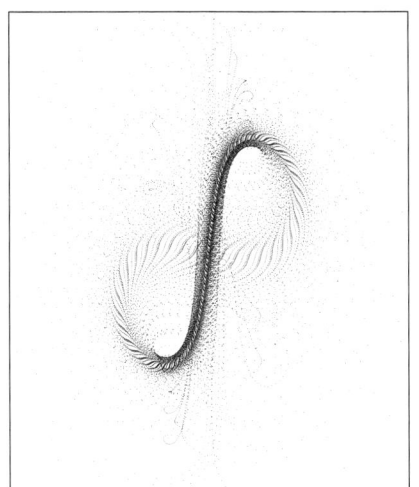 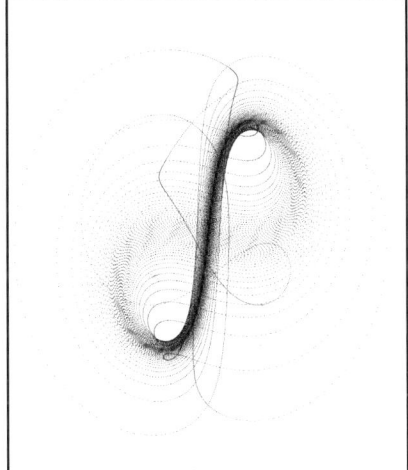

Chaotic and Periodic Solutions

Figure 8.5. A nearly periodic solution, with $\Delta t = 0.01$, on the right, and a chaotic attractor with $\Delta t = 0.001$ on the left, both for the Nosé-Hoover conducting oscillator with maximum temperature gradient $\epsilon = 0.4$. Both views show the square: $\{-4 < p, \zeta < +4\}$. See text for more details.

The (p, ζ) projections of two numerical "solutions" appear in Figure 8.5. The righthand projection, with $\Delta t = 0.01$ is, to a very good approximation, a (rather long) periodic orbit, while the lefthand projection, with a significantly smaller timestep, $\Delta t = 0.001$, the optimum choice for double precision, is not, and is typical of a strange attractor. Both solutions began with initial condition

$$(q \, , \, p \, , \, \zeta) = (0.0 \, , \, 5.0 \, , \, 0.0) \, .$$

Every 100th point is plotted for the last half of the runs. One would likely conclude that the higher precision of the calculation with $\Delta t = 0.001$ gives a better "solution", the strange attractor, shown at the left.

However, if a (q, p, ζ) point *on* the righthand periodic orbit (the last point from the righthand calculation will do) is used as an *initial condition* for the smaller timestep (0.001), that timestep too follows the periodic orbit (at least for many millions of timesteps). From the numerical standpoint it appears that the higher-dimensional oscillator problems are better behaved than is the Nosé-Hoover oscillator and we will turn to such a case next. See Section 5.9.2 for another such problem. The main message of Figure 8.5 is that a careful study of the conducting Nosé-Hoover oscillator would be rewarding and is definitely warranted. Let us turn now to the algorithms which characterize chaos through the Lyapunov spectrum.

8.10.2 *Doubly-Thermostated Oscillator; Lyapunov Spectra*

Lyapunov's exponential instability, the growth of a small perturbation as $\simeq e^{+\lambda t}$, can be estimated quite easily; take *two* initial conditions for the system under investigation, separated in the state space (usually phase space, though configuration space or momentum space can be used) by a tiny perturbation on the order of the computational roundoff error. Measure the separation of the two solutions [usually $\Delta r = \sqrt{\sum(\Delta q^2 + \Delta p^2)}$] as a function of time in order to estimate λ. Sophistication and resolution can be enhanced by "rescaling" the separation at intervals of Δt and estimating the growth rate (the Lyapunov exponent λ) by analyzing the mean separation ratio before and after the scaling;

$$\lambda \equiv \langle \ln(\Delta r_{\text{before}}/\Delta r_{\text{after}}) \rangle / \Delta t \, .$$

Even better, a *continuous* rescaling algorithm can be achieved by treating the separation Δr as a constraint, enforced by a Lagrange multiplier (whose average value turns out to be the largest Lyapunov exponent λ_1). These two approaches, periodic rescaling, and continuous rescaling, can be applied to small but finite separations or to infinitesimal ones. Thus there are *four* different approaches to computing the Lyapunov spectrum.

A four-variable problem is easy to program but sufficiently complex to illustrate all of the numerical steps involved. The doubly-thermostated oscillator, with relatively simple equations of motion, is a good choice for study, with an æsthetic attractor. The equations of motion are
$$\{\dot q = p;\ \dot p = -q - \zeta p - \xi p^3;\ \dot\zeta = p^2 - T;\ \dot\xi = p^4 - 3p^2 T\}\ ;$$
where again the temperature of the oscillator is coordinate-dependent:
$$T(q) \equiv 1 + \epsilon \tanh(q)\ .$$
The friction coefficient ζ drives the oscillator's kinetic temperature toward the instantaneous target value $T(q)$. Similarly, ξ controls the fourth moment, $\langle p^4 \rangle$. Time averages of the $\dot\zeta$ and $\dot\xi$ equations give nonequilibrium generalizations of the equilibrium velocity moments:
$$\langle p^2 \rangle \longrightarrow \langle T \rangle;\ \langle p^4 \rangle \longrightarrow \langle 3p^2 T \rangle\ .$$
The equilibrium distribution is a four-dimensional Gaussian,
$$4\pi^2 T f_{\rm eq} = e^{-(q^2+p^2)/2T} e^{-(\zeta^2+\xi^2)/2}\ .$$
The *nonequilibrium* distribution depends upon the maximum temperature gradient ϵ. Figure 8.6 shows the time development of the (ζ,ξ) projection of $f_{\rm neq}$ for a very similar problem detailed in Reference [83]. The projected trajectory is a one-dimensional line in the two-dimensional (ζ,ξ) space.

The fractal "information dimension" D_I can be estimated from the small-bin dependence of Gibbs' entropy on bin size ($\Delta \longrightarrow 0$):
$$-(S_{\rm Gibbs}/k) \equiv \langle \ln f_{\rm neq} \rangle \equiv \sum f_{\rm neq} \ln f_{\rm neq} / \sum f_{\rm neq} \propto (D_I - D_E) \ln \Delta\ ,$$
where D_E is the dimensionality of the fractal's embedding space. Here accumulated trajectory points eventually fill in a 2.56-dimensional multifractal object in the four-dimensional (q,p,ζ,ξ) embedding space.

Measuring the Lyapunov spectrum begins with a trajectory inspection. Trajectories can be checked by noticing that the quadratic form,
$$E = \tfrac{1}{2}(q^2 + p^2 + \zeta^2 + \xi^2)\ ,$$
resembles an energy shell in (q,p,ζ,ξ) space. The quadratic's time derivative leads to a "constant of the motion", a useful check on the accuracy of the trajectory. The time derivative is
$$\dot E = q\dot q + p\dot p + \zeta\dot\zeta + \xi\dot\xi =$$
$$q(p) + p(-q - \zeta p - \xi p^3) + \zeta(p^2 - T) + \xi(p^4 - 3p^2 T) =$$
$$-(\zeta + 3p^2 \xi)T(q) \longrightarrow$$
$$\tfrac{1}{2}(q^2 + p^2 + \zeta^2 + \xi^2) + \int_0^t [\zeta(t') + 3p^2(t')\xi(t')]T(t')dt' =$$
$$\tfrac{1}{2}(q^2 + p^2 + \zeta^2 + \xi^2)_{t=0} = {\rm constant}\ .$$

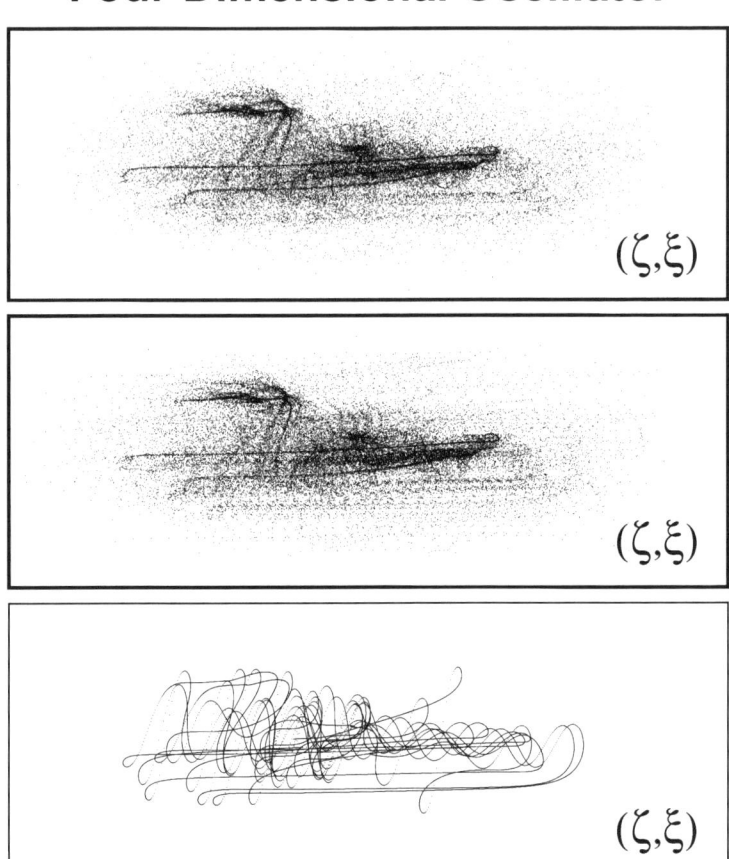

Figure 8.6. Three trajectory-sampling intervals, 0.005, 0.500, and 50.000, are used here to display the projection of the (q, p, ζ, ξ) oscillator dynamics into (ζ, ξ) space. The smallest interval shows the one-dimensional nature of the trajectory. Each trajectory includes 40,000 points with the longest sampling time corresponding to the longest trajectory. The largest interval (top), of order $(1/\lambda_1)$, clearly indicates the multifractal nature of the trajectory, which has an information dimension of 2.56. The "correlation dimension" in (ζ, ξ) space corresponds to the limiting variation of the number of points within a small distance δ, $D_C = 1.77 \longrightarrow \#_{\text{pairs}} \propto \delta^{1.77}$.

The constant of the motion can be checked by adding a fifth equation, for yyp(5) ⟶ yy(5), to the trajectory calculation in the righthand side subroutine:

```
subroutine rhs(yy,yyp)
implicit double precision(a-h,o-z)
dimension yy(5),yyp(5)
common eps
q = yy(1)
p = yy(2)
z = yy(3)
x = yy(4)
T = 1.0d00 + eps*dtanh(q)
yyp(1) = p
yyp(2) = -q - z*p - x*p*p*p
yyp(3) = p*p - T
yyp(4) = p*p*p*p - 3.0d00*p*p*T
yyp(5) = z*T + 3.0d00*x*p*p*T
return
end
```

The simplest approach to computing the Lyapunov spectrum is to advance four nearby satellite trajectories, separated from the (q, p, ς, ξ) reference trajectory by $(\delta_1, \delta_2, \delta_3, \delta_4)$. The righthand side subroutine then acquires an additional sixteen differential equations:

```
q = yy(1)
p = yy(2)
z = yy(3)
x = yy(4)
T = 1.0d00 + eps*dtanh(q)
Tprime = eps/dcosh(q)**2
do i = 1,4
dq(i) = yy(4*i+1)
dp(i) = yy(4*i+2)
dz(i) = yy(4*i+3)
```

```
dx(i) = yy(4*i+4)
enddo
dq1dot = dp(1)
dq2dot = dp(2)
dq3dot = dp(3)
dq4dot = dp(4)
dp1dot = -dq(1) - z*dp(1) - p*dz(1) - p*p*p*dx(1)
dp1dot = dp1dot - 3.0d00*p*p*x*dp(1)
dp2dot = -dq(2) - z*dp(2) - p*dz(2) - p*p*p*dx(2)
dp2dot = dp2dot - 3.0d00*p*p*x*dp(2)
dp3dot = -dq(3) - z*dp(3) - p*dz(3) - p*p*p*dx(3)
dp3dot = dp3dot - 3.0d00*p*p*x*dp(3)
dp4dot = -dq(4) - z*dp(4) - p*dz(4) - p*p*p*dx(4)
dp4dot = dp4dot - 3.0d00*p*p*x*dp(4)
dz1dot = 2.0d00*p*dp(1) - Tprime*dq(1)
dz2dot = 2.0d00*p*dp(2) - Tprime*dq(2)
dz3dot = 2.0d00*p*dp(3) - Tprime*dq(3)
dz4dot = 2.0d00*p*dp(4) - Tprime*dq(4)
dx1dot = 4.0d00*p*p*p*dp(1) - 6.0d00*T*p*dp(1)
dx1dot = dx1dot - 3.0d00*Tprime*p*p*dq(1)
dx2dot = 4.0d00*p*p*p*dp(2) - 6.0d00*T*p*dp(2)
dx2dot = dx2dot - 3.0d00*Tprime*p*p*dq(2)
dx3dot = 4.0d00*p*p*p*dp(3) - 6.0d00*T*p*dp(3)
dx3dot = dx3dot - 3.0d00*Tprime*p*p*dq(3)
dx4dot = 4.0d00*p*p*p*dp(4) - 6.0d00*T*p*dp(4)
dx4dot = dx4dot - 3.0d00*Tprime*p*p*dq(4)
yyp( 1) = p
yyp( 2) = - q - z*p - x*p*p*p
yyp( 3) = p*p - T
yyp( 4) = p*p*p*p - 3.0d00*p*p*T
yyp( 5) = dq1dot
yyp( 6) = dp1dot
yyp( 7) = dz1dot
yyp( 8) = dx1dot
yyp( 9) = dq2dot
yyp(10) = dp2dot
yyp(11) = dz2dot
yyp(12) = dx2dot
yyp(13) = dq3dot
```

```
yyp(14) = dp3dot
yyp(15) = dz3dot
yyp(16) = dx3dot
yyp(17) = dq4dot
yyp(18) = dp4dot
yyp(19) = dz4dot
yyp(20) = dx4dot
return
end
```

If, as in the sample code above, the equations of motion are linearized, the *length* of the offset vectors is arbitrary. We maintain it equal to unity for simplicity and impose that constraint after each timestep Δt. In the main code the four "offset vectors" $\{\delta q, \delta p, \delta \zeta, \delta \xi\}$ are made orthonormal by the "Gram-Schmidt [orthonormalization] procedure". The Lyapunov exponents are computed as averages, updated after each rescaling. The rescaling process considers each of the offset vectors in turn. First, the length of δ_1 is scaled to unity, giving the local value of the first Lyapunov exponent. Next, the projection of δ_2 on δ_1 is removed and the remaining δ_2 is rescaled to get the local value of $\lambda_2(t)$. Third, the projections of δ_3 on (δ_1, δ_2) are removed, with the remaining δ_3 rescaled to get λ_3, and so on. The Fortran coding is as follows:

```
dot11 = dq(1)*dq(1)+dp(1)*dp(1)+dz(1)*dz(1)+dx(1)*dx(1)
dq(1) = dq(1)/dsqrt(dot11)
dp(1) = dp(1)/dsqrt(dot11)
dz(1) = dz(1)/dsqrt(dot11)
dx(1) = dx(1)/dsqrt(dot11)
dot12 = dq(1)*dq(2)+dp(1)*dp(2)+dz(1)*dz(2)+dx(1)*dx(2)
dq(2) = dq(2) - dq(1)*dot12
dp(2) = dp(2) - dp(1)*dot12
dz(2) = dz(2) - dz(1)*dot12
dx(2) = dx(2) - dx(1)*dot12
dot22 = dq(2)*dq(2)+dp(2)*dp(2)+dz(2)*dz(2)+dx(2)*dx(2)
dq(2) = dq(2)/dsqrt(dot22)
dp(2) = dp(2)/dsqrt(dot22)
```

```
dz(2) = dz(2)/dsqrt(dot22)
dx(2) = dx(2)/dsqrt(dot22)
dot13 = dq(1)*dq(3)+dp(1)*dp(3)+dz(1)*dz(3)+dx(1)*dx(3)
dq(3) = dq(3) - dq(1)*dot13
dp(3) = dp(3) - dp(1)*dot13
dz(3) = dz(3) - dz(1)*dot13
dx(3) = dx(3) - dx(1)*dot13
dot23 = dq(2)*dq(3)+dp(2)*dp(3)+dz(2)*dz(3)+dx(2)*dx(3)
dq(3) = dq(3) - dq(2)*dot23
dp(3) = dp(3) - dp(2)*dot23
dz(3) = dz(3) - dz(2)*dot23
dx(3) = dx(3) - dx(2)*dot23
dot33 = dq(3)*dq(3)+dp(3)*dp(3)+dz(3)*dz(3)+dx(3)*dx(3)
dq(3) = dq(3)/dsqrt(dot33)
dp(3) = dp(3)/dsqrt(dot33)
dz(3) = dz(3)/dsqrt(dot33)
dx(3) = dx(3)/dsqrt(dot33)
dot14 = dq(1)*dq(4)+dp(1)*dp(4)+dz(1)*dz(4)+dx(1)*dx(4)
dq(4) = dq(4) - dq(1)*dot14
dp(4) = dp(4) - dp(1)*dot14
dz(4) = dz(4) - dz(1)*dot14
dx(4) = dx(4) - dx(1)*dot14
dot24 = dq(2)*dq(4)+dp(2)*dp(4)+dz(2)*dz(4)+dx(2)*dx(4)
dq(4) = dq(4) - dq(2)*dot24
dp(4) = dp(4) - dp(2)*dot24
dz(4) = dz(4) - dz(2)*dot24
dx(4) = dx(4) - dx(2)*dot24
dot34 = dq(3)*dq(4)+dp(3)*dp(4)+dz(3)*dz(4)+dx(3)*dx(4)
dq(4) = dq(4) - dq(3)*dot34
dp(4) = dp(4) - dp(3)*dot34
dz(4) = dz(4) - dz(3)*dot34
dx(4) = dx(4) - dx(3)*dot34
dot44 = dq(4)*dq(4)+dp(4)*dp(4)+dz(4)*dz(4)+dx(4)*dx(4)
dq(4) = dq(4)/dsqrt(dot44)
dp(4) = dp(4)/dsqrt(dot44)
dz(4) = dz(4)/dsqrt(dot44)
dx(4) = dx(4)/dsqrt(dot44)
```

The local Lyapunov exponents can then be added to cumulative sums:

```
sum1 = sum1 + (dlog(dot11)/2.0d00*dt)
sum2 = sum2 + (dlog(dot22)/2.0d00*dt)
sum3 = sum3 + (dlog(dot33)/2.0d00*dt)
sum4 = sum4 + (dlog(dot44)/2.0d00*dt)
```

which are divided by the number of iterations, at the run's end, to get the global Lyapunov exponents:

```
slam1 = sum1/itmax
slam2 = sum2/itmax
slam3 = sum3/itmax
slam4 = sum4/itmax
```

A variation on this algorithm uses small offset vectors—a length of 0.000001 is typical—rather than infinitesimal ones. In that case, the coordinates of each satellite trajectory can be advanced in time *without* linearizing the equations of motion. For example, we illustrate here the propagation of the first satellite trajectory in rhs $[\delta_2 = \text{q(2)},\text{p(2)},\text{z(2)},\text{x(2)}]$ through one timestep dt:

```
dq2dt = p(2)
dp2dt = - q(2) - z(2)*p(2) - x(2)*p(2)*p(2)*p(2)
dz2dt = p(2)*p(2) - T(2)
dx2dt = p(2)*p(2)*p(2)*p(2) - 3.0d00*p(2)*p(2)*T(2)
```

The satellite propagations are followed in the main code by applying Gram-Schmidt orthonormalization to the four unit vectors describing the satellite offsets,
$$(\delta_2, \delta_3, \delta_4, \delta_5) \times 10^6 \ .$$

Another more-elegant (but more time-consuming) procedure is to compute the array of Lagrange multipliers constraining the offsets in rhs:

```
slam12 = dq(1)*dq2dot+dp(1)*dp2dot+dz(1)*dz2dot+dx(1)*dx2dot
slam12 = slam12 + dq(2)*dq1dot + dp(2)*dp1dot + dz(2)*dz1dot + dx(2)*dx1dot
slam13 = dq(1)*dq3dot+dp(1)*dp3dot+dz(1)*dz3dot+dx(1)*dx3dot
slam13 = slam13 + dq(3)*dq1dot + dp(3)*dp1dot + dz(3)*dz1dot + dx(3)*dx1dot
slam14 = dq(1)*dq4dot+dp(1)*dp4dot+dz(1)*dz4dot+dx(1)*dx4dot
slam14 = slam14 + dq(4)*dq1dot + dp(4)*dp1dot + dz(4)*dz1dot + dx(4)*dx1dot
slam23 = dq(2)*dq3dot+dp(2)*dp3dot+dz(2)*dz3dot+dx(2)*dx3dot
slam23 = slam23 + dq(3)*dq2dot + dp(3)*dp2dot + dz(3)*dz2dot + dx(3)*dx2dot
slam24 = dq(2)*dq4dot + dp(2)*dp4dot + dz(2)*dz4dot+ dx(2)*dx4dot
slam24 = slam24 + dq(4)*dq2dot + dp(4)*dp2dot + dz(4)*dz2dot + dx(4)*dx2dot
slam34 = dq(3)*dq4dot + dp(3)*dp4dot + dz(3)*dz4dot+ dx(3)*dx4dot
slam34 = slam34 + dq(4)*dq3dot + dp(4)*dp3dot + dz(4)*dz3dot + dx(4)*dx3dot
slam11=dq(1)*dq1dot+dp(1)*dp1dot+dz(1)*dz1dot + dx(1)*dx1dot
slam22=dq(2)*dq2dot+dp(2)*dp2dot+dz(2)*dz2dot + dx(2)*dx2dot
slam33=dq(3)*dq3dot+dp(3)*dp3dot+dz(3)*dz3dot + dx(3)*dx3dot
slam44=dq(4)*dq4dot+dp(4)*dp4dot+dz(4)*dz4dot + dx(4)*dx4dot
sum = slam11 + slam22 + slam33 + slam44
```

The sum of the Lagrange multipliers is calculated above as a check, in view of the relation

$$\lambda_1 + \lambda_2 + \lambda_3 + \lambda_4 = (\dot{\otimes}/\otimes) \equiv -\zeta - 3p^2\xi \ .$$

The Lagrange-multiplier contributions are then added on to the offset vector equations of motion before returning to the main code:

```
dq1dot = dq1dot - slam11*dq(1)
dp1dot = dp1dot - slam11*dp(1)
dz1dot = dz1dot - slam11*dz(1)
dx1dot = dx1dot - slam11*dx(1)
dq2dot = dq2dot - slam12*dq(1) - slam22*dq(2)
dp2dot = dp2dot - slam12*dp(1) - slam22*dp(2)
dz2dot = dz2dot - slam12*dz(1) - slam22*dz(2)
dx2dot = dx2dot - slam12*dx(1) - slam22*dx(2)
dq3dot = dq3dot - slam13*dq(1) - slam23*dq(2) - slam33*dq(3)
dp3dot = dp3dot - slam13*dp(1) - slam23*dp(2) - slam33*dp(3)
dz3dot = dz3dot - slam13*dz(1) - slam23*dz(2) - slam33*dz(3)
```

```
dx3dot = dx3dot - slam13*dx(1) - slam23*dx(2) - slam33*dx(3)
dq4dot = dq4dot - slam14*dq(1) - slam24*dq(2) - slam34*dq(3) - slam44*dq(4)
dp4dot = dp4dot - slam14*dp(1) - slam24*dp(2) - slam34*dp(3) - slam44*dp(4)
dq4dot = dz4dot - slam14*dz(1) - slam24*dz(2) - slam34*dz(3) - slam44*dz(4)
dx4dot = dx4dot - slam14*dx(1) - slam24*dx(2) - slam34*dx(3) - slam44*dx(4)
```

The mean values of the multipliers are accumulated in the **rhs** integration to get yy(21),yy(22),yy(23),yy(24):

```
yyp(21) = slam11
yyp(22) = slam22
yyp(23) = slam33
yyp(24) = slam44
```

In following this Lagrange-multiplier idea it is necessary to use an occasional Gram-Schmidt rescaling in the main code to compensate for accumulated roundoff errors in the Runge-Kutta integration.

As benchmarks for the reader to check his own oscillator problem we can cite results from our papers, "Local Gram-Schmidt and Covariant Lyapunov Vectors and Exponents for Three Harmonic Oscillator Problems" and "The Second Law of Thermodynamics and MultiFractal Distribution Functions: Bin Counting, Pair Correlations, and the Kaplan-Yorke Conjecture", the latter written with Posch and Codelli. With $\epsilon = 0$, the equilibrium case, the symmetric Lyapunov spectrum is:

$$\{\lambda\}_{\text{eq}} = \{+0.066, +0.000, -0.000, -0.066\} \ .$$

With $\epsilon = 1$, corresponding to the strange attractor of Figure 8.6, the spectrum is:

$$\{\lambda\}_{\text{neq}} = \{+0.073, +0.000, -0.091, -0.411\} \ ,$$

and sums to

$$\langle (\dot{\otimes}/\otimes) \rangle = -\langle \zeta + 3p^2 \xi \rangle = -0.430 \ .$$

The covariant vectors mentioned in the references are linear combinations of the forward and backward Gram-Schmidt vectors considered here, and are restricted to follow the linearized motion equations:

$$\{\dot{\delta} \equiv D \cdot \delta\} \ .$$

The covariant vectors, for a particular coordinate system, are a "unique" set of time-dependent offset vectors. By contrast there are two different Gram-Schmidt sets, one forward and one backward in the oscillator case just treated.

Some of the literature suggests that the covariant vectors and spectra are superior to their Gram-Schmidt relatives. With respect, we disagree. The first and last of the covariant vectors are *identical* to the first forward and first backward Gram-Schmidt vectors and so provide no new information. Second, the covariant vectors, like the Gram-Schmidt vectors, are *not* invariant to phase-space transformations, though it is often stated that they are "norm-independent". Like the Gram-Schmidt vectors their directions, and the magnitudes of the corresponding local exponents, depend upon the coordinate system chosen for their evaluation, as is evident from the scaled harmonic oscillator example mentioned on page 287. Finally, the Gram-Schmidt vectors are intimately related to the time rate of change of phase volume, $\dot{\otimes}/\otimes \equiv \sum \lambda$, while the covariant vectors are not so simply related to phase-space deformation. For these three reasons we choose to use the original Gram-Schmidt vectors and exponents to describe Lyapunov instability.

8.10.3 *Lyapunov Spectra for a Chaotic Double Pendulum*

Next consider the chaotic planar Hooke's-Law double pendulum problem, with both masses unity, but with variable pendulum lengths, in a unit gravitational field. The phase space for the problem, either Cartesian or polar, is *eight*-dimensional. We arbitrarily choose the force constants in the two Hooke's-Law springs equal to four. The equations of motion are simply related to those of the single pendulum considered in Section 1.9.3. Here is the Cartesian version:

$$\dot{x}_1 = p_{x1};\ \dot{y}_1 = p_{y1};\ \dot{x}_2 = p_{x2};\ \dot{y}_2 = p_{y2}\ ;$$

$$\dot{p}_{x1} = 4x_1[(1/r_1) - 1] + 4(x_1 - x_2)[(1/r_{12}) - 1]\ ;$$

$$\dot{p}_{y1} = 4y_1[(1/r_1) - 1] + 4(y_1 - y_2)[(1/r_{12}) - 1] - 1\ ;$$

$$\dot{p}_{x2} = 4(x_2 - x_1)[(1/r_{12}) - 1]\ ;$$

$$\dot{p}_{y2} = 4(y_2 - y_1)[(1/r_{12}) - 1] - 1\ .$$

Whether or not a particular solution is chaotic depends solely on the initial conditions. Small displacements from the least-energy configuration,

$$\{x_1 = 0;\ x_2 = 0;\ y_1 = -(3/2);\ y_2 = -(11/4)\};\ [\text{equilibrium}]\ ,$$

give four normal vibrational modes and no chaos. High energies eliminate the influence of the gravitational field. Again no chaos results.

The initial horizontal unstressed configuration shown in Figure 8.7, with no initial kinetic energy, *is* chaotic. The distributions of the *local* values of the four largest Lyapunov exponents, $\{-5 < \lambda < +5\}$, calculated in both the Cartesian phase space, $\{x, y, p_x, p_y\}$, and the polar phase space, $\{r, \theta, p_r, p_\theta\}$, are shown in the Figure. In both cases the averaged exponents are $\{\pm 0.17,\ \pm 0.08,\ \pm 0.04,\ 0.00,\ 0.00\}$. The constant-energy condition corresponds to one vanishing exponent. The second paired zero exponent corresponds to motion parallel to the trajectory. Note that the corresponding instantaneous exponents do not vanish. The local exponents *do* nevertheless satisfy the exact instantaneous pairing, $\{\lambda_i + \lambda_{9-i} \equiv 0\}$.

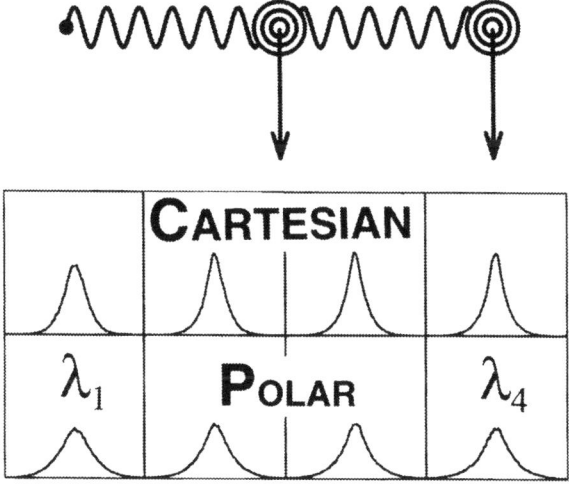

Figure 8.7. Distributions of double Hookean pendula spectra in Cartesian and Polar phase spaces. Gravity accelerates both masses downward. Distributions $\{\text{prob}(\lambda \geq 0)\}$ of the four most-positive exponents mirror the distributions $\{\text{prob}(\lambda \leq 0)\}$ for the four most-negative exponents, which are not shown here. For each exponent the range of values shown is $-5 < \lambda < +5$.

8.10.4 Coarse-Grained Galton Board Entropy

Here we illustrate the connection between fractal distributions and the divergence of Gibbs' entropy for the Galton Board problem treated in Section 5.9.1. The two-dimensional Poincaré section cataloging all possible collisions (see Figure 5.8 on page 189 for an example) is relatively easy to visualize. Further, the dynamics is simple and the motion is ergodic for small to moderate driving fields. Enhancing the resolution of the Poincaré section displays the inexorable approach of the Gibbs' entropy to $-\infty$. In the equilibrium case, with zero accelerating field, the probability density is Gibbs' uniform microcanonical distribution:

$$f_{\text{eq}}(0 < \alpha < \pi, -1 < \sin(\beta) < +1) = 1/(2\pi) ,$$

corresponding to the equal probability of the direction of motion and the location of scattering collisions. A nonvanishing accelerating field changes the uniform density in an intrinsically singular way—the mean value of the excess entropy, $(S/k) = -[\ln(f/f_{\text{eq}})]$ diverges. This can be seen by spanning the unit square $[(\alpha/\pi), (\sin(\beta)/2)]$ with a regular grid of square boxes, and computing the dependence of the entropy on the box size. As the boxes become small, this same calculation gives also the information dimension, as explained in Sections 2.12.1 and 8.5:

$$D_{\text{I}} = \langle \ln(\text{prob}) \rangle_\epsilon / \ln(\epsilon) .$$

In the equilibrium case, with a uniform distribution, all the box probabilities are simply equal to the box size, ϵ^2, and the information dimension is the same as the box-counting dimension, $D_{\text{I}}(\text{eq}) = \langle \ln(\epsilon^2) \rangle / \ln(\epsilon) \equiv 2$.

Because the nonequilibrium fractal phase-space distribution contains information on *all* scales, and is singular everywhere, the effective extension in phase vanishes. Gibbs' entropy S_{G} for such a nonequilibrium system diverges, and so is not a meaningful quantity. Figure 8.8 shows the divergence of Gibbs' entropy with decreasing box-size ϵ.

$$D_{\text{I}} = \langle \ln(f\epsilon^2) \rangle_{\text{CG}} / \ln(\epsilon) = \langle \ln \text{prob} \rangle_{\text{CG}} / \ln(\epsilon) = -[S_{\text{G}}/k \ln(\epsilon)] + 2 .$$

Here "coarse-grained" averages are indicated by a subscript $_{\text{CG}}$.

The information dimension just calculated (see the Figure caption) agrees perfectly with Dellago's independent calculation of the Kaplan-Yorke dimension from the Lyapunov spectrum. See his 1995 PhD thesis (University of Vienna) for details. Gary Morriss has also investigated this same system, and all three sets of results are in good agreement.

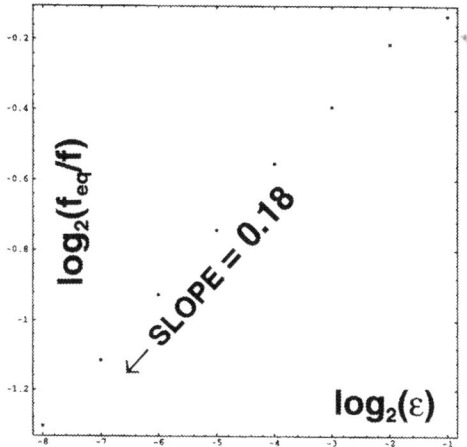

Figure 8.8. Dependence of Gibbs' entropy on the box size ϵ. The linear relation, $(S_G/k) = +0.18 \ln \epsilon < 0$, indicates an information dimension of 1.82. Details are in Bill's paper in the 1998 *Journal of Chemical Physics*.

8.10.5 *Color Conductivity*

Finally, we consider the "color conductivity" problem, illustrating the shift of Lyapunov exponents to more negative values in a nonequilibrium manybody system. In the time-reversible *non*equilibrium problems considered here this shift of Lyapunov exponents toward more negative values is directly related to phase-space contraction. In the Galton Board and Color Conductivity problems the phase-space contraction rate is *identical* to the external rate of thermodynamic dissipation. Klages, Rateitschak, and Nicolis have emphasized that this identity need not hold in more complicated cases. In *every* case, the resulting fractal distributions *do* explain, quantitatively, the general rarity of nonequilibrium phase-space states relative to their much more numerous equilibrium relatives.

The Galton Board problem follows the motion of a single particle through a lattice of fixed scatterers. This one-body problem is isomorphic to a *two*-body problem, with periodic boundaries, in which the two particles are accelerated, in opposite directions, by a fixed external field. In this form it is natural to generalize the two-particle version of the Galton Board problem to N particles, with $(N/2)$ of each type, accelerated in opposite directions. This "Color Conductivity" problem is the many-body analog of the Galton Board. In this many-particle case the equations of motion have the form:

$$\{\dot{x} = (p_x/m);\ \dot{y} = (p_y/m);\ \dot{p}_x = F_x \pm E - \zeta p_x;\ \dot{p}_y = F_y - \zeta p_y\}\ ,$$

where we arbitrarily choose the field E in the x direction. It is natural, and usual, to choose the friction coefficient ζ to thermostat the total kinetic energy, making K a constant of the motion:

$$(d/dt)[\sum_{\text{white}} (p^2/2m) + \sum_{\text{black}} (p^2/2m)] \equiv 0\ .$$

Generally, the friction coefficient ζ can be chosen to stabilize the kinetic or internal energy of the system, or it can simply be chosen equal to a convenient constant. This latter choice corresponds to a pervasive cold *stochastic* thermostat, with $T_{\text{bath}} = 0$. In either case the current leads to a well-defined conductivity, the "color conductivity", given by the ratio of the mean velocity in the field direction to the field strength. For simplicity, we minimize numerical errors by using a short-ranged purely-repulsive potential, with a characteristic energy ϵ and three continuous derivatives at the cutoff distance, $r = \sigma$:

$$\phi(r < \sigma) \equiv 100\epsilon[1 - (r/\sigma)^2]^4\ .$$

The length σ is an effective "collision diameter". At the lower of two field values, $E = 0.25(\epsilon/\sigma)$, the large-system steady distribution of the two particle "colors" turns out to be spatially homogeneous. The smooth dependence of the results on system size indicates that, like shear viscosity, the color conductivity in two dimensions need not exhibit any hydrodynamic instability.

At the higher field value, $0.50(\epsilon/\sigma)$, the situation is different, with the two colors separating and the "fluid" *freezing*. It is clear from these data that a nonequilibrium phase transition, analogous to the coexisting morphologies seen in the Rayleigh-Bénard simulations of Sections 4.11.2 and 7.8, separates the two field strengths. The Lyapunov exponents for the higher-field systems are significantly smaller, indicating collective motion, with a corresponding reduction in mixing activity in the phase space.

An unexpected and significant finding emerged from an analysis of these Lyapunov instability studies. We found that those particles which make the *largest* contribution to the maximum exponent, λ_1, tend to be localized in space. Figures 8.9 and 8.10, for $N = 25,600$, illustrate typical cases, following the 1998 paper by Hoover, Boercker, and Posch.

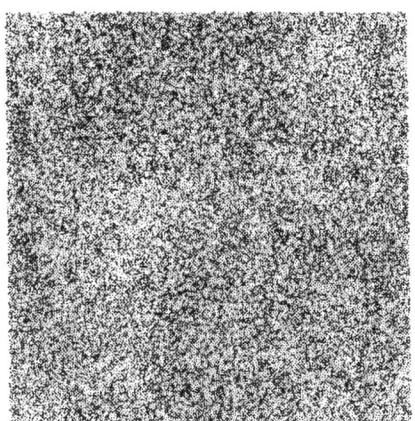 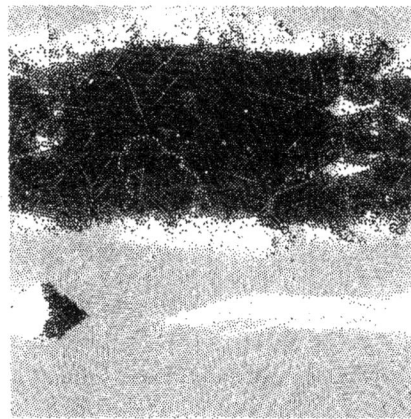

Figure 8.9. The distributions of the two particle "colors" are shown for the two different field strengths discussed in the text. At the higher field strength, shown on the right, the two-component fluid has frozen into crystallites—the distribution is inhomogeneous, with similarly-colored particles clumped together.

Figure 8.9 shows the homogeneous mixed arrangement of the two "colors" which prevails at the lower field strength. Figure 8.10 shows, as larger dots, those particles which make an above-average contribution to the local Lyapunov exponent λ_1, first in coordinate space, and then in momentum space. The correlated clumps of "important particles" which result are nearly the same for the two representations. We found no particular properties of the "important" particles, such as temperature, energy, or stress, which were correlated with the instability. The very smallest nonzero Lyapunov exponents exhibit a much more regular behavior, with a spatial periodicity, like sound waves. This problem has been thoroughly investigated by Posch's group in Vienna.

Numerical studies of the Lyapunov exponents also reveal, as suggested by Green and Kubo's work, that the distribution has a (fractal) "information dimension", D_I, which lies below the equilibrium embedding dimension, D_E, by a "dimensionality loss" ΔD *quadratic* in the field strength. Because the sum total of the Lyapunov exponents *must* necessarily be negative, and of order $(E/\sigma)^2$ for small fields, the dimensionality loss can be estimated:

$$\Delta D \equiv D_E - D_I - \langle \dot{S}_{\text{external}}/(k\lambda_1) \rangle \ .$$

For the fluid example problem illustrated in Figure 8.9 the loss is about 170, where the total phase-space dimensionality is 4 x 25,600 = 102,400. The external dissipation rate due to the thermostat forces is equal to the rate at which energy is extracted from the system divided by the thermostat temperature.

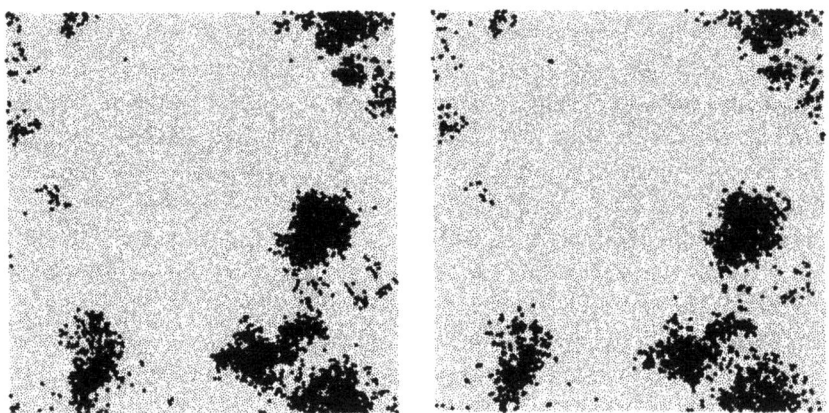

Figure 8.10. Localization of chaos. In the homogeneous lower-field case illustrated on the left side of Figure 8.9, those particles making the above-average contributions to the largest *instantaneous* Lyapunov exponent are emphasized here by using larger dots. The coordinate-space contributors are emphasized on the left; the momentum-space contributors are emphasized on the right.

8.11 Summary

The chaos—localized exponential instability—underlying the fractal attractors characteristic of time-reversible nonequilibrium steady states provides a clear interpretation of the dissipation underlying the Second Law of Thermodynamics. As extraneous information is destroyed by heat transfer, $T\dot{S}_{\text{external}}$, the logarithmic extension in phase volume vanishes. Gibbs' entropy approaches $-\infty$ as the phase-space probability density diverges. The stability of the resulting singular set of states—the set is a strange and multifractal attractor—is to be contrasted with the *instability* of its time-reversed image, the repeller. The Lyapunov Spectrum provides a detailed description of the time-reversible phase-space flow—*from* the unobservable repeller to the inevitable attractor—as well as Kaplan and Yorke's connection to the fractal geometry of the attractor and to the time-averaged rate of external entropy increase. The thermomechanical form of the Second Law of Thermodynamics links the external and Gibbs entropies to the collapse of the phase-space distribution function, as given by the long-time-averaged friction coefficients and the Lyapunov spectrum:

$$\langle \dot{S}_{\text{external}}/k \rangle = -\langle \dot{S}_{\text{Gibbs}}/k \rangle =$$

$$\langle d \ln f_N / dt \rangle = -\langle \dot{\otimes}/\otimes \rangle = \langle \sum \zeta \rangle = -\sum \lambda > 0 \ .$$

$$f_{\text{CG}} \propto \epsilon^{D_I - D_E} \simeq \epsilon^{-\dot{S}_{\text{ext}}/\lambda k} \longleftrightarrow \otimes_{\text{CG}} \propto \epsilon^{D_E - D_I} \simeq \epsilon^{+\dot{S}_{\text{ext}}/\lambda k} \ .$$

Once again:

> **Long-time-averaged time-reversible nonequilibrium steady-state flows invariably generate external entropy and correspond to fine-grained contracting flows from a fractal repeller to its mirror-image strange attractor.**

8.11.1 *Notes and References*

Sprott's *Chaos and Time Series Analysis*, Schröder's *Fractals, Chaos, and Power Laws: Minutes from an Infinite Paradise*, plus Farmer, Ott, and Yorke's "The Dimension of Fractal Attractors" provide helpful and entertaining reading on the basic definitions, along with plenty of examples.

The classic paper on the numerical evaluation of Lyapunov spectra is Benettin, Galgani, Giorgilli, and Strelcyn's: "Lyapunov Exponents for Smooth Dynamical Systems and for Hamiltonian Systems; a Method for Computing All of Them". The Lagrange-Multiplier approach was discovered independently, in 1987, by Goldhirsch, Sulem, and Orszag in "Stability and Lyapunov Stability of Dynamical Systems: A Differential Approach and a Numerical Method" and by Harald Posch and Bill, in "Direct Measurement of Equilibrium and Nonequilibrium Lyapunov Spectra". The connection with the Second Law of Thermodynamics was stressed by Holian, Hoover, and Posch in "Resolution of Loschmidt's Paradox: The Origin of Irreversible Behavior in Reversible Atomistic Dynamics".

The covariant exponents and vectors were popularized by Ginelli, Poggi, Turchi, Chaté, Livi, and Politi in "Characterizing Dynamics with Covariant Lyapunov Vectors" (2007). A four-dimensional oscillator problem with some æsthetic aspects and relevant to the Kaplan-Yorke conjecture is in our 2007 work with Codelli and Posch, "The Second Law of Thermodynamics and Multifractal Distribution Functions: Bin Counting, Pair Correlations, and the Kaplan-Yorke Conjecture".

For references to recent work on Lyapunov "modes" for Hard-Particle systems see "Covariant Lyapunov Vectors for Rigid Disk Systems" by Bosetti and Posch, as well as Harald Posch's webpage.

Chapter 9

Resolving the Reversibility Paradox

*It's a Naïve Domestic Burgundy
without any Breeding, But I think
You'll be Amused by its presumption.*

James Thurber

9.1 Introduction

Our exploration of time reversibility from the perspective of computer simulation and chaos has provided us with insights into the breaking of time symmetry which were not available to Boltzmann or Gibbs or Maxwell or Poincaré or to their successors: Green, Kubo, and Onsager. Insights gleaned from careful computer simulations have also provided grist for the mathematicians' mills. The main results are already quite clear, despite the lack of formal proofs. Simulations have clarified the formation and significance of time-reversible ergodic multifractal phase-space structures. Those structures, which arise from dissipative chaos, provide a natural link between an underlying time-reversible microscopic dynamics and the one-way irreversibility of the macroscopic Second Law of Thermodynamics. The various approaches to understanding are necessarily consistent with this view, but they are often expressed in quite different languages, sometimes vague or contrived. By searching through libraries, and now the internet, attending topical conferences, and exchanging ideas with our colleagues, we have sought to assimilate these different routes to understanding with our own. We are both of us very grateful to those many who have helped.

Understanding evolving views, but always with a perspective grounded in the past, is the continuous pursuit of a moving target. Only a partial understanding can result. Our own points of view reflect in part those of our predecessors and contemporaries. Though consensus is necessarily incomplete in research, our goals here in this book are first, to emphasize common features, and second, to point out useful directions in which new insights leading to a more complete understanding might be found. Wherever disagreements or uncertainties remain, we reïterate that the most powerful and useful means to understanding are based on the study of relatively simple computational models. This approach, gaining understanding through the analysis of "computer experiments", is far from being exhausted.

9.2 Irreversibility from Boltzmann's Kinetic Theory

Even today, an understanding of Boltzmann's views is complicated by the sheer bulk of the work he generated during the lifelong evolution of his ideas. His language and notation are also unfamiliar—he often uses $\{q\}$ for momenta and $\{p\}$ for coordinates, for example. Steve Brush, Carlo Cercignani, Ezechiel Cohen, Martin Curd, and Giovanni Gallavotti have all provided useful guideposts to the chronology and development of his work.

In the early 1960s, while Steve was still at the Livermore Laboratory, he made Bill a very welcome gift—his translation of Boltzmann's gas-theory lectures. Boltzmann was always interested in clarifying and exploring precise detailed consequences of well-defined mechanical models and was, for this reason, unable to do much with liquids and solids. Even air, with its missing vibrational heat capacity, was already too much for Boltzmann's simple classical mechanical models. Kinetic theory was nonetheless a superb model for the low-density transport coefficients. It was a welcome supplement to continuum mechanics, which can only assume and accept constitutive properties.

Boltzmann's irreversible equation for the time evolution of the single-particle distribution function $f_1(r, v, t)$, due to its statistical assumption for the two-body collision frequencies, provides a direct, though approximate, link between two-body atomistic mechanics and the accepted manybody flow equations of continuum mechanics. Despite its approximate nature, the Boltzmann equation really does *apply* to (classical models of) dilute gases.

The exponential amplification of perturbations justifies his simplification for the relative probability of finding two particles in a small volume element:
$$f_2(r_1,r_2,v_1,v_2,t)\delta(r_1-r_2) \longrightarrow f_1(r_1,v_1,t)f_1(r_2,v_2,t)\delta(r_1-r_2) \ .$$
For dilute gases, solutions of the Boltzmann equation provide useful recipes for computing the diffusion coefficient, the shear viscosity, and the heat conductivity. The Boltzmann equation provides an understanding too of the chaotic collisional mechanism underlying the Maxwell-Boltzmann distribution. Best of all, a rigorous consequence of the (approximate) Boltzmann equation for evolving $f(r,v,t)$ is an expression for the time-dependence of the corresponding single-particle "Boltzmann entropy":
$$S_B(t) \equiv -Nk\langle \ln f_1\rangle = -Nk\int\int drdv f_1 \ln f_1 \ .$$
According to the Boltzmann equation the one-body entropy rigorously obeys the usual macroscopic thermodynamic form of the Second Law, *without* the thermal fluctuations that necessarily accompany microscopic dynamical simulations, and without the time reversibility too. For any isolated system, away from equilibrium, the Boltzmann equation predicts that the Boltzmann entropy S_B *invariably* increases with time. This strong result, called the H Theorem, is a direct consequence of the plausible, but approximate, nature of Boltzmann's derivation. Boltzmann replaced Liouville's exact and *time-reversible* evolution equation, $\dot{f}_N = 0$, with his own *irreversible* one, $\dot{f}_1 = (\partial f/\partial t)_{\text{collisions}}$, "Boltzmann's Equation".

Let us check on the time-reversibility of Boltzmann's Equation by imagining the consequences of reversing all the velocities in a many-body system at some particular time. To explore the consequences, take a distribution $f(r,v,t)$ and change the signs of all of the particle velocities, $\{+v\} \to \{-v\}$. Next, reverse the time ordering of points along all the particle trajectories at the chosen reversal time, $+\Delta t \to -\Delta t$. Evidently these two operations *change* the sign of the *lefthand* side of the Boltzmann equation, $+\dot{f}_1(+v) = -\dot{f}_1(-v)$. But the collision term, $(\partial f/\partial t)_{\text{collisions}}$, on the *righthand* side is completely unchanged by the time-reversal operation. Thus the Boltzmann equation is *not* time-reversible. Uncorrelated collisions *always* bring a velocity distribution closer to local equilibrium.

To see this in detail, notice, following Boltzmann, that each time-reversed two-body collision contributing to the righthand side can be replaced by its "reversed inverse" collision, with both the relative separation and the velocity changing sign $(+r_{ij}, +v_{ij}) \to (\ r_{ij}, -v_{ij})$. (The *macroscopic* location of *any* colliding particle pair is unchanged during their collision, but the *microscopic* geometry of each colliding pair is inverted.) For all "forward-collision" velocity pairs $\{(+v_i, +v_j) \to (+v'_i, +v'_j)\}$ inverting both the relative separation vectors and the velocities gives the time-reversed and spatially-inverted collisions $\{(-v_i, -v_j) \to (-v'_i, -v'_j)\}$. See Figure 9.1 for an example collision.

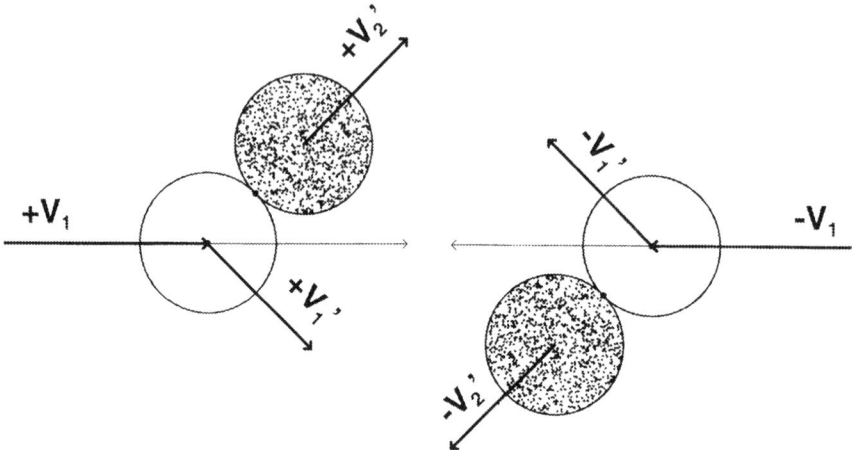

Figure 9.1. The collision on the left: $(+v_1, +v_2) \to (+v'_1, +v'_2)$ corresponds exactly [to see this just rotate the drawing 180 degrees] to the time-reversed and spatially-inverted collision on the right: $(-v_1, -v_2) \to (-v'_1, -v'_2)$.

Thus the collision term on the righthand side, evaluated for the reversed velocities, is unchanged:

$$(\partial f[+v]/\partial t)_{\text{collision}} = (\partial f[-v]/\partial t)_{\text{collision}} .$$

With the lefthand side changing sign, and the righthand side not, we see that the Boltzmann equation *is* intrinsically irreversible. This lack of time reversibility only became clear as objections to the irreversibility were raised. Attempts to rationalize this irreversibility, in the face of Zermélo's and Loschmidt's objections to it, engaged Boltzmann's efforts for decades.

Boltzmann finally attempted to avoid both Zermélo's recurrence objection and Loschmidt's reversibility objection by imagining a built-in cosmological irreversibility based on an unlikely low-entropy fluctuation, or initial condition, with a resemblance to our current "Big-Bang" "understanding" of how things began. A cosmological resolution of the reversibility paradox was appealing to Boltzmann, and later to the Ehrenfests, and still has an avid following today. Our own view agrees with that articulated by Prigogine: the basic small-scale and relatively-rapid processes we associate with dissipation and irreversibility—the linear nonequilibrium flows of mass, momentum, and energy which respond to gradients—deserve relatively *simple* explanations, local in time and space. Green and Kubo have already provided such simple explanations, by carrying Gibbs' equilibrium approach one step further. Their linear-response theory of transport is the result. Green and Kubo's transport theory *is* irreversible.

The long-standing paradox that reversible equations of motion and irreversible behavior coexist can be addressed in a variety of ways. It is straightforward to apply perturbation theory to the statistical ensembles of Boltzmann and Gibbs, obtaining Green and Kubo's linear-response theory of transport. This approach is restricted to infinitesimal *linear* deviations from equilibrium. Boltzmann's equation applies arbitrarily far from equilibrium, but is restricted to low-density gases. Boltzmann "showed" that such systems invariably and irreversibly progress from less likely to more likely states, with the "state" concept loosely defined. His approximate "proof" was restricted to gases. Assume the system state can be described through the one-body distribution function f_1, and further that the system is *isolated*. Then, with a statistical irreversible assumption for the collision law, the corresponding Boltzmann entropy, $S_B = -H_B = -Nk\langle \ln f_1 \rangle$, obeys the H-Theorem—the entropy increases monotonically to reach the equilibrium value. Boltzmann certainly suggested that similar ideas should apply to liquids and solids, but in quite vague terms.

Computational thermomechanics is able to treat liquids and solids just as easily as gases, even far from equilibrium. The accuracy of this approach is primarily limited by computer time. Analytic approaches, like Boltzmann's, are a different approach to understanding transport processes. Computer simulations help us to see, understand, and surmount, many of the difficulties that can frustrate and defeat analytic approaches.

In dense fluids both coordinates and momenta have distributions qualitatively different to the equilibrium product distribution. Outside the linear-response theory, there is no useful way to approximate the many-body distribution with a product of one-body functions. Interparticle spatial correlations become dominant at high density.

We can estimate the substantial errors incurred by a *nonequilibrium* product approximation by considering the simpler equilibrium case. At equilibrium, the "excess" nonideal part of the dense-fluid entropy follows from the Mayers' virial series:

$$(\Delta S/Nk) = [S(N,V,T) - S_{\text{ideal}}(N,V,T)]/Nk =$$

$$-\sum_{n>1}[B_n + (dB_n/d\ln T)]\rho^{n-1}/(n-1) \ .$$

Here the $\{B_n(T)\}$ are "virial coefficients" giving the density expansion of the pressure:

$$(PV/NkT) = 1 + \sum_{n>1} B_n \rho^{n-1} \ .$$

For hard spheres the ten-term virial-expansion data displayed in Figure 3.5, page 107, are in convincing good agreement with computer simulations throughout the fluid phase. Both approaches show that the excess entropy (relative to an ideal gas at the same density and temperature) decreases with increasing density, reaching a value near $\Delta S_{\text{excess}} = -6Nk$ at freezing (in three dimensions—the two-dimensional excess is about $-4Nk$). This fluid-phase entropy decrease is in rough conformity to the behavior of real "simple liquids", like liquid argon. This decreased entropy means that the phase-space volume for freezing spheres is smaller than that from an ideal-gas phase-space estimate by a factor of $e^{6N} \simeq 400^N$.

We see today that the isolated low-density systems which were necessarily Boltzmann's primary interest differ from those constrained by or driven at their boundaries in two important ways: (i) steadily-driven systems eventually come to occupy attractors which have zero measure relative to the equilibrium distribution ; (ii) the entropy of such fractal states is not at all well-defined. Both these differences are consequences of the strong correlations built up by nonequilibrium steady states. Evidently the equilibrium distribution is not a promising start for describing nonequilibrium systems, except in the linear-response regime of Green, Kubo, and Onsager.

Today *algorithms* for fast computers have replaced *analysis* as our primary means of "understanding". The difficulty of calculating many-body collisions can be overcome with molecular dynamics. Distributions can be characterized and dissipation can be related to the Lyapunov spectrum. There are also related numerical methods for solving the Boltzmann equation. Today these numerical algorithmic approaches have largely replaced the outmoded analytic approaches of an earlier century. We describe a numerical approach to Boltzmann's equation in the next Section and illustrate the corresponding computer algorithm toward the end of this Chapter.

9.3 Boltzmann's Equation Today

Boltzmann's kinetic theory achieved at least four separate goals: (i) it described the "approach to equilibrium" through the H Theorem ; (ii) it provided a derivation for the resulting equilibrium (Maxwell-Boltzmann) velocity distribution ; (iii) it provided verifiable estimates for the dependence of the low-density transport coefficients on density, temperature, and the interatomic forces ; and (iv) it gave us a means for following far-from-equilibrium time evolutions of dilute gases. Today, computer simulation provides simpler alternative approaches to these same four goals, and without the restriction to dilute gases. Simulation makes it possible to follow the relaxation effects described by the H Theorem, to characterize the spatial and velocity distributions and to measure the transport coefficients. In addition, kinetic theory relates these transport coefficients to the shifts of the Lyapunov exponents in far-from-equilibrium gases and to the equilibrium correlation functions emphasized by Green-Kubo linear-response theory.

Computers have likewise revolutionized the study of dilute gases. An important computational development, largely due to Græme Bird, the "Direct Simulation Monte Carlo" method, provides the best means to solve the Boltzmann equation numerically, for discrete statistical samples of typical particles. Because the number of particles studied is finite, typically ranging from millions to billions, fluctuations which are absent in the original Boltzmann-equation approach complicate the "Monte Carlo" solutions. This many-particle method imposes a grid of cells on the system under study, with a typical cell containing about a dozen particles, each with its own time-dependent location and velocity.

Following an evolving gas in a *six*-dimensional phase-space is already too much for a serious numerical calculation, but a *three*-dimensional purely spatial grid is possible. Bird's approach eliminates the three μ-space dimensions corresponding to the velocities, making low-density gas simulations feasible. The system is divided into *spatial* cells with a size approximating the mean free path, and this cell structure is used to select collision partners.

The numerical time evolution of the system combines two separate operations during each timestep Δt: (i) a "free-streaming motion", for the time Δt, during which *all* the particles move without collisions, followed by (ii) a separate collisional update, modifying the velocities of *some* pairs of particles, those chosen for collision "during" the time interval Δt. There is an explicit example of this algorithm toward the end of this Chapter.

The collision probability for nearby pairs is assumed to be proportional to their relative speed. "Colliding" particle pairs need not be closer together than the width of a computational cell. Thus the method is inaccurate for cells larger than the mean free path between collisions. The collision parameters for pairs of particles chosen for collision in each cell are determined statistically. In two dimensions, for a fixed relative velocity, the possible collisional outcomes are tabulated according to the "impact parameter" b shown in Figure 9.2. All values of b up to the range $\sigma > b$ of the chosen forces need to be included. The necessary table, or statistical algorithm, gives the "scattering angle" χ as a function of the impact parameter b and the relative speed $|v|$. From it the post-collisional velocities can be computed.

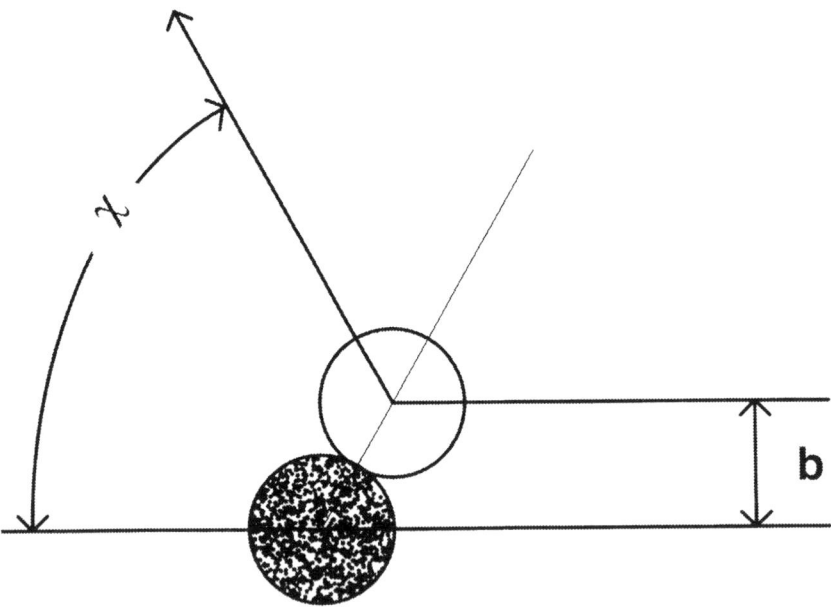

Figure 9.2. The closest distance of approach, in the absence of a collision, defines the *impact parameter* b. The change in relative velocity defines the *scattering angle* χ. The coordinates of the lefthand particle define the origin. In this coordinate system the motion corresponds to a "light" particle, with mass $(m/2)$, scattered by an infinitely-massive particle at the origin. In this example the light particle moves from right to left.

In three dimensions all the possible pre-collisional orientations, for two colliding particles with a relative velocity v and an impact parameter less than σ, fill a three-dimensional cylinder of radius σ and length $|v|\Delta t$. In two dimensions, as shown in the Figure, these orientations occupy a strip of width 2σ. Thus in three dimensions an extra angle is required. In either case the probability for a pair of particles to collide is proportional to their relative velocity. The collision probability is *assumed* to be independent of the particle positions within the cell.

Because this statistical method of selecting collisions randomly, looking up the outcomes in a table, is so much simpler than following the exact N-body dynamics, the *number* of particles which can be treated exceeds that for conventional molecular dynamics by a factor of ten to one hundred. We illustrate the simplicity of the algorithm in Section 9.16, applying it to the problem of a maximally strong—threefold compression—dilute-gas shock-wave in two space dimensions. Because the shockwave properties depend only on the coordinate in the direction of shock propagation, the programming is no more complex than is one-dimensional molecular dynamics.

Bird's Direct Simulation Monte Carlo can provide convincing flow information, at the atomic level, with only a little programming effort. There are two or three disadvantages. It is evident that both angular momentum conservation and the H Theorem are casualties of this numerical solution method. Additionally, fluctuations give rise to the same difficulties, or opportunities, as in molecular dynamics.

9.4 Gibbs' Statistical Mechanics

As experience with statistical calculations was gained, based on Gibbs' systematic formulation of statistical mechanics, it became clear that the equilibrium thermodynamic entropy simply reflects the number of accessible microstates,

$$S_{\text{Thermodynamic}} \equiv S_{\text{Gibbs}} \equiv -k \langle \ln f_N \rangle = k \ln \Omega_N .$$

In classical statistical mechanics, this result describes the entropy of liquids and solids, as well as gases, both dense and dilute. Gibbs' result expresses equilibrium thermodynamics in terms of the number of available microstates. An alternative approach to entropy is its direct thermodynamic evaluation by integration of the measured equation of state: $TdS = dE + PdV$. Molecular dynamics or Monte Carlo simulations can be used to make the required "measurements" of energy and pressure as functions of temperature and volume. The difficulties in applying Gibbs' statistical mechanics to real problems are *computational*, just as are the difficulties in applying Hamilton's motion equations.

Hamilton's motion equations point out the difficulty in extending statistical mechanics to nonequilibrium problems. Liouville's incompressible theorem,

$$\dot{f}_N \equiv (df_N/dt) \equiv (\partial f/\partial t) + \sum \dot{q}(\partial f/\partial q) + \sum \dot{p}(\partial f/\partial p) \equiv -f_N(\dot{\otimes}/\otimes) \equiv 0 \,,$$

shows that the phase-space probability density doesn't change with time. On the other hand, following the time-development of the "fine-grained" probability density f_N in a *driven* nonHamiltonian steady state shows that f_N continually *increases* over time. Hamilton's equations of motion can never give any of the *nonequilibrium* entropy increases described by the Second Law of Thermodynamics.

With thermomechanics, and the realization that $\langle \dot{f}_N \rangle > 0$ corresponds to the *fractal* rarity of nonequilibrium states, the use of Gibbs' entropy away from equilibrium—a concept even Gibbs found puzzling—became even less plausible. Chaotic dynamics furnishes a singular phase-space description of nonequilibrium distribution functions which the traditional thermodynamic investigations had not even suggested.

For the N-body distribution function, Loschmidt and Zermélo's reversal and recurrence objections to Boltzmann's proof of the H Theorem remain as powerful as ever. Isolated systems can tend toward equilibrium only briefly. Gibbs' classic text *Elementary Principles in Statistical Mechanics* begins with an echo of his 1884 talk, a careful exposition of Liouville's incompressible theorem, the basis for Gibbs' microcanonical ensemble. To make this base compatible with the Second Law Gibbs then had to introduce the "coarse-graining" of Section 8.9, page 297. This picture requires spanning phase space with infinitesimal (hyper)cubes of side length ϵ. Then Gibbs, quite plausibly and correctly, argued that an initially compact localized phase-space density would spread out and eventually contribute density to all accessible hypercubes. This useful intuitive picture of phase-space mixing, though certainly correct, does not pass muster at the *rigor mortis* level, as Krylov emphasized repeatedly in his unfinished book.

A more detailed and compelling rationale for Gibbs' coarse-graining became apparent two generations later, with the recognition that Lyapunov instability and chaos automatically provide the necessary mixing to validate the ensemble viewpoint and to render the notion of a complete deterministic trajectory illusory. Though curious, Gibbs remained leery of exploring in print the details required to construct and describe nonequilibrium systems. Such details necessarily include realistic boundary conditions and detailed descriptions at the atomic level, and so offer no convenient general analog to Boltzmann's kinetic equation for dilute gases.

After the Second World War computer simulations began to be carried out for many-body systems. In those cases using "realistic" force laws, simulations validated Gibbs' simplification of equilibrium physics, showing the equivalence of molecular dynamics' time averages and Monte Carlo's phase averages. Accurate equilibrium thermodynamic properties could be calculated either way. The agreement showed that there was no reason to doubt the validity of Gibbs' statistical approach. For nonlinear nonequilibrium physics there is no simple analytic ensemble approach like Gibbs'. But his canonical distribution did lead naturally to the calculation of the *linear* transport coefficients using first-order perturbation theory.

That "linear-response theory" or "Green-Kubo theory" computes the ensemble-averaged response (such as a shear stress or a heat flux) to an imposed macroscopic gradient (a transverse velocity gradient in the case of shear stress—a temperature gradient in the case of heat flux). The mass, momentum, and energy flows which result are all linear in the imposed gradients, providing theoretical bases for Fick's law of diffusion, Newtonian viscosity, and Fourier's heat conduction. Each of these nonequilibrium flows is "dissipative". The diffusive currents are eventually converted to heat. Without adding a thermostat to Gibbs' approach, to extract the heat generated by the perturbing gradient, there is no way to attain a true stationary state. An isolated system decays to equilibrium. A driven insulated system gradually gets hotter, as the inexorable heating continues. But because such irreversible heating is *quadratic* in the responsible gradient there is no special difficulty in obtaining the small-gradient linear transport coefficients for driven systems. Their calculated or simulated values can then be used in numerical macroscopic simulations of more-complicated nonequilibrium flows.

Of course, *linear* transport theory is not the full story. When materials are irreversibly compressed, shockwaves can result. Shockwaves are the consequence of the general increase of sound speed with increasing density. Compressive pressure waves steepen. The technical importance of shock waves to understanding high-speed flight, blast waves, high-pressure thermodynamics, and chemical kinetics, has been responsible for a host of numerical studies. Some of them have been based on continuum mechanics. Some have been based on the Boltzmann equation. Both types of studies date back to the early days of computers. Shockwave simulations in dense fluids, using molecular dynamics, were described in Section 5.7 (page 181) and Chapter 6 (page 199). *All* such shock waves are intrinsically irreversible nonlinear transitions—from an initial cold state, to a hotter, denser, higher-pressure state. The cold and hot system boundaries are relatively simple equilibrium states. This simplicity made possible the many shockwave simulations carried out since the 1940s.

9.5 Jaynes' Information Theory

Any *general* approach to *non*equilibrium states needs to handle both linear and nonlinear processes. Ed Jaynes embraced Gibbs' entropy as the basis for his own "information theory". Gibbs had already commented on the analogy between entropy and information in his boxed notes at the Yale University library. This *general* approach to constrained *non*equilibrium systems requires an extension of Gibbs' phase-space entropy from its natural isobaric isothermal equilibrium habitat, to far-from-equilibrium states. Jaynes suggested (i) maximizing the entropy subject to the restrictions that (ii) all essential nonequilibrium properties of the investigated system be included as constraints. For example, in addition to composition, energy, and volume, a homogeneous nonequilibrium system could additionally be constrained to have a specified stress component σ_{xy} or a specified heat-flux-vector component Q_x. Such nonequilibrium requirements can be imposed by using Lagrange multipliers. The multipliers led formally to new generalized Gibbs-like ensembles with nonequilibrium probability densities varying exponentially in the additional constraints.

But we know now that true nonequilibrium distributions are infinitely more subtle than this. Imposing a single stationary constraint requires, in addition, Lagrange multipliers for *all* the time derivatives of that constraint (such as $\dot{\sigma}_{xy}, \ddot{\sigma}_{xy}, \ldots$). Taking Jaynes' Lagrange-multiplier approach seriously would accordingly suggest adding another exponential controlling $\dot{\sigma}_{xy}$, and then another, constraining $\ddot{\sigma}_{xy}$, and so on. Presumably this infinite chain of restrictions implicit in Jaynes' work corresponds also to the practical translation of Zubarev's ideas for formulating nonequilibrium ensembles.

Because purely Hamiltonian equations of motion do not satisfy any useful nonequilibrium constraints—such as constant shear stress or heat flux—it appears that Jaynes' or Zubarev's extensions of Gibbs' ideas to nonequilibrium flows are so cumbersome as to be fundamentally flawed. Rather than simply including thermostats, Jaynes' approach requires the evaluation of an *infinite* number of Lagrange multipliers.

Author Bill studied the simplest possible problem, thermostated two-body color conductivity, from Jaynes' point of view, in order to quantify the errors in the information-theory approach. At low density the thermostated two-body problem can be solved by assuming a statistical distribution of collisions, just as Boltzmann did. The stationary distribution of velocities which results depends on the direction of motion relative to the accelerating field direction. This distribution, $f(0 < \theta < \pi)$, is compared to Jaynes' approximate maximum-entropy Lagrange-multiplier approach below:

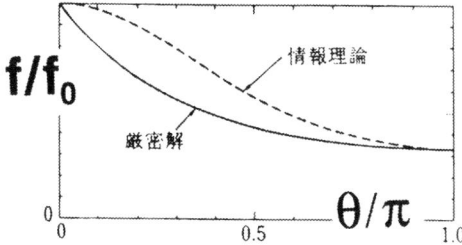

Figure 9.3. Comparison of velocity distributions (f/f_0) for two-body low-density color conductivity. Jaynes' information theory (above) does not agree with the correct solution (below) from Boltzmann's equation for $f(\theta)$ where θ is measured relative to the field direction.

The hope that Jaynes' generalization of Gibbs' statistical approach *could* describe nonequilibrium systems has a shaky basis. How could one reasonably expect to characterize nonequilibrium systems without ever describing the kinetic mechanisms necessary for maintaining them? *Dynamics is required* for an understanding of nonequilibrium systems. Think again of the Rayleigh-Bénard problem in only two space dimensions. For fixed boundary conditions this problem can have a multiplicity of different solutions, each with its own time-averaged thermodynamic and hydrodynamic properties, and each one stable for as long as one cares to wait. Evidently the maximum-entropy approach is quite useless here. The initial conditions and the chosen algorithm together determine which of the many stationary, periodic, and chaotic solutions will be observed.

For many years the only significant progress made in the direction of understanding nonequilibrium states beyond what Boltzmann had done and what continuum mechanics had explained was, we believe, Green and Kubo's linear-response theory, a development important enough to warrant a Section of its own.

9.6 Green and Kubo's Linear Response Theory

The simplest version of Linear Response theory begins with Gibbs' canonical phase-space distribution for a system with Hamiltonian $\mathcal{H}_0$:

$$f_0 \propto e^{-\mathcal{H}_0/kT} \ .$$

In the event that the Hamiltonian is perturbed slightly, the corresponding linear change in Gibbs' canonical distribution function can be evaluated by formal perturbation theory:

$$\mathcal{H}_0 \to \mathcal{H}_0 + \Delta\mathcal{H}; \ f_0 \to f_0 e^{-\Delta\mathcal{H}/kT} \simeq f_0[1 - (\Delta\mathcal{H}/kT)] \ ,$$

where terms of order $(\Delta\mathcal{H}/kT)^2$ are ignored. For example, the Doll's-Tensor Hamiltonian for periodic shear flow with the strain rate $\dot\epsilon = (\partial u_x/\partial y)$, which we mentioned in Section 3.9 and applied in Section 5.8 – page 184,

$$\mathcal{H}_{\text{shear}} = \mathcal{H}_0 + \dot\epsilon \sum y p_x \ ,$$

provides equations of motion describing a flow field with the stream velocity

$$\langle \dot x(y) \rangle = \langle (\partial\mathcal{H}/\partial p_x) \rangle = \langle (p_x/m) + \dot\epsilon y \rangle = \dot\epsilon y \ ,$$

and periodic shearing boundary conditions. The *nonequilibrium* Hamiltonian $\mathcal{H}_{\text{neq}}(t)$ is necessarily constant. The energy perturbation $\Delta\mathcal{H}(t)$ induced by the strain rate acting for a particular time interval $(0\ldots t)$ is simply the stored energy, the work $V\int_0^t \sigma\dot\epsilon\, dt'$ done by the shear strain:

$$\mathcal{H}_{\text{neq}}(t) = \mathcal{H}_{\text{eq}}(t) - \dot\epsilon V \int_0^t \sigma_{xy} dt' = \text{constant} \longrightarrow$$

$$f_{\text{neq}}(t) \propto e^{-\mathcal{H}/kT} \equiv f_{\text{eq}}(t) e^{+\dot\epsilon V \int_0^t \sigma_{xy} dt'/kT} \ .$$

When the corresponding probability density perturbation, $\Delta f = f_0(\Delta\mathcal{H}/kT)$, is used to calculate the mean shear stress at time t,

$$\langle \sigma_{xy} \rangle = \int\ldots\int \prod(dqdp) \Delta f \sigma_{xy}(t) =$$

$$+\dot\epsilon(V/kT)\int\ldots\int\prod(dqdp) f_0 \int_0^t \sigma_{xy}(t') dt' \sigma_{xy}(t) \ ,$$

we see that for short times the additional shear stress grows linearly in the time and is proportional to the equilibrium fluctuation $\langle\sigma_{xy}^2\rangle_{\text{eq}}$. We expect that the large-t limiting shear stress will approach the macroscopic value, $(\eta\dot\epsilon)$. In that case the limiting shear viscosity coefficient $\eta = (\sigma_{xy}/\dot\epsilon)_\infty$ takes on the limiting form of an *equilibrium* stress-stress autocorrelation integral:

$$\eta = (V/kT)\int_0^\infty \langle\sigma_{xy}(0)\sigma_{xy}(t)\rangle_{\text{eq}}dt \ .$$

In the small-strain-rate limit, this "autocorrelation" average can be computed numerically, using the unperturbed equilibrium distribution.

Perturbations which induce mass and energy currents similarly provide Green and Kubo's linear-response formulas for diffusion and heat conductivity. They also provide microscopic motion equations for the simulations of these nonequilibrium flows. These expressions link microscopic mechanics, properly ensemble-averaged, to the macroscopic phenomenological linear transport laws which characterize dissipative systems. This approach to the understanding of nonequilibrium systems begins with Gibbs' statistical mechanics and is perfectly general. It requires only an *equilibrium* time history. It is still true that any *practical* applications of Green-Kubo theory require computational molecular dynamics for a numerical evaluation of the transport coefficients.

The most important lessons from linear-response theory are that the irreversibility described by the Second Law requires *averaging* and an allowance for *time delay*. The sampling time in the autocorrelation integrals has to be long enough for the linear response to stabilize. From the pedagogical standpoint Green-Kubo theory is important in showing that Newtonian viscosity and Fourier's heat conduction are exact consequences of Gibbs' statistical mechanics. Specific nonequilibrium problems require more analysis. They require boundaries and thermostats incorporating the additional constraints and driving forces which distinguish nonequilibrium molecular dynamics—thermomechanics—from ordinary Newtonian mechanics. From the formal standpoint, progress beyond linear-response theory requires confronting the fine-grained fractal information content of the phase-space distribution functions which defeated Jaynes and Zubarev. Thermomechanics provides a practical means for discarding extraneous information through a computational coarse-graining, in which no more than sixteen digits need to be retained.

9.7 Thermomechanics

Steve Smale and David Ruelle each suggested relatively early, though not so early as Krylov, that the classical mechanics of Newton, Lagrange, and Hamilton would have to be generalized so as to treat and discuss nonequilibrium systems in thermal contact with their surroundings. For engineers and experimental physicists, treating nonequilibrium surroundings as sources of feedback and control forces is routine. More theoretically-inclined physicists and mathematicians have been much less likely to include this useful approach in their computational toolkits. Smale emphasized that the exponential divergence of linear trajectory separations within a bounded space—such as an energy surface in phase space—must ultimately lead to nonlinear bending and folding. The "Smale Horseshoe" mapping idealizing this deformation embodies the idea of a swirling shear flow familiar from watching clouds or mixing paint.

Molecular dynamics began with Fermi's numerical work at Los Alamos, which was continued there by Wood, and taken up by Alder and Wainwright at Livermore, and independently by Vineyard at Brookhaven. As a result, computer simulation began to augment, and sometimes replace, the older theories of both equilibrium and nonequilibrium systems. In formulating algorithms suited to fast computers, it was necessary to develop computational analogs for thermal variables, like pressure and temperature. This could have been done indirectly, by fixing the energy of selected degrees of freedom constituting an ergostat, but the parallel with laboratory practice suggested instead thermostats, based on the ideal-gas temperature scale appropriate to classical mechanics, and barostats, based on Clausius' virial theorem. These ideas led to modifications of the equations of motion in which the moments of the velocity distribution were controlled, either instantaneously, by using constraints, or on a time-averaged basis, by using integral feedback. Either way, new forces, and temperature with them, were added to Newton's equations of motion.

About 1970, computer simulations of *driven* steady-state systems, such as shear flows and heat flows, began to be carried out. These thermomechanical simulations of driven flows, using "nonequilibrium molecular dynamics", showed an overall agreement with the phenomenological transport laws, reproducing viscosities and conductivities well despite relatively large fluctuations and uncertainties. The presence of gradients, generated with real boundaries, in nonequilibrium simulations, introduced a much larger number dependence than the $(1/N)$ effects associated with spatially-periodic equilibrium conditions. Though it was not explicitly stated in the earliest work, it was soon realized that these nonequilibrium simulations used *time-reversible* motion equations and yet still obtained *irreversible* behavior. Thus the new nonequilibrium computer simulations revealed the *same* time-symmetry breaking and illustrated the *same* related paradox that had puzzled and stimulated Boltzmann, Loschmidt, and Zermélo—How *could* time-reversible motion equations yield irreversible behavior?

It is tempting to take seriously Loschmidt's argument that a time-reversible mechanics cannot lead to a future differing from the past. Stated this way, more recent investigations have shown this point of view to be naïve, though this is only known for sure in the case of time-reversible dissipative systems with phase-space compressibility, $(\langle \dot{\otimes} \rangle < 0)$. Such motion equations are *formally* reversible. Any solution of them can played back (with the coordinates appearing in reversed order, and with the momenta, and friction coefficients changed in sign). Although satisfying the *same* motion equations, all such played-back nonequilibrium solutions are unobservable and correspond to repellers. This is completely unlike the situation with classical Newtonian mechanics. So long as the boundary conditions are equilibrium ones, there are no known *stationary* Newtonian situations in which time symmetry is broken.

Of course some *transient* or *time-periodic* Newtonian problems are *effectively* irreversible—the free expansion of a cold crystal, to form a gas, is irreversible because the initial conditions correspond to zero measure in the corresponding phase space. The Newtonian shockwave problem of Chapter 6, followed by a confined free expansion, is another example, and was also discussed in Section 3.11.3, page 108. In that case the Lyapunov instability of the motion forward in time is quite different to that of its time-reversed backward twin. This shockwave problem, with its symmetry-breaking, is an excellent example of an effectively-irreversible problem solved with purely time-reversible Newtonian mechanics.

Generally, initial conditions cannot account for irreversibility. Zermélo and Loschmidt were both right—apart from zero-measure initial conditions like the perfect cold crystal isolated Newtonian dynamics cannot give long-time irreversibility. Green and Kubo's linear-response theory offers a more promising way out of the reversibility paradox. They found that the *average* response of equilibrium ensembles to small perturbations generating mass, momentum, and energy currents *can* be described by the diffusion, viscosity, and heat conductivity coefficients. It seems perfectly acceptable to take these results over to the continuum form of the conservation laws, from which the irreversibility described by the Second Law follows. The essential step is to replace time-reversible trajectory dynamics by an ensemble average of that dynamics along with a well-chosen perturbation.

9.8 The Delay Times Separating Causes from their Effects

It has been clear since Maxwell that the microscopic model of kinetic theory is not quite consistent with the macroscopic linear laws:

$$Q = -\kappa \nabla T;\ \sigma_{xy} = \eta[(\partial u_x/\partial y) + (\partial u_y/\partial x)]\ .$$

Both the time asymmetry of these "laws" and the collisional delay times before they can apply show needs for the laws' extensions and generalizations. In deriving the Green-Kubo transport coefficients the autocorrelation integrals trace the transient growth of shear stress or heat flux in response to a suddenly imposed strain rate or temperature gradient.

In molecular dynamics reversing the time instantaneously reverses also the *direction* of the heat flux (which has terms both linear and cubic in the velocities) but leaves the temperature gradient (which is quadratic in the velocities) unchanged. In molecular dynamics reversing the time reverses also the strain rate (linear in the velocities) but leaves the stress (quadratic) unchanged. These contradictions show that the two descriptions are incompatible in their details. Fourier's Law and Newtonian viscosity both suggest an instantaneous response to imposed velocity or temperature gradients, while a little reflection suggests instead that transport involves a diffusive relaxation process which reduces propagation by a scattering process described by a mean free path.

This more detailed look at transport phenomena shows that the macroscopic linear laws are oversimplifications. They omit the collisional delay times which separate causes from their effects. Maxwell estimated the time required for air to move in response to an unbalanced stress—or equivalently, the time for the stress to respond to deformation. This time, "Maxwell's relaxation time", is the ratio of the shear viscosity to the bulk modulus, about two hundred picoseconds for air. The same idea applies to the time lag of heat flux in response to an imposed temperature gradient.

Unless these delays are taken into account in the constitutive models, the macroscopic flow equations suggest *supersonic* propagation of stress and heat waves according to the linear-transport diffusion equation. For example, the diffusion of shear stress in the x direction, where the kinematic

viscosity is $\nu = (\eta/\rho)$, follows from:
$$(\partial \sigma/\partial t) = \nu \nabla^2 \sigma \longrightarrow \sigma(x,t) \propto (e^{-x^2/4\nu t}/\sqrt{4\pi \nu t}) ,$$
suggesting that the effect of a stress initially localized at the origin extends to *all* x a short time t later. Evidently some delay needs to be incorporated into the diffusion equation for it to make physical sense.

A simpler example, with an explicit delay, is the acceleration of a mass m, subject to a viscous drag force $-(mv/\tau)$ as well as to an external field E, turned on at $t = 0$ with the mass at rest. The corresponding solution of the motion equation is:
$$m\dot{v} = E - (mv/\tau) \longrightarrow v = (E\tau/m)[1 - e^{-t/\tau}] .$$
Maxwell and Cattaneo both used linear relaxation equations to describe delay, Maxwell for stress and Cattaneo for heat flux:
$$\sigma + \tau\dot{\sigma} = \eta\dot{\epsilon} \text{ or } Q + \tau\dot{Q} = -\kappa\nabla T .$$
These Maxwell-Cattaneo equations provide a phenomenological bridge joining the conventional linear laws with kinetic relaxation.

Steady-state shockwave simulations produce direct evidence for the time delay. As we detailed in Chapter 6, stationary shockwaves can be generated by introducing a rightmoving particle flux at the left boundary and extracting the resulting slower hotter denser flux at the right boundary. When the mass input flux is properly tuned to match the output the left and right mass, momentum, and energy fluxes all match:
$$\{\rho u_x;\ P_{xx} + \rho u_x^2;\ \rho u_x[e + (P_{xx}/\rho) + (u_x^2/2)] + Q_x\} \text{ all constant} .$$
Close examination of the results shows that the stress and heat flux do lag behind the velocity and temperature gradients which drive them.

9.9 A Fluctuation Theorem

Short-time fluctuations can provide unlikely wrong-way *reversals* of the linear laws, not just time delays. These unlikely events can be related to Landau and Lifshitz' fluctuation theory and to Green and Kubo's response theory. The added features are the concepts of Lyapunov instability, periodic orbits, and phase-space attractor-repeller pairs. This combination can provide a quantitative understanding of short-term (but not too short!) observations *opposite* to the direction expected from the Second Law of Thermodynamics. These highly-unlikely fluctuations are limited in size to a few multiples of Boltzmann's constant in the case of Gibbs' entropy and to a few multiples of kT in terms of work and heat.

An approximate "fluctuation theorem", described by Cohen, Evans, and Morriss in 1993, gives the relative probability of observing trajectories forward in time to their time-reversed backward twins in terms of the (external) entropy produced:

$$\ln\left[\frac{\text{prob}_f(+\sigma\delta\tau/k)}{\text{prob}_b(-\sigma\delta\tau/k)}\right]_{\delta\tau} = \ln\left[\frac{\text{prob}_f(+\sigma)}{\text{prob}_b(-\sigma)}\right]_{\delta\tau} = +\langle\sigma\delta\tau/k\rangle\ .$$

Here the trajectories forward or backward in time produce or absorb an entropy $S = \sigma\delta\tau$ during the observation time interval $\delta\tau$. σ is the "entropy production rate". Apart from a factor of two this simple result looks very like Gibbs' association of probability with the exponential of $(\Delta S/k)$.

Evidently the "theorem" cannot apply for short times. As the observation time window goes to zero, $\delta\tau \to 0$, it is evident that the probabilities of forward and backward trajectories are *not* at all equal. Like the Central Limit Theorem for Gaussian probability densities, the Fluctuation Theorem only holds for sufficiently long times.

The derivation of the theorem is simplest if we blur the distinction between orbits and distributions. Imagine that the stationary phase-space distribution contains a long attractive periodic orbit and includes also a repulsive twin corresponding to the time-reversed orbit. τ is the orbit period. Consider now the rates at which probability flows between the neighborhoods of these two periodic orbits, attractive and repulsive. The mean probability on the attractive orbit is f_A and that on the repulsive orbit is f_R. The stationary flow rate away from the attractor necessarily balances the flow rate toward the attractor from the repeller. Both these escape rates can be expressed in terms of the *positive* Lyapunov exponents. These exponents describe the effect of the *expanding* flow directions in the neighborhoods of the periodic orbits:

$$f_A \exp[\sum_{\lambda_A > 0} \lambda_A \tau] = f_R \exp[\sum_{\lambda_R > 0} \lambda_R \tau]\ .$$

Evidently the ratio of the forward-to-backward probabilities, (f_A/f_R) is given by the *complete* sum of Lyapunov exponents:

$$(f_A/f_R) \equiv \exp[\sum_{\lambda_A>0} \lambda_A \tau]/\exp[\sum_{\lambda_R>0} \lambda_R \tau] = \exp[\sum \lambda_A] = e^{\langle\dot{S}\rangle(\tau/k)}\ .$$

This simplification, which *is* the Cohen-Evans-Morriss Fluctuation Theorem, follows from the symmetry of the Lyapunov exponents — for every positive exponent on the attractor there is a corresponding negative exponent of equal magnitude on the repeller, and *vice versa*.

The attractor is certainly equivalent to a *sufficiently long* periodic orbit. We can increase the length arbitrarily by increasing with the number of significant figures kept in the calculation. For a nonequilibrium Galton Board problem first solved by Thomas Gilbert, we generated periodic orbits using from 3-digit to 12-digit accuracy. The number of collisions for these ten orbits varied from 774, for three digits, to 793,951,594 for 12, with the 12-digit periodic orbit requiring 447,064,397,614 Runge-Kutta timesteps of length 0.0005.

These data can be used to test (for skeptics), or to illustrate (for believers), the Fluctuation Theorem. The relative probabilities of entropy production and consumption can be computed by sampling exhaustively *all* the windows of a fixed length $\delta\tau$ on each of the periodic orbits. The numerical data are shown in Figure 9.4 for our 12-digit orbit. The Figure shows that the ratio of probability logarithms *is approximately linear* for sampling windows in the range $\{2 \leq \delta\tau \leq 50\}$. The ratio for the shortest window shown is definitely *non*linear. The window length of 50 is about as long as is practical with current workstations. We found no instances, in the entire orbit of nearly a billion collisions, in which the longer window length, $\delta\tau = 100$, contained a negative entropy production.

The form of the Fluctuation Theorem's probability ratio can be compared to the same ratio taken from the Central Limit Theorem, which also applies for sufficiently long sampling times $\delta\tau$. The Central Limit Theorem predicts a *Gaussian* distribution of entropy production rates:

$$\ln\left[\frac{\text{prob}_f(+\sigma)}{\text{prob}_b(-\sigma)}\right]_{\delta\tau} = -\frac{(+\sigma - \langle\sigma\rangle)^2}{2\Sigma^2} + \frac{(-\sigma - \langle\sigma\rangle)^2}{2\Sigma^2} = \frac{2\sigma\langle\sigma\rangle}{\Sigma^2}.$$

Here Σ is the Gaussian distribution's standard deviation. The distribution of probabilities for the longest feasible sampling time reveals substantial discrepancies with the Gaussian form. We conclude that the Fluctuation Theorem complements the Central Limit Theorem by providing a semiquantitative description of wrong-way trajectories' probabilities. This is a relatively rare circumstance: a nonlinear result valid even far from equilibrium. To the extent that the Theorem is *true* we can say we "understand" the Second Law of Thermodynamics as applying for sufficiently long time-averaging windows, a time of 100 in the case of the Galton Board example.

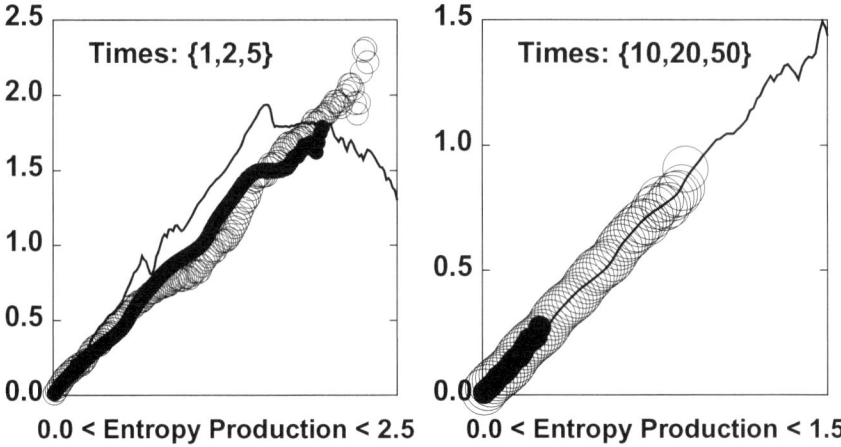

Figure 9.4. $\ln[\text{prob}(+\sigma\delta\tau/k)/\text{prob}(-\sigma\delta\tau/k)]$ as a function of the entropy production rate (σ/k) averaged over the six time intervals mentioned in the text. For sufficiently long times the unit slope is consistent with the prediction of the Fluctuation Theorem.

9.10 Are Initial Conditions Relevant?

Another possible, but highly-artificial, way out of the reversibility paradox is to choose special initial conditions which *agree* with thermodynamic behavior, and to rule out any which do not. This approach seems to be a common refuge of philosophers who are unfamiliar with the insights furnished by computer simulation. The ultimate hard-to-prove modification of this reliance on initial conditions is to regress back to the Big Bang as the initial condition. This provides a low-entropy initial condition for the evolving parts of the Universe we have the luck and skill to observe. A second, more down-to-earth approach focuses instead on simple and idealized conservative systems.

Lawrence Sklar is a philosopher with a gift for explaining physics. Jean Bricmont is a staunch articulate defender of truth and simplicity in research. Both these men have argued that *the* initial conditions are important to any understanding of irreversibility. This point of view results from the desire to "understand" irreversibility by analyzing isolated systems obeying Newtonian or Hamiltonian dynamics. If one imagines that the hypothetical state of an isolated system is *exactly* known, then that state is *in principle* linked to others at all previous and all future times. This point of view suffers from the need to imagine an initial condition so precise as to contain the entire system trajectory for all previous and future times. This infinitely-detailed "view" of the system, including its entire history, parallels Price's "Block Universe", discussed below. We view the information content of such constructs as so outrageously unreasonable as to invalidate the entire argument. Given chaos, we see no way to reasonably study, or even contemplate, irreversibility based on the evolution of a *precisely-known* isolated system. "Exact" trajectories are a casualty of Lyapunov instability.

Huw Price is likewise a philosopher, and also feels that *the* "initial conditions" (of the universe!) are the crucial thing to understand. Price champions the view that time is not so different from space (where the positive and negative directions are arbitrary choices). Price adopts the "Block Universe" point of view, in which the "state of the universe" is laid out in both space and time, including all of space and all of time, both past and future. With this task accomplished, he then sees no special difference between positive and negative directions, for either type of variable. He seems to miss entirely the necessity for defining, or accepting, a "subjective" time in order to *formulate* the underlying differential equations for a physical theory. Price believes that our quest for an understanding of dissipation and irreversibility is both futile and outmoded.

> *For all their intrinsic interest, the new methods of nonlinear dynamics do not throw new light on the asymmetry of thermodynamics. Writers who suggest otherwise have failed to grasp the real puzzle of thermodynamics—Why is entropy low in the past?—and to see that no symmetric theory could possibly yield the kind of conclusions they claim to draw.*

Price feels that the tendency of systems to approach equilibrium, the most likely state, is pretty well understood, and that the "Big Question" is how the Universe happened to start out in such an unlikely state. It seems to us that this "big question" is one of the "meaningless questions" (ill-posed questions, for which the path to an answer is inconceivable) about which Paul Schmidt warned Bill in his undergraduate philosophy class at Oberlin. It has become very clear to us, based on examples of the kind worked out in this book, that a simple understanding of irreversible processes can be based on the inherent relative stability of processes obeying the Second Law of Thermodynamics and the much greater *instability* of the time-reversed unobservable repeller versions of these same processes.

Thermodynamically irreversible processes involve the dissipation of energy and the growth of entropy. Continuum mechanics can incorporate a variety of processes, stationary or not. From the formal standpoint any process which conserves mass can be converted into a time-periodic process *if* it is possible for work to be done and heat to be extracted or added to the final state in such a way as to regain the initial state. Thus studying a time-periodic process is enough. In laboratory experiments it is an article of faith that the results depend only upon the *macroscopic* initial state, and not upon details at the atomic level. This same attitude has been adopted by those whose experimental medium is the fast computer. A clever choice of initial conditions can often reduce the time necessary to attaining a desired stationary state, but can never prevent it. The evolution algorithm, including boundary conditions, constraints, and driving forces, should be chosen so as to (i) cause the results *not* to depend upon the details of the initial state and to (ii) make analyses of the simulation as straightforward as possible.

The choice of boundary types is quite arbitrary. For example, a system can be contained within a boundary composed of solid or fluid particles interacting with relatively arbitrary force laws and subject to arbitrary mechanical and thermal constraints. Thermal constraints can be imposed by differential or integral feedback or by stochastic choices of velocities or accelerations. Among all these arbitrary choices the simplest forms of feedback lead to the simplest analyses.

Purely-theoretical analyses lack the efficiency, scope, and relevance of numerical steady-state simulation studies. *At* equilibrium, interesting fluctuations are quite rare. Choosing an initial state far from equilibrium results in early-time behavior dominated by the chosen initial conditions, and late-time behavior dominated by equilibrium fluctuations. It is arguably simpler, and certainly more educational, to follow the experimentalists, designing algorithms incorporating boundaries, constraints, and driving forces so as to study particular nonequilibrium states.

It seems to us that an emphasis on the importance of initial conditions is doomed to failure for the two reasons cited by Zermélo and Loschmidt in their criticisms of Boltzmann's work on irreversibility. Eventually, all isolated systems *will* return [Zermélo], and any forward-running trajectory *is* equally-likely to be found running backward [Loschmidt]. The presence of chaos also implies, as best articulated by Joseph Ford and Ilya Prigogine, that the very concept of *exact* initial conditions can have no operational significance. Thus, a meaningful discussion of physical correlations and evolution requires that the results *not* depend on all the unobservable details of the initial conditions.

Because the usual Hamiltonian systems are fundamentally inadequate for treating the very simplest nonequilibrium systems, steady states, and because the details of the initial conditions are quickly lost, by contraction below any observable scale, reliance on these initial conditions is a highly unproductive and artificial point of view. It needs to be realized that chaos makes computer simulation our *primary* source of knowledge for predicting the future and connecting that future to its past. It is just as wrong to imagine that particular computer algorithms are the "cause" of the outcome as it is to ascribe macroscopic results to particular microscopic initial conditions. Both initial conditions and algorithms need to be irrelevant to the macroscopic outcome. The checks of *any* reasonable computational approach are precisely those of laboratory experiments, reproducibility, intelligibility, and consonance with experience.

At the 1982 Boulder meeting John Barker made the case for explicit *nonequilibrium* simulations as follows:

> *It seems to me that to constrain the molecular dynamist to use only Green-Kubo is somewhat analogous to constraining the experimentalist to measure the viscosity by studying only the decay of spontaneously arising velocity fluctuations, e.g., by light scattering. This makes for time consuming and expensive experiments. In actual fact, of course, we permit the experimentalist to put his hands in the apparatus and impose a finite shear. I do not see why we should not allow the molecular dynamist to do the same. Naturally, we should expect him to use all necessary corrections and extrapolations. If this requires a theory of its own, so be it! We will learn something this way.*

9.11 Constrained Hamiltonian Ensembles

Bob Dorfman, Pierre Gaspard, and Gregoire Nicolis have developed their own novel boundary-based approach to "bridging the gap" between microscopic mechanics and continuum hydrodynamics. They consider the loss of the mass, momentum, and energy escaping from a region within which the usual conservative mechanics applies. For a sufficiently large region, the long-time-averaged *escape rate* satisfies a macroscopic diffusion equation, establishing a connection with standard macroscopic hydrodynamics. This very simple physical idea is the basis of their "escape-rate theory".

The particle escaping from a finite Galton Board, in Figure 1.3 on page 26, is a small-scale example of this approach. Escape-rate theory leads to interesting mathematical conclusions. As time goes on, the particles remaining within the Newtonian region dwindle, so that their phase-space states become more-and-more special, eventually reducing to multifractal repeller distributions somewhat similar to those found in nonequilibrium steady states.

The escape-rate approach seems not to be intended to facilitate simulations or to attempt practical applications of nonequilibrium systems. It is instead an intellectual and relatively formal attempt to understand the physical meaning of the great (fractal) rarity of nonequilibrium states in terms of standard Newtonian mechanics. It is significant that this approach leads also to a fractal explanation of the dissipative states underlying the Second Law of Thermodynamics. Otherwise, the complex style of the approach, and the lack of significant applications beyond Green-Kubo theory, limit the appeal of the work.

Escape-rate theory bears a family resemblance to periodic-orbit theory. Rather than attack the problem of nonequilibrium states headon, by adding and analyzing boundary conditions, driving forces, and thermostats, an entire *equilibrium* ensemble is considered. Then *almost all* of it is gradually discarded, bit by bit, until only the relevant fractal part remains. This gradual pruning away of the equilibrium distribution also characterizes Jaynes' approach, with an infinite number of Lagrange multipliers, and the periodic-orbit approach, with an infinite number of unstable orbits. In all these cases the desired states are those few remaining after all others, the vast majority, have been discarded. The pruning process, corresponding to the loss of information is certainly an important part of irreversibility. But it is far better to allow the irrelevant information to depart quietly, through the negative Lyapunov exponents, rather than taking on the relatively arduous task of *analyzing* the details of that loss. Gödel guarantees that such a detailed analysis will ultimately have to fail.

9.12 Anosov Systems and Sinai-Ruelle-Bowen Measures

David Ruelle has discussed the possibility that the useful properties exhibited by certain oversimplified and quite rare dynamical systems, termed "Anosov systems", have counterparts in the more usual thermostated systems studied with nonequilibrium simulation methods. Anosov systems are oversimplifications, like square clouds or spherical chickens.

There is one "realistic" example Robert MacKay kindly pointed out to us. To emhasize its physical nature he had a physical model with him that fitted into an attaché case. The physical model closely corresponds to the "triple-linkage" problem of Figure 9.5. This model, analyzed by MacKay and Tim Hunt, his student, consists of three coplanar rotating disks, linked together by off-center rods.

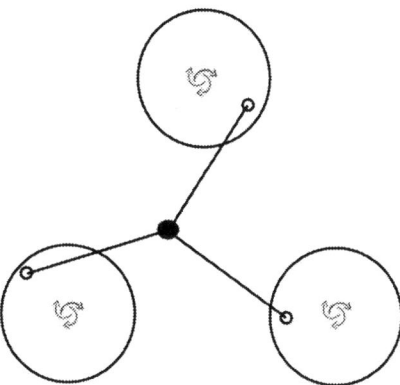

Figure 9.5. An Anosov System from the work of Bob MacKay and Tim Hunt. The configuration of the system can be described by the three angles together with the coordinates of the central point. These five coordinates and their five time derivatives, give a ten-dimensional description. The three length constraints, and *their* time derivatives, plus the constraint of constant energy reduce the motion from ten dimensions down to a manageable three-dimensional space, where chaos can occur.

Another Anosov system, not so well suited to an attaché case, is a point mass constrained to travel a "surface of negative curvature", a surface, like the inner part of a doughnut, which *always* looks locally like a saddle. The defining property of Anosov systems is that they are "uniformly hyperbolic", with expanding and contracting saddle-like manifolds corresponding to their positive and negative Lyapunov exponents. Because nonequilibrium dissipative conditions correspond to an overall *negative* shift of the Lyapunov exponents the number of negative exponents typically exceeds the number of positive ones for large systems with a continuous spectrum. This rather common feature of time-reversible dissipation is evidently incompatible with the Anosov picture. In typical molecular dynamics simulations, just as in Lorenz' Attractor trajectories, there is an incessant sign changing and reordering of the local Lyapunov exponents, associated with the wildly chaotic rotation of the "offset vectors" $\{\delta\}$ described in Section 8.4.

Although realistic differential equations cannot be expected to display the Anosov behavior, there are examples for maps, where discontinuities abet chaos. The Baker Map is the best illustration of consistent hyperbolic behavior, with a smoothly expanding unstable manifold and a stable contracting manifold along which the long-time-averaged distribution becomes fractal. Though smooth differential equations with the Anosov property are quite rare, it is admittedly stimulating to imagine and discuss them, with the avowed goal of gaining a better understanding of physics.

To date, there have been two "results" of the Anosov effort, both retrospective. Both these theoretical results seem to us to have more to do with time reversibility than with the unrealistically-smooth and near-vanishingly-rare manifolds characterizing Anosov systems. The first result is a proof that "conjugate pairing" holds in nonequilibrium steady states. Conjugate pairing means that the sums of corresponding pairs of Lyapunov exponents—pairs which would always sum to zero at equilibrium—are likewise the same, for *all* pairs, away from equilibrium:
$$\lambda_1 + \lambda_N = \lambda_2 + \lambda_{N-1} = \lambda_3 + \lambda_{N-2} = \cdots = \lambda_{N/2} + \lambda_{(N/2)+1} \ .$$
So far as we know, such pairing results are invariably restricted to spatially homogeneous systems. Similar results hold also for non-Anosov, but homogeneous, Hamiltonian mechanics, with frictional damping added, *if* all the Cartesian momenta are damped with the same constant friction coefficient.

The second retrospective result obtained from the Anosov studies is the Evans-Morriss-Cohen Fluctuation Theorem, valid at long times according to the detailed-balance periodic-orbit argument of Section 9.9. Both these retrospective "results"—the Lyapunov-exponent pairing and fluctuation theorem—have been proven rigorously *for reversible Anosov systems*, despite the specialized rarity of such systems.

These same "results" were actually given earlier by Denis Evans and several of his coworkers, for more general circumstances and through more elementary arguments. Because the simple geometric argument of Section 8.8 shows that nonequilibrium attractors are actually generated by *any* stable, time-reversible, steady dynamics, the applicability of the Anosov proofs is limited to very special cases.

The oversimplified phase-space distribution associated with a dissipative Anosov system would evidently be somewhat intermediate between the volume-preserving flows of Hamiltonian mechanics and the attractor-repeller pairs of time-reversible nonequilibrium molecular dynamics. Sinai, Ruelle, and Bowen established the properties of such "SRB" distributions and discussed the Lyapunov spectra for such hypothetical hyperbolic Anosov systems. The static hyperbolic structure gives a Lyapunov spectrum with a very simple structure, with fixed numbers of positive and negative local exponents. If the dynamics were also time-reversible then these two numbers would necessarily be equal. Analogs of the established relations were observed earlier in sufficiently simple computer simulations, typically homogeneous simulations with periodic boundary conditions.

Theoretical constructs, such as "measures", should be viewed with a healthy suspicion until algorithms for evaluating them are supplied. The chaos inherent in *interesting* differential equations guarantees that our only access to the "strange sets" which constitute attractors and repellers will be representative time series from dynamical simulations. In no way can we construct, or even conceive of constructing, a Sinai-Ruelle-Bowen measure for an interesting system. On the other hand for very small attractors we *can* make rudimentary low-dimensional checks for ergodicity. For more interesting attractors, with dimensionalities of a few hundred, we can also obtain the entire Lyapunov spectrum and so estimate the attractors' Kaplan-Yorke dimensions. Occam urged us to avoid the extraneous. Bridgman cautioned us to avoid adopting concepts which are not susceptible to an operational investigation. It is not surprising that Bridgman singled out set theory for special criticism.

More recently, Gallavotti and Cohen have emphasized the "nice" properties of Anosov systems. Rather than finding realistic Anosov examples, they have instead promoted their "Chaotic Hypothesis": if a system behaved "like" a [wildly-unphysical but well-understood] time-reversible Anosov system, there would be simple and appealing consequences, of exactly the kind mentioned above. Whether or not speculations concerning such hypothetical Anosov sytems are an aid or a hindrance to understanding seems to be an æsthetic question.

9.13 Trajectories *versus* Distribution Functions

Ilya Prigogine devoted most of his lifelong research effort to supporting and promoting the study of "complex systems". In the 1980s he was a champion of the significance of Hopf's Baker Map. In the 1990s he made strong claims for the merits of a formal approach which is nearly disconnected from the analysis of experiments or simulations. John Dougherty has kindly attempted to provide a perspective on Prigogine's work, critically comparing it to the information theory of Jaynes and Zubarev. See also Zetie's perceptive review. Prigogine finally adopted the very reasonable view that trajectories are themselves not a proper description of dynamics. His reasoning is quite consistent with Joseph Ford's longstanding observation—the information content of a chaotic trajectory is inaccessible in principle. This is because the information required to reverse such a trajectory grows in exact proportion to the trajectory's length. For this reason Prigogine chose instead to emphasize the definition and evolution of distribution functions. He argued, quite justifiably, that distribution functions are the more faithful representation of the inherent uncertainty of physical theories.

Prigogine went on, claiming to find that the time evolution of distribution functions can be expressed in terms of two one-way "semigroup" structures. This gives a time-symmetry breaking analogous to the Second Law of Thermodynamics. Perhaps it is no coincidence that his point of view is not so different to that which comes directly from computer simulations. We guess that his symmetry-breaking "semigroup" approach corresponds to neglecting the unobservable repellers, and selecting the inevitable attractors, which latter are by now so familiar from simulation work with time-reversible thermostated systems.

The time-reversible deterministic computer experiments, beginning with the Galton Board study in 1986, showed that the trajectories *can* be followed stably in only one of the two time directions, *despite* time-reversibility. Further, the typical ergodicity of the computer simulations indicates that the trajectory and distribution-function approaches provide *identical* coarse-grained solutions. Prigogine's "rigged Hilbert spaces" for quantum systems probably correspond to the constrained fractal distributions which simulations generate for stationary nonequilibrium systems.

9.14 Are Maps Relevant?

Both the equilibrium and the dissipative Baker Maps were instructive aids to understanding in the early days of chaos. Elaborations of these maps continue to be studied today, particularly as models of spatially-extended systems. Chaotic ergodicity in maps requires only two variables, while differential equations require three, or possibly four, for a continuous situation. The dissipative but time-reversible Baker Map embodies many of the same paradoxical qualities as do its many-body dynamical analogs. The breaking of time symmetry is particularly transparent for maps.

The long recurrence times in numerical work can be modeled accurately by stochastic jumps between Ω microstates. Even that simplest of chaotic continuous systems, the Galton Board, has so many distinct microstates that an exhaustive exploration with 16-digit double precision arithmetic lies near, and probably beyond, the margin of what can be done today.

The message delivered by the maps is clear enough: stochastic models are sufficient for a qualitative explanation of typical computational results for deterministic systems. This message implies that Gibbs' and Boltzmann's entropy analyses are generally good enough for describing even very simple systems.

The "enclosed shockwave + free expansion" problem of Chapters 3 and 6 is a good example of irreversible behavior obtained with Newtonian-Hamiltonian motion equations. Here we again imagine two cold blocks of material, one moving right and the other to the left, and occupying the two halves of the periodic unit square, x negative and $\dot{x}$ positive for one block—x positive and $\dot{x}$ negative for the other. As the blocks make contact a pair of shockwaves is launched along the contact line at $x = 0$. A little later the blocks are slowed to a stop by the propagating shockwaves. If the initial block velocities are properly chosen the density can double. In any case the initial kinetic energy is converted to internal energy. Next, in the final phase of the motion, the hot fluid expands twofold, again filling the square, but this time as a quiescent homogeneous warm fluid.

All this highly-complex motion converting the two cold blocks to a warm homogeneous fluid is in principle time-reversible. In fact the system's past history, backward in time and prior to the initial condition, could be made an *exact* mirror-image of its forward motion! We explored a bit-reversible segment of this problem, forward and backward in time, in Chapter 6, page 236. There we found an "Arrow of Time", with the forward Lyapunov vectors quite different to their more complex backward relatives.

Similarly, but much more simply, the equilibrium Baker Map converts a square, half black and half white, to a uniform grey, with the scale of the black/white mixing diminishing exponentially with time. If the phase space is finite, as with finite-precision arithmetic, then the motion necessarily repeats, with the repeat time increasing exponentially with the square root of the precision of the calculation. These problems have a family resemblance to familiar illustrations of Poincaré recurrence: a half-full container of gas expanding isoenergetically to fill its container, or a coin-flipping problem beginning with 100 heads and no tails, waiting for a repetition of that unlikely initial condition. The stochastic and deterministic problems are alike in that recurrence times, with Ω states, are of order $\sqrt{\Omega}$, just as in the "Birthday Problem". [With 23 people and random birthdays the odds are favorable that two have the same Birthday.]

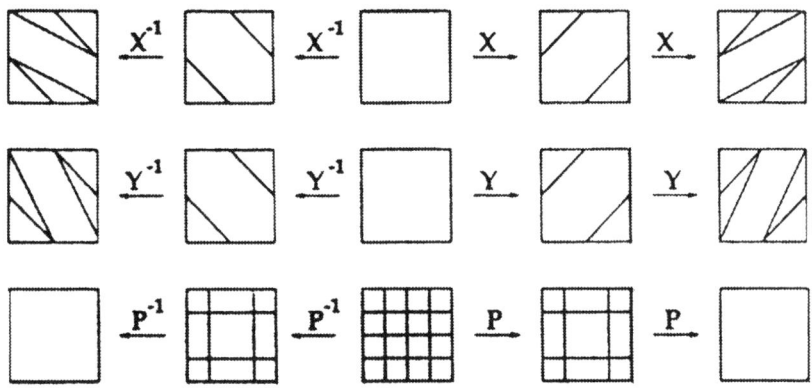

Figure 9.6. Time-reversible shear (X, Y) and reflection P maps in two space dimensions. Each of these three operations is "time-reversible" if the vertical y coordinate changes sign in the reversed mapping.

There is evidently an irresistible temptation to make maps more and more elaborate, to approach the complexity of continuum simulations by increasing the number and complexity of spatially distinct regions. *Time-reversible* maps must obey the defining relations:

$$M(x, y) = (x', y') \longleftrightarrow M(x', -y') = (x, -y) ,$$

where the "coordinate" x is unchanged by time reversal while the "momentum" y changes sign. Shears and reflections are the simplest examples of time-reversible maps. See Figure 9.6. *Any symmetric combination* of

time-reversible maps, such as
$$M_{12321} \equiv M_1 \times M_2 \times M_3 \times M_2 \times M_1 \;,$$
has also the time-reversibility property.

Maps tell us that symmetry breaking results in the net destruction of information, with an overall *decrease* in area corresponding to a concentration of probability density. The strange attractors which maps generate display a continuous *loss* of information. By using concatenated *chains* or *arrays* of maps, the shrinking can also diffuse to the interior of the chain as the result of boundary interactions. See, for example, the work by Vollmer, Tél, and Breymann.

The shrinking characterizing an attractor means that initially important information becomes gradually less so. At the same time the underlying chaos in a *strange* attractor generates expansion and the emphasis of initially unimportant information. Both properties, destruction of initially *relevant* information, and increasing importance of initially irrelevant information, characterize nonequilibrium dissipative systems, and are faithfully caricatured by maps. Thus maps *are* useful and relevant to an understanding of deterministic time-reversible dissipative systems. To a large extent this role has already been filled by the Baker Maps. Maps *are* indeed a generally useful tool for investigating topological effects in a few dimensions and in investigating the behavior of spatially-extended aggregates.

9.15 Irreversibility ⟵ Time-Reversible Motion Equations

It is often stated that "time-reversible equations cannot lead to irreversible behavior". Though seemingly obvious, this statement is *false*. Bill learned this in 1986, by looking at Bill Moran's computer simulations of the isokinetic Galton Board problem described in Section 5.9.1, page 189. Independent of the detailed initial conditions, the Galton Board invariably generated the *same* fractal distributions and the *same irreversible* dissipative behavior, despite the time reversibility of the underlying equations of motion. Because the dynamics of the Board can be displayed in two dimensions, by cataloging collisions according to two characteristic angles, its evolution is easy to visualize. This simplicity is shared with the evolving planar distributions described by two-dimensional maps.

The irreversibility exhibited by the Galton Board is this: invariably, for any initial conditions, stable chaotic flow states give rise to a current extracting energy from the field and dissipating it in the form of heat. The heat is extracted by a time-reversible thermostat variable ζ. The time-reversed flow, satisfying the same motion equations, corresponds to an unstable unobservable multifractal repeller, with heat converted into work and with ζ changed in sign.

If one were to take the point of view that reversing time is simply replacing $+\Delta t$ by $-\Delta t$ in integrating the equations of motion then the dissipative flow just described would be occurring for decreasing time. But this artificial mathematical way of "reversing time" is unacceptable, for it discards the connection between velocities and coordinates, $\{\dot{r} \equiv +v\}$, in favor of $\{\dot{r} \equiv -v\}$. The physicist accepts as basic an intuitive flow of time, including its "subjective" or "psychological" "arrow of time", in distinguishing the past from the future and in making it possible to write kinetic differential equations of motion. With this distinction in place, and with time flowing forward, he then can describe the changes resulting from time-reversible laws. Dissipation is the generic result, as is evident from the Galton Board example.

The dissipative Baker Map is an even simpler, though more idealized, example of the irreversible behavior hidden within time-reversible motion equations. Heat flow is a more physical example and involves no macroscopic mass currents. Analyses of heat conduction through a Newtonian region, sandwiched between two time-reversible heat reservoirs, one hot and one cold, provide convincing demonstrations of time symmetry lost through reversible dissipation. Such an example was considered in Section 5.9.3.

The *cause* of time-symmetry breaking is the systematic destruction of the *information* required to reverse a trajectory. Information corresponds to phase-space localization so that any contracting action in the phase space gives additional information. But, analogous to a coarse-grained cell in phase space, computers operate with a fixed wordlength, and have no way to accomodate all of this additional information. The loss of information which results becomes clear only gradually, for at any stage in a long trajectory, it is possible to reverse the trajectory reasonably accurately for several collision times. On the other hand, the distribution function obtained from a long-time-averaged trajectory is *not* symmetric in time, and invariably shows a net dissipative current parallel to the field direction. This current was already explained by Green and Kubo.

Because time reversal has no effect at all on the equations of motion, any attempt to find the past history of the system, at increasingly negative times, generates a "reversed" trajectory identical to the forward trajectory. When analyzed in the "forward" time direction, the reversed sequence of trajectories becomes a fractal repeller, differing from the attractor only in the signs of the velocities. This same symmetry applies to a heat transfer problem where the flow of heat in the "reversed" trajectory is from cold to hot, leading to a *negative-conductivity* repeller differing from the positive-conductivity attractor only in the sign of the heat flux component parallel to the temperature gradient. A reversed shear flow is slightly more complicated. In the formal time-reversed motion the strain rate changes sign while the shear stress does not. The quotient would give a negative shear viscosity, emphasizing the instability of the corresponding repeller. A time-reversed mass flow likewise corresponds to a *negative* diffusion coefficient.

9.16 Boltzmann-Equation Shockwave-Structure Algorithm

We illustrate a simple shockwave solution by considering the maximum (threefold) compression of a cold monatomic ideal gas, with the two-dimensional equation of state $PV = Ne = NkT$. The shock compression of a cold two-dimensional ideal gas is necessarily threefold, as follows from the Hugoniot relation:

$$\Delta E = \Delta V (P_c + P_h)/2 \longrightarrow$$

$$E_h - E_c = E_h = P_h V_h = (V_c - V_h)(P_h + P_c)/2 = P_h(V_c - V_h)/2 \longrightarrow$$

$$V_h = (V_c/3) \ .$$

Conditions satisfying the Hugoniot energy-conservation equation are:

$$u : [3 \to 1]; \ \rho : [1 \to 3]; \ P : [0 \to 6]; \ e : [0 \to 2] \ ,$$

where the velocity u is measured in the frame comoving with the wave. In an alternative frame, fixed on the hot material, a cold gas moving with a velocity of 2 stagnates against the wall and the kinetic energy per unit mass, $2^2/2 = 2$, is converted into internal energy. The mass, momentum, and energy fluxes are all constant in the frame fixed on the shockwave:

$$\rho u_x = 3; \ P_{xx} + \rho u_x^2 = 9; \ \rho u_x [e + (P_{xx}/\rho) + (u_x^2/2)] + Q_x = (27/2) \ .$$

Rather than simulate the shockwave with molecular dynamics we solve the Boltzmann equation for this problem by using a Monte Carlo technique appropriate to the low-density limit. This is a real simplification. Low-density molecular dynamics is slowed by keeping track of the long free paths which separate infrequent collisions. On the other hand, the zero-density Boltzmann equation solutions can be separated into two simple parts: (1) *streaming* in coordinate space through a computational mesh, punctuated by statistical *jumps* in velocity space, jumps due to collisions. The collisions are probabilistic, rather than deterministic, corresponding computationally to the use of random numbers to generate collisions. The distribution of impact parameters (distance of closest approach) is random in two dimensions while the collision probability for pairs close enough to collide is directly proportional to the speed at which the two particles approach one another.

An algorithm based on these ideas is often called "Direct Simulation Monte Carlo", and has been developed and championed by Græme Bird. For simplicity here we visualize hard-disk particles of unit mass and diameter. The disk "diameter" is also the minimum distance between two colliding disks' centers. The collision rate for two particles with relative velocity v in a two-dimensional periodic "volume" $V \gg 1$ is $2|v|/V$ so that the mean free path (the distance traveled between collisions, and also the expected width of the shockwave region) is of order (V/N). In the numerical work here we alternately advance all of the particles for a streaming timestep of $dt = 0.1$ and then consider the likelihood of collisions for particles close enough to collide during the time dt. In the Boltzmann equation colliding particles share a small volume so that the particle size vanishes in comparison to the mean free path. In our algorithm pairs of particles selected for collision lie within a small interval dx much less than the mean free path. We choose $dx = 0.01$ in our numerical example.

The steady-state solution also depends upon the range of the weight function chosen for spatial averages and the number of particles introduced during dt. Thus there are five characteristic lengths in the problem, the length of the system, 100 in the case below; the mean free path, of order unity; the width of the region containing pairs of colliding particles $dx = 0.01$, the range of the weight function $h = 0.20$, and the distance traveled in a timestep, of order 0.1. Provided that the spatial and collisional resolution scales h and dx are both much less than the mean free path (which is in turn less than the system length, the results are insensitive to these parameters. Because individual runs can be carried out in a few minutes this is an excellent pedagogical kinetic-theory problem for studying the sensitivity of numerical results to chosen computational parameters.

The initial conditions we choose here, for a system of 100,000 particles, include 25,000 to the left of the origin, moving to the right at speed 3 and zero temperature, and 75,000 "hot" particles to the right of origin, with

mean speed 1 and initial kinetic temperature
$$\langle v_x^2 \rangle = \langle v_y^2 \rangle = (kT/m) = 2 \ .$$
We set the (arbitrary) density scale by choosing a cold inflow density of unity (so that the hot exit density corresponds to the maximum threefold compression, $\rho_h = 3$). The "shockfront" is at the coordinate origin $x = 0$ and the mass, momentum, and energy fluxes are as follows:
$$\rho u_x = 3; \ P_{xx} + \rho u_x^2 = 9; \ \rho u_x [e + (P_{xx}/\rho) + (u_x^2/2)] + Q_x = (27/2) \ .$$
Both the pressure and the internal energy are zero for the incoming cold "gas". The particle coordinates and the Gaussian distribution for the hot particle velocities are generated as follows:

```
Temp = 2.0d00
intx = 0
inty = 0
do j = 1,25000
x(j) = -50.0d00 + (j - 0.5d00)*50.0d00/25000.0d00
vx(j) = 3.0d00
vy(j) = 0.0d00
enddo
do j = 1,75000
x(j+25000) = (j - 0.5d00)*50.0d00/75000.0d00
5 try = 10.0d00*(rand(intx,inty) - 0.5d00)
exp = dexp(-try*try/(2.0d00*Temp))
if(rand(intx,inty).gt.exp) go to 5
vx(j) = 1.0d00 + try
6 try = 10.0d00*(rand(intx,inty) - 0.5d00)
exp = dexp(-try*try/(2.0d00*Temp))
if(rand(intx,inty).gt.exp) go to 6
vy(j) = try
enddo
```

The timestep chosen here is $dt = 0.1$. During each step each particle is first moved according to its x velocity component:
$$x(t + dt) = x(t) + v_x dt \ .$$

The y coordinates are not recorded as the statistical collisions make no use of them. Next, particles with an x separation of 0.01 or less collide with a probability proportional to

```
vij = dsqrt((vx(i) - vx(j))**2 + (vy(i) - vy(j))**2) .
```

The velocities of a colliding pair of particles are changed by a collision subroutine which selects impact parameters for collisions randomly:

```
subroutine collision(vx1,vy1,vx2,vy2,intx,inty)
vx1b = vx1
vy1b = vy1
vx2b = vx2
vy2b = vy2
emu = 2.0d00*rand(intx,inty) - 1.0d00
theta = dacos(emu)
if(rand(intx,inty).gt.0.5d00) theta = - theta
vxs = (vx1b + vx2b)/2.0d00
vxd = (vx1b - vx2b)/2.0d00
vys = (vy1b + vy2b)/2.0d00
vyd = (vy1b - vy2b)/2.0d00
choose rotation of difference velocity through angle theta
vxr = vxd*dcos(theta) - vyd*dsin(theta)
vyr = vxd*dsin(theta) + vyd*dcos(theta)
choose rotation of difference velocity through angle theta
vx1 = vxs + vxr
vx2 = vxs - vxr
vy1 = vys + vyr
vy2 = vys - vyr
return
end
```

It is easy to check that this routine generates typical hard-disk collisions and that it conserves momentum and energy at each collision. After a few hundred timesteps a stationary state emerges. The profiles can be calculated using Lucy's one-dimensional weight function (with h set equal to 0.20 in the main program and dx defined positive, as $|dx|$):

```
function w(dx,h)
implicit double precision(a-h,o-z)
z = dx/h
w = (5.0d00/(4*h))*(1.0d00 + 3.0d00*z)*(1.0d00 - z)**3
return
end
```

Note that Lucy's one-dimensional weight function is properly normalized:

$$(5/4)\int_{-1}^{+1}(1 - 6z^2 + 8|z|^3 - 3z^4)dz \equiv 1 \ .$$

As the calculation proceeds, averages are calculated at **ng** grid points. First, the needed sums are initialized at zero for the entire grid:

```
do ig = 1,ng
xg(ig) = (ig - 0.5d00*(1 + ng))*100.0d00/ng
vg(ig) = 0.0d00
rhog(ig) = 0.0d00
rhovxg(ig) = 0.0d00
rhovxvxg(ig) = 0.0d00
rhovyvyg(ig) = 0.0d00
rhovxvxvxg(ig) = 0.0d00
rhovxvyvyg(ig) = 0.0d00
enddo
```

Then, for all particles within h of a grid point, the particle contributions needed for computing the stream velocity, the kinetic temperatures, and the heatflux vector are accumulated:

```
do ig = 1,ng
do i = 1,N
dx = dsqrt((xg(ig) - x(i))**2)
if(dx.lt.h) then
rhog(ig) = rhog(ig) + w(dx,h)
rhovxg(ig) = rhovxg(ig) + w(dx,h)*vx(i)
rhovxvxg(ig) = rhovxvxg(ig) + w(dx,h)*vx(i)*vx(i)
```

```
rhovyvyg(ig) = rhovyvyg(ig) + w(dx,h)*vy(i)*vy(i)
rhovxvxvxg(ig) = rhovxvxvxg(ig) + w(dx,h)*vx(i)*vx(i)*vx(i)
rhovxvyvyg(ig) = rhovxvyvyg(ig) + w(dx,h)*vx(i)*vy(i)*vy(i)
endif
enddo
enddo
do ig = 1,ng
vg(ig) = vg(ig)/rhog(ig)
enddo
```

These grid values are summed up over sufficiently many timesteps (300 were enough for Figure 9.7) to provide smooth time-averaged profiles. That Figure shows typical results for the velocity, pressure tensor, shear stress, temperature tensor, and heat flux profiles. The maximum in the kinetic temperature indicates a failure in Fourier's law, as the negative slope for T_{xx} corresponds to a negative heat conductivity. Two other failures of linear transport theory can be seen in the profiles. The shear stress $(P_{yy} - P_{xx})/2$ lags behind the strain rate (du/dx), just as Maxwell suggested, as does also the heat flux, Q_x, lagging behind the temperature gradient.

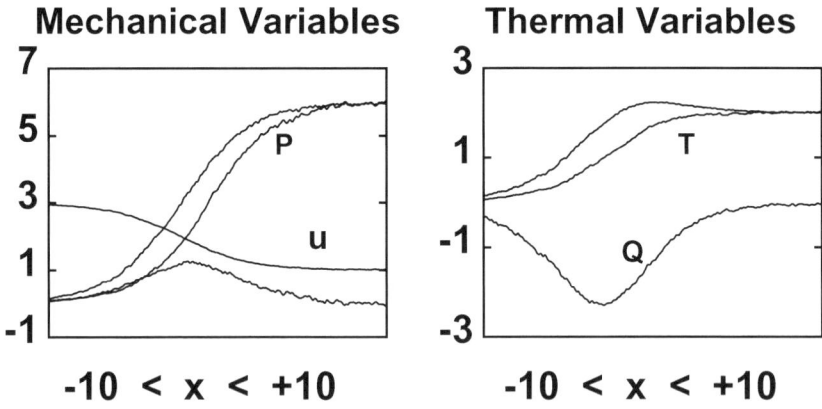

Figure 9.7. Solution of the Boltzmann Equation for threefold compression of a two-dimensional zero-density zero-temperature ideal gas. Mechanical variables shown in the figure are velocity, the diagonal components of the pressure tensor, and their difference, $P_{xx} - P_{yy}$. Thermal variables shown in the figure are T_{xx}, T_{yy}, and the (negative) heat flux vector Q_x.

The shockwave problem, like the Rayleigh-Bénard problem, is simple, and well-suited to numerical work, because the boundary conditions require little effort. In the shockwave problem illustrated here, fresh particles are injected at each timestep, 150 of them, equally spaced in x and located between -50.0 and -49.7. The righthand moving thermal boundary is most easily implemented (and stabilized) by refreshing the velocities of particles near it, just after moving each particle to its new x coordinate:

```
do i = 1,N
x(i) = x(i) + vx(i)*dt
if(x(i).gt.45.0d00) then
7 try = 20.0d00*(rand(intx,inty) - 0.5d00)
exp = dexp(-try*try/(2.0d00*Temp))
if(rand(intx,inty).gt.exp) go to 7
vx(i) = 1.0d00 + try
8 try = 20.0d00*(rand(intx,inty) - 0.5d00)
exp = dexp(-try*try/(2.0d00*Temp))
if(rand(intx,inty).gt.exp) go to 8
vy(i) = try
endif
enddo
```

Figure 9.7 showed profiles computed from 300 timesteps (after discarding the first 300). The time lags between the stress and strain rate and between the heat flux and the temperature are characteristic of results for dense fluids. The lags are another symptom of the difference between time-reversible particle mechanics (simulated here by collisions) and the irreversible macroscopic "Laws" which lack that time symmetry.

9.17 Summary

It is evident that the difficult conceptual questions underlying the reversibility paradox and its resolutions cannot be usefully addressed in completely general terms. Examples are required. Despite this, the desire at least to understand, if not to control, the future, has led philosophers to ponder the meanings of time, probability, and irreversibility. Quantum mechanics provided philosophers with the additional unsettling need to consider probability. To us, it seems presumptuous of them to attempt a resolution of the argument between Bohr and Einstein, but they remind us that this subject is not yet closed.

Maxwell recognized that Lyapunov instability destroys the apparent determinism of physics. Joseph Ford emphasized the same point. The impossibility of precisely specifying initial conditions for a chaotic problem seems lost on those who stress the importance of initial conditions. Any "good" computer simulation, meaning a useful algorithm, is certainly better controlled than most any laboratory experiment, and loses knowledge of the initial conditions in a time of order a few collision times. In computer simulations relevant to macroscopic physics, both the initial conditions and the chosen algorithm must be irrelevant to the macroscopic outcome.

A literary, completely theoretical approach to physics questions is unconvincing. Physics is grounded in observation with limited resources. Because computer simulations can be reliable sources of observation, they provide an essential adjunct to our understanding of physics. The exposure to philosophical ideas required for the preparation of this book led to the inclusion of many of the worked-out numerical examples. By providing tests of new ideas, examples can help to keep the literature relatively free of published errors.

From the standpoint of physics it is quite clear that Boltzmann, followed by Green and Kubo, formulated the phenomenological transport laws governing mass, momentum, and energy flows as equilibrium ensemble averages. The necessity of averaging, ignoring the transients and the fluctuations, is obvious in all of this work. The result is a verification of the simplest phenomenological nonequilibrium constitutive relations. These can then be incorporated in the macroscopic simulations to provide results in fine agreement with experiment. They can also be used to extend the treatment of equilibrium thermodynamic systems to nonequilibrium systems, with a basis for both in microscopic mechanics.

It is interesting to contrast the sensitive chaotic instability of microscopic trajectories to the relentless predictable stability of macroscopic dissipation. The microscopic equations of motion generate wildly different solutions with only slightly different computers, or compilers, or computer languages. But the resulting transport coefficients and dissipative behavior are not at all sensitive to these effects.

The analyses of stability in the neighborhood of a nonequilibrium trajectory deserve more investigation. The particles to which the motion is most sensitive (or least sensitive) can be determined in three different ways: (i) analysis of the usual forward-in-time Lyapunov vectors to determine the individual particle contributions; (ii) analysis of the maximum (or minimum) growth rate directions in phase space using singular value decompostion of the dynamical matrix; (iii) analysis in the reversed time direction, of Lyapunov vectors. All three of these analyses can be carried out precisely, using bit-reversible algorithms, for patently irreversible but purely Newtonian problems, fracture, inelastic collisions of macroscopic masses, shockwaves, to name a few. This approach is bound to bear fruit in bolstering our understanding of Second Law irreversibility from time-reversible mechanics.

The interventionist approaches necessary to steady-state nonequilibrium simulations were certainly foreseen by Ruelle and Smale. Krylov's agonized search for irreversibility could have ended with the discovery of linear-response theory, but there was then, and perhaps still is, more to do. The concept of a Lyapunov-unstable fractal is a key ingredient of our current understanding of reversibility. It was unknown in Boltzmann's day. The simple picture that flow proceeds from a terribly-unlikely past to an inevitable future is consistent with our everyday experience. To the extent that our knowledge of the current situation is incomplete we could explore the future of a corresponding ensemble, or of a typical initial condition. When the two provide similar answers we can conclude that a physical description is possible.

The mathematics of fractal sets remains as uncomfortably incomplete and paradoxical as it was in Bridgman's day. We heartily recommend a look at his 1934 "Second Look at Mengenlehre". How *can* the set of even integers be the "same size" as the set of all integers? How can it be that the number of points in an orange is equal to the number of points in *two* oranges? How can it be that there are fewer rationals than irrationals, while the binary grid composed of numbers with equal fractions of zeros and ones has exactly the same dimensionality as does the continuum (while any other fraction gives a lower dimensionality)? How can it be that *exactly the same points* make up *both* the inevitable attractor *and* the unobservable repeller?

9.17.1 *Notes and References*

Boltzmann's work is profusely documented. His most famous result, the H Theorem, was validated by Alder and Wainwright's early molecular dynamics simulations, as summarized in their "Molecular Dynamics by Electronic

Computers" in 1956. The reversibility problem Boltzmann confronted was a primary motivation for the study of many simple nonequilibrium few-particle models as well as the many-body kinetic theory and molecular dynamics investigations that followed.

Other potentially promising work seems to be headed toward a dead end. Jaynes' began his lifelong entropy-based explorations with his 1957 papers on "Information Theory and Statistical Mechanics". The results obtained so far, like Zubarev's efforts toward describing nonequilibrium distributions, in *Nonequilibrium Statistical Thermodynamics*, have had little influence on computer simulation.

Some of the serious mathematicians' contributions *have* had an impact on computer simulation. Smale's famous horseshoe cartoons of strange attractor bending and folding in his 1967 work, "Differentiable Dynamical Systems" and Eckmann and Ruelle's 1985 long review, "Ergodic Theory of Chaos and Strange Attractors", are examples of mathematicians' helping us better to understand nonequilibrium phase-space dynamics. The instability studies that are the staple fare of contemporary research owe their underlying concepts to the mathematicians.

Loschmidt's paradox (irreversibility from time-reversible dynamics) has been a major motivation for our research over the past 25 years. Bill's 1987 work with Brad Holian and Harald Posch, "Resolution of Loschmidt's Paradox: The Origin of Irreversible Behavior in Reversible Atomistic Dynamics" was an early realization of the repeller-attractor structure of particle-based nonequilibrium simulations.

There remains an abiding gap between the interesting results from molecular dynamics and what can be proved rigorously. Nevertheless the two approaches complement and aid one another. The interest in thermostated systems has been gratifying and future progress seems assured. See Pierre Gaspard's 1998 work, *Chaos, Scattering and Statistical Mechanics*, and Rainer Klages' *Microscopic Chaos, Fractals, and Transport in Nonequilibrium Statistical Mechanics* (2007) for an indication of today's mathematical research frontier.

Chapter 10

Afterword—a Research Perspective

Ah, make the most of what we yet may spend
Before we too into the dust descend;
Dust into dust, and under dust, to lie,
Sans wine, sans song, sans singer, and sans end.

Omar Khayyám, as translated by Edward FitzGerald

10.1 Introduction

In this final Chapter we take stock of our progress in defining and confronting the reversibility paradox from the modern perspective of chaos and computer simulation. We summarize the way in which our current knowledge resolves the long-standing problem of "understanding" irreversibility, and we point out why some are still unsatisfied with this explanation. Our views, and the challenges to them, change with passing time. We discuss some of our own experiences with innovation and change. Computers provide a mechanism for change through the exploration of example problems. We stress the necessity and importance of well-chosen numerical examples to an understanding of chaos and irreversibility. We also point out difficulties associated with an overly-mathematical approach to these questions.

Finally, we assess those real problems, still in need of clarification, which are so nearly feasible now as to constitute the research of the near-term future.

10.2 What do We Know?

Boltzmann, Einstein, Gibbs, and Maxwell were all well aware that the Second Law of Thermodynamics holds only in an averaged sense. The statistical probabilistic point of view, so useful and familiar to us today, was forced on Boltzmann by his critics. By now we have added new knowledge, with the help of chaos and computers, to the integrated experience of the past. Nonequilibrium molecular dynamics (1972), Nosé-Hoover thermomechanics (1985), Moran's multifractal Galton Board simulations (1986), the investigations of nonequilibrium Lyapunov spectra, and the studies of strong shockwaves which followed—all of these advances led us to our current state of knowledge. Today we see that even the smallest chaotic systems, when time-averaged, have behavior and properties fully consistent with the Second Law of Thermodynamics.

Chaos theory shows exactly *how* the macroscopic Second Law results from microscopic chaotic dissipative processes. Processes obeying the Second Law are stable, corresponding to attractors—they *destroy* the information which would be required to reverse them. This information flows *from* the system under investigation *to* its surroundings, where it is lost, as "heat". Processes violating the Second Law are unstable, corresponding to repeller states. Repellers are *only* observed formally, and artificially, by playing recorded attractor trajectories backward. They require an external store of relevant historical information. Chaos theory has led us from the blind alley of large conservative systems to the understanding of deterministic dissipation through thermomechanics. As forecast by many seers, this explanation of the Second Law *does* require an extension of mechanics to the maintenance of nonequilibrium states. It thereby avoids the silly emphases on dilute gases, "initial conditions", and isolated systems which have plagued this subject for so long.

Were it *not* for chaos initial conditions *would* be important. But chaos limits our descriptive abilities. Ilya Prigogine and Joe Ford correctly stated that chaotic trajectories are unobservable, beyond computation. The acceptance of chaos makes it possible to identify and exclude meaningless constructions, such as the precise trajectories of the Block Universe. Chaos makes it clear that irreversibility is not at all restricted to infinite systems. Irreversibility can be seen in very small samples, in long-time-averaged nonequilibrium steady states involving only one or two degrees of freedom. It follows that we must examine subsystems, "open systems" interacting with reservoirs, in order to understand real-world irreversibility.

Mechanical thermostats are essential to the simulation and understanding of open systems. Without them we could not so easily relate mechanics to thermodynamics and the real world around us. Time-reversible mechanical thermostats are a natural outgrowth of computer simulation. The time-reversibility property, pervasive in physics, turned out to simplify the implementation and interpretation of thermomechanics, and to lead naturally, through Liouville's Theorems, to an understanding of irreversibility.

At equilibrium, Gibbs' entropy $-k\langle \ln f_N \rangle$ and Shannon's "information" $\langle \ln f_N \rangle$ provide essentially the same picture. Maximizing entropy corresponds to minimal information. *At* equilibrium no understanding of dynamical processes or mechanisms is necessary. But, *away* from equilibrium the continual creation and destruction of information cannot balance. The imbalance is quantified by the information dimension of the strange attractors and by their Lyapunov instability. This is the lesson of the Second Law of Thermodynamics.

The inaccessibility and completely negligible measure of repeller states away from equilibrium, despite the formal reversibility of the equations of motion, makes this point clear in detail. Nonequilibrium core states occupy a vanishingly small fraction of the total states available at equilibrium. The difference between the embedding dimensionality of equilibrium states D_E, and the nonequilibrium information dimension D_I reflects both the disappearance of phase volume and the divergence of the probability density $f(q, p, \zeta)$ as the coarse-graining length ϵ is reduced toward zero:

$$\langle \otimes_{\text{noneq}} / \otimes_{\text{eq}} \rangle \propto \langle (f_{\text{eq}}/f_{\text{noneq}}) \rangle_\epsilon \propto \epsilon^{+\Delta D} \propto \epsilon^{(D_E - D_I)} .$$

This same relationship can alternatively be expressed in terms of the Lyapunov spectrum $\{\lambda\}$ and the external entropy dissipation rate $\dot{S}$. Kaplan and Yorke pointed out that the attractor information dimension can be estimated as the number of terms included in the partial sum $\lambda_1 + \lambda_2 + \ldots$ at the point where the sum changes sign, from positive to negative. With Nosé-Hoover dynamics the *full* sum is identically equal to $-(\dot{S}_{\text{external}}/k)$. Thus the nonequilibrium reduction in attractor dimension is approximately $(\dot{S}_{\text{external}}/\lambda_1 k)$, the number of terms which must be omitted to get a partial sum of zero:

$$\langle(\otimes_{\text{noneq}}/\otimes_{\text{eq}})\rangle \propto \langle(f_{\text{eq}}/f_{\text{noneq}})\rangle_\epsilon \simeq \epsilon^{(+\dot{S}_{\text{ext}}/\lambda_1 k)} \ .$$

Fractals are a universal feature of irreversible flows. That formally time-reversible nonequilibrium simulations invariably yield *fractals*, with detail at all scales, *was* a surprise. There have been *many* discoveries along the way. Ruelle had an early inkling of this, based on Anosov systems. Alder, Gass, and Wainwright's comparisons of Boltzmann's and Enskog's kinetic-theory predictions with the results of molecular dynamics simulations, in 1970, emphasized the importance of chaos to irreversibility. Moran's work on the Galton Board showed fractals without any doubt. Prigogine found them in Hopf's Baker Map. Prigogine discussed semigroups, similar to the repeller-attractor pairs evident in the early computer studies. Much more recently Dorfman, Gaspard, and Nicolis showed how fractal structures arise when ordinary Hamiltonian systems are allowed to decay through escape at open-system boundaries.

It is wonderful to see now the profuse expression of Michael Barnsley's idea, *Fractals Everywhere*, published by Academic Press in 1988. The fractal explanation of irreversibility is by now accepted and in common use, joining together the two ends of the spectrum connecting Boltzmann and Prigogine, and including all those many of us in between.

10.3 Why Reversibility is Still a Problem

Squaring nature's irreversibility with reversible equations of motion has been discussed, sometimes heatedly, for more than a century. Why has this problem been with us so long? Partly, it is its difficult nature. We have no unified theory applicable to far-from-equilibrium systems. See Brad Holian's published lecture contrasting the "beauty" of present day formalisms with the "ugly reality" of fractal distributions. Unlike equilibrium states, which are well-defined and unique once the state variables are given, the simplest nonequilibrium states—steady states—owe their existence and their complexity to *boundary conditions*. More complicated states can also depend on time.

A common reaction to this intrinsic difficulty is denial. One can entirely ignore the need for boundaries, constraints, and driving, by choosing to study the "approach to equilibrium", beginning with special nonequilibrium initial conditions. Though this choice avoids a characterization of nonequilibrium states, it retains the useful equilibrium property of incompressible phase-space flows obeying standard Newtonian mechanics. This approach exaggerates the importance of initial conditions. It would baffle those concerned with real laboratory experiments where the irrelevant details of the initial state are never precisely known. Gibbs' equilibrium ensembles require no special boundary conditions and are time independent. At equilibrium the macroscopic properties require no time averages and no dynamical considerations. Nonequilibrium states are more difficult.

For how much longer will "irreversibility" be an interesting subject? The intrinsic interest of the subject guarantees continuing discussions of three kinds: (i) those intended to further the design and control of irreversible processes for practical needs, (ii) those intended to clarify the issue on intellectual grounds, and (iii) those intended to win fame and fortune for the participants. All three motivations will follow mankind into the unforeseeable future. The design and control of irreversibility for practical ends is in good shape despite the continuing need for clarification and explanation.

There is an understandable tendency for armchair philosophers to seek excessive generality. Their reach exceeds their grasp. Rather than confining their ruminations to well-posed problems, they prefer to visualize a "final solution", a "theory of everything". This urge to generalize can lead to oversimplification. It is startling to see discussions of physical theories which steadfastly ignore the fact that these are theories of *time evolution*. Change comes very slowly, reflecting Joseph Ford's favorite extract from Leo Tolstoy's work:

> *I know that most men, including those at ease with problems of the highest complexity, can seldom accept even the simplest and most obvious truth if it be such as would oblige them to admit the falsity of conclusions which they have delighted in explaining to colleagues, which they have proudly taught to others, and which they have woven, thread by thread, into the fabric of their lives.*

To what extent is Gibbs' ensemble viewpoint useful away from equilibrium? Because the phase-space probability density for simple nonequilibrium states is fractal, it is plain that the ensemble view is *not* very useful. The independence of results to initial conditions, for stationary or time-periodic nonequilibrium states, implies that the dynamics of a single system—Gibbs' "time ensemble"— provides just as much information as does the dynamics of an ensemble. So the utility of ensembles away from equilibrium must be decided on grounds of æsthetics or efficiency.

There certainly would be little more to prove if nature's underlying motion equations were themselves irreversible. Such a view seems totally unnecessary and is currently quite rare. Because there seems to be no compelling physical evidence for it, Occam would suggest avoiding this assumption of intrinsic irreversibility. For the more usual isolated mechanical systems, with many degrees of freedom and with phase volume conserved, the only irreversibility possible is the approximate type described by Boltzmann, the shedding of unlikely initial conditions. Isolated systems contain within themselves the paradoxical possibilities of return and reversal. With reversible nonequilibrium equations of motion, can there be irreversible behavior? The answer to this question is undeniably "yes", but for reasons which are difficult to understand, even for the simplest case, which is the reversible dissipative Baker Map.

Any study of irreversibility could be complicated further by introducing (somehow) gravitational, relativistic, and quantum effects. But these complexities seem totally unnecessary. To understand the dissipation associated with viscous and conducting flows, despite time reversibility, is today relatively simple. Our view is this: the theoretical advances made since Boltzmann's discussion of dilute-gas irreversibility come from set theory, dynamical systems theory, an understanding of chaos, and the extension of Gibbs' ensemble theory to include linear response. A practical and believable understanding required, above all, digital computers and computer simulation. Beyond denial and dogmatism, a part of the reason for the long life of the reversibility problem lies in the complexity of *any* approximations or simplifications which could make possible a confrontation of the microscopic and macroscopic points of view *beyond* linear response theory: (i) one needs a clearly formulated mathematical theory, amenable to simulation and precise calculation, so that results can be obtained; (ii) a useful microscopic theory must necessarily contain precise, even if arbitrary, analogs of the relatively-vague macroscopic concepts of temperature and entropy; (iii) it must be admitted that numerical solutions, subject to Lyapunov instability, cannot attain precision for long times, so that a probabilistic description of the trajectory-based theory needs to be discussed. It is more difficult to pose meaningless questions and to pursue outmoded approaches if we keep in mind the need to achieve consistency among all three routes to knowledge: experiment, theory, and computer simulation.

10.4 Change and Innovation

Understanding irreversible flows has undergone qualitative change, from the smooth distributions of Maxwell, Boltzmann, and Gibbs to fractals. From simple analytic solutions of mechanical problems to the chaotic simulations of today. Maxwell and Boltzmann's work changed our views forever. They introduced probabilistic ideas. Despite the approximate and restricted nature of Boltzmann's gas-phase explanation of irreversibility, through the H Theorem, Lebowitz has promoted his own view that Boltzmann's understanding of irreversibility was essentially complete, hampered only by an inability to work out phase-space integrals for condensed phases. This generous view credits Boltzmann with much more than he ever did, or would, claim. Boltzmann certainly did reach a nearly-complete understanding of the approach to equilibrium of dilute gases, but he could do little outside that area. It is certainly true that his fundamental relation $S_{\text{eq}} = k \ln \Omega$, linking the *equilibrium* entropy to the number of phase-space states, is valid for *equilibrium* liquids and solids. His probabilistic collisional assumptions, highly plausible (and even correct!) *for gases*, provided a basis for the phenomenological irreversible macroscopic flow equations.

But it is clear that Boltzmann had no idea at all about the existence of fractal distributions. Boltzmann certainly had no idea that simple differential equations of motion can lead to the complex irreversible fractal structures revealed by thermomechanical computer simulations. His accomplishments are monumental enough as it is that it seems wholly inappropriate and unnecessary to credit him with more than what he actually did. The compressible Navier-Stokes equations, including Newton's viscosity and Fourier's heat conduction, as well as some interesting higher-order effects, and specialized modifications of the continuum equations to describe plasticity and fracture, are still quite adequate for most engineering problems. They are not adequate for problems in which fluctuations are important, for which Bird's modern numerical approach, Direct Simulation Monte Carlo, represents a distinct improvement over the analytic Boltzmann equation.

The passage of time leads inexorably to new concepts, tools, and vantage points from which past views seem limited, and of little value. Krylov's critical work, as well as the Ehrenfests' encyclopedia article, so valuable in its early summary of the accomplishments and the problems of kinetic theory, now hold little interest for a physicist, for whom Schrödinger's quantum mechanics, Maxwell and Boltzmann's kinetic theory, and Gibbs' statistical mechanics all coexist, though within slightly different areas of usefulness.

The creative and possessive impulses are active in every generation of researchers. Paraphrasing Niels Bohr: "Physics progresses one death at a time." Man has a deep-seated need for novelty, the replacing of old ideas with his own creations, which he is then loathe to relinquish. This natural evolution leads, over time, to the continual development of extremely esoteric methods for analyzing simple problems. Schrödinger even imagined concealing a microscopic source of randomness, modifying classical mechanics to introduce irreversibility. Prigogine followed this lead for quantum systems. Systematic, highly-complex, barren formulations can be developed too. For example, a long trajectory, in a bounded space, necessarily nearly repeats, and so can be expressed in terms of periodic orbits in that space. Operators which advance distributions forward in time have eigenfunctions. These can be followed into the well-named complex plane.

Sometimes change is the result of an illusory quest for novelty. It is quite possible to pursue blind alleys in physics, roads through an imaginary landscape, which lead nowhere. Anosov systems, conjugate-pairing theorems, and fluctuation theorems represent real progress only to the extent that they have interesting analogs in the real world. So far there is no indication that something like exponent pairing really holds for a system with realistic nonequilibrium boundary conditions.

There is no doubt that trajectories and distributions propagate in different ways. A phase-space trajectory follows a completely-hypothetical one-dimensional track. It gains no complexity with passing time. Nevertheless, a Lyapunov unstable trajectory (unless it is bit-reversible!) cannot be reversed by any means other than storing the results of the forward motion, and using these to direct the results of the reversed motion. The difficulties are even more severe in reversing a *non*equilibrium trajectory, for the sum of Lyapunov exponents is *positive* in the time-reversed motion, not just zero. *Any* reasonable distribution spreads, due to Lyapunov instability, eventually filling out a coarse-grained probability density with no real memory or clue as to the initial conditions. This disparity between trajectories and probability densities has led some to think the two approaches are profoundly different. Prigogine has asserted that trajectories themselves have no reality. Distributions *must* be considered.

Because distributions can be nothing more than the result of averaging a long trajectory this distinction is a bit puzzling. From the computational standpoint trajectories are obtained from *ordinary* differential equations. Summed-up coarse-grained trajectories then provide coarse-grained distributions. Trajectories are in fact the *only way* to get these distributions. Though formally these distributions satisfy a *partial* differential flow equation there is no practical way to implement that equation in interesting cases. The phase-space gradient operation ∇f is apparently ill-defined for typical fractals. In the end, the distinction between one-dimensional trajectories and the resulting fractal probability densities is certainly a red herring, as must be any explanation that distinguishes strongly (i) phenomenological reality, (ii) theoretical analysis, and (iii) computer simulation. There is no doubt that these three are separate pathways to a common partial understanding of reality and that including relativity, gravitation, quantum mechanics, or cosmology can enrich this understanding while adding nothing fundamentally new to our picture of irreversibility.

At the Livermore Radiation Laboratory of the early 1960s Bill soon discovered hurdles along the path of innovation. Francis Ree and he had to get Edward Teller's permission to use the computer time necessary for an accurate calculation of the hard-sphere solid entropy using constrained ensembles. Fortunately, Teller quickly agreed and that work was soon published. It wasn't until 1986 that Bill found that publishing new ideas could be as difficult as generating and exploring them. When Bill Moran's solutions of the nonequilibrium Galton Board problem revealed multifractal objects, both of them were truly surprised. Though Bill already knew what fractals were, from his Nosé-Hoover oscillator work in Vienna with Harald Posch and Franz Vesely, he had not expected to see them in nonequilibrium simulations, and had even proposed to the National Science Foundation investigating of nonequilibrium distribution functions using Fourier analysis!

An editor at Physical Review Letters, George BasBas, rejected a manuscript describing the new fractal findings. Bill sought him out, at a New York City cocktail party hosted by the American Physical Society, in the naïve hope that he would have some interest in understanding the novel physics. But BasBas was steadfastly uninterested and relentlessly uninformed. Finally, by permuting the order of the authors' names and changing the title of the manuscript, Hoover, Posch, and Holian were able, on their *third* attempt, to publish it in Physical Review Letters. Meanwhile Bill had submitted another manuscript to the Journal of Statistical Physics, which was honoring Ilya Prigogine with a special issue devoted to the question of reversibility. Despite the relevance of the paper, Joel Lebowitz rejected it. Today it is hard to believe that these early papers, which still seem to us to be clearly enough written, and certainly interesting, required extraordinary efforts to publish. But stories of this kind are legion. Jaynes' and Ruelle's are both well known.

Without the heroes of the past we would have had no option but to reïnvent their discoveries. The luxury of building on the successful ideas from the past, and forgetting the rest, is what continues to provide the excitement and sense of progress still present today. From the long-term point of view the all too common priority disputes and struggles to differentiate minor results appear quite ludicrous.

It seems increasingly rare today to find people willing to admit their mistakes. During Bill's graduate student days at Ann Arbor things were different. When his advisor, Andy De Rocco, called Bob Zwanzig to report an error Bill had found, it took only a few seconds for Zwanzig to understand, and accept correction. Today even hours of conversation followed up with months of correspondence can be ineffective in resolving relatively simple questions. For an example we remember very well, work backward from Bill's Physics Letters A manuscript of mid-1999.

10.5 Role of Examples

Examples are essential to progress on conceptual problems in physics. As it became apparent that fractal distributions underlie nonequilibrium problems occasional naysayers insisted that this view was mistaken. The simplest fractal examples are generated by maps. Maps and cellular automata [as well as digital computers!] replace the space-time continuum with discrete states. This simplifies state counting and can certainly return substantial conceptual rewards. The Baker map has both equilibrium and nonequilibrium forms. These forms can be concatenated into chains yielding large-system fractal distributions. Such a chain of maps is simpler and more interesting than Lorenz' continuous model, which can also be concatenated but which lacks time reversibility and has no equilibrium form.

With hindsight, it was possible to embed the successful thermomechanical simulation techniques into traditional mechanics. Whether or not the embedding is simpler than choosing *ad hoc* boundary, constraint, and driving forces is an æsthetic choice. The essential point is this: the perceived need to simulate the irreversible flows found in the laboratory led naturally to the time-reversible algorithms. These algorithms were only later linked to the structure of mechanics. Numerical solutions have provided valuable insights. Without them it would truly be impossible to improve upon Boltzmann's understanding. With them, the fields of Nonlinear Dynamics and Chaos have developed, providing new ideas, described with a rich vocabulary, and accessible to scientists and engineers. The notions of mixing and Lyapunov instability, coupled with Cantor's set theory and Mandelbrot's fractals, are essential to a faithful description of the strange attractors accompanying dissipation.

The Galton Board is a particularly good model for the study of fractals and Lyapunov instability. It has both equilibrium and nonequilibrium versions. Its generalization to a many-body flow—the color conductivity problem—revealed the uniform shifts of Lyapunov exponent pairs called "conjugate pairing". Certainly it is worthwhile to complicate mechanics sufficiently to include the irreversibility of thermodynamics within it, as a consequence. This approach breathes life into Gibbs' ensemble theory, and allows us to apply some of his ideas away from equilibrium. Thermostats are also useful for *equilibrium* thermodynamics, for they provide mechanisms for generating Gibbs' ensembles from dynamical trajectories. Generalizing the concept of a mechanical system to a *thermomechanical* system, including reversible boundary interactions and constraints, provides a natural chaos-based irreversibility: the collapse to a relatively-stable strange attractor.

This collapse, corresponding to information loss, or, equivalently, to the extreme rarity of nonequilibrium states, is the inevitable consequence of the thermomechanical approach. Beyond the reward of establishing a mechanism for irreversibility, this approach also avoids the completely hypothetical, misleading, and unrealistic large-system limit. The *same* time-reversible dissipative behavior can be seen in systems both large and small. The simple dissipative Baker Map and Galton Board are the prototypical examples, for discrete and continuous time, respectively. In these cases, as is typical, a repeller-attractor pair forms, linking two ergodic fractal objects in a way which breaks time symmetry through the dynamical sensitivity—Lyapunov instability—called chaos.

All of the important features of interesting nonequilibrium systems can be seen in simple example problems. Lorenz' Attractor is mixing and chaotic. The Galton Board adds reversibility and ergodicity to these properties. The thermostated oscillator, with a temperature gradient, shows that phase-space compressibility is not always identical to the external entropy production. Shockwaves exhibit a time-symmetry breaking *despite* their purely Hamiltonian nature. These examples are our favorites. Each scientist has his own. This is how we learn.

10.6 Role of Chaos and Fractals

It has taken quite a while to see that chaos, fractal geometry, and time-reversible ergodicity, with mixing, are all vital to an understanding of the reversibility paradox. Though Maxwell, Boltzmann, and Poincaré recognized the importance of "sensitivity to initial conditions", it remained to computer simulation to show the *need* for this property in order to obtain results *independent* of initial conditions.

The variety of theoretical approaches to irreversibility which have all developed since the seminal discussions at Howard Hanley's 1982 Boulder Conference brings to mind the earlier profusion of *ad hoc* integral equations in the 1950s. Eventually, with the development of molecular dynamics and Monte Carlo simulations, the equilibrium "structure of simple liquids" became a solved problem. The need and the advocates for the obsolete approaches disappeared. Likewise, the fractal ergodic nature of time-reversible *non*equilibrium states is now well-known. We can anticipate that the periodic orbits, maps, Anosov systems, and other exotic tools for "understanding" what is now a solved problem will likewise gradually disappear.

Chaos makes it possible to access *all* the states while fractals quantify the global *rarity* of nonequilibrium states. It is not easy to avoid the need for evolution equations in order to find an instantaneous analog of this property, but perhaps the quest is not impossible. Computer simulation will continue to play its useful role, providing the answers to the new questions, just as it has for the old.

10.7 Role of Mathematics

The unreasonable applicability of mathematics to physics expresses two roles—the ability of computation to simulate real phenomena, and to stimulate mathematical advances. From the viewpoint of a computational physicist, or even a skilled theorist like Feynman, the mathematics appears to lag well behind our understanding. It was with some difficulty, in 1986, that Bill came to understand the importance of time-reversible deterministic simulations, with nonequilibrium boundary conditions, to a fractal explanation of the Second Law of Thermodynamics. At that time Boltzmann's approach, understanding infinite systems, with Hamiltonian mechanics and without boundaries, was the only approach under serious study by the mathematicians. The need for something new had been articulated, but *little* had been done.

What will be the role of mathematics in continuing these advances? Perhaps it will widen their appeal and accelerate exploration and understanding? For some reason mathematics, while no doubt inspired by the computer results, is presently so restricted and complex as to be of little help in understanding or in making further progress. Quite often the mathematical treatment requires the acceptance of unrealistic assumptions, special unphysical cases, or nonuniform convergence. These difficulties suggest that the underlying phenomena are not being properly treated, that the models are oversimplified.

But the numerical results *can* be understood, and are not really *so* complicated. The existence of fractals is an obvious consequence for any flow which is simultaneously chaotic and dissipative. When the first computer simulation results were obtained—fractal attractor-repeller pairs, shifted Lyapunov spectra, decreased dimensionality in phase space (as opposed to just decreased volume)—it was relatively difficult to publish them. Only about ten years later did most physicists working in the field accept some version of the new ideas, usually on the basis of their own specialized models, with more or less restrictive assumptions, and expressed in their own jargon. As we have stated in hundreds of seminar and conference talks, time reversibility, coupled with the requirement of boundedness, implies that only flows with long-time-averaged contraction can be observed. This is the symmetry breaking underlying the Second Law. "Explanations", based on Sinai-Ruelle-Bowen measures which lie completely outside the domain of operational definitions and practical techniques, are relatively specialized exercises for the experts. It certainly would be useful to have results for *instantaneous* nonequilibrium fluctuations, but this modest goal seems still to be far beyond the capability of the various theoretical approaches.

The Second Law of Thermodynamics is perfectly general, and, when properly averaged, applies to all nonequilibrium systems. On the other hand, it seems to be necessary to examine each individual special case in order to make useful predictions of constitutive behavior. Are there any guidelines or general rules, analogous to Gibbs' ensembles, which would simplify the computational task of predicting nonequilibrium behavior? Is there any way to discover the fractal nature of a distribution function by examining only a small part of the system? These questions lead us to a consideration of the puzzles which still remain.

10.8 Remaining Puzzles

Taking stock, we have seen that chaos, when (i) constrained, (ii) dissipative, (iii) stationary, and (iv) time-reversible, guarantees not only the irrelevance of initial conditions, but also the destruction and loss of the information which would be required to reverse a dissipative trajectory. From the computational standpoint this understanding is convincing, complete, and verifiable. Simulation *is* a sufficiently faithful representation of irreversible phenomena. It explains our inability to recapture the past as a destruction of information, an automatic book burning built into chaotic dynamics.

Nonequilibrium thermostats have finally taken their rightful place in traditional classical mechanics as well. Gauss' Principle of Least Constraint can be used to derive the isokinetic or isoenergetic equations of motion for a driven system. Energy surfaces for Dettmann's Hamiltonian obey the Nosé-Hoover equations of motion. This derivation is not possible for situations, such as prototypical heat flow, involving more than one thermostat. More work needs to be done. Hamilton's Principle of Least Action, specially emphasized by Feynman, provided Dettmann, Morriss, and Choquard with another good starting point for thermostats.

Perhaps there is no lasting need to seek a better understanding of the fundamental reversibility paradox than that available through an analysis of the dissipative Baker Map. More-elaborate *chains* of maps have been studied by Breymann, Gaspard, Tél, and Vollmer. This work is but the beginning few links in a logical chain coupling simple maps to complex numerical solutions of the complete partial differential equations of continuum mechanics. A uniform probability, subject to the action of the dissipative Baker map, spreads out as an exact analog of a biased random walk. An initially uniform probability density approaches the multifractal distribution shown in Chapter 2, with the average value of the information entropy,

$$S_{\text{information}} \equiv -\sum \text{prob} \ln \text{prob} = -\langle \ln \text{prob} \rangle \; ,$$

where the {prob} are the bin probabilities, decreasing by $(1/3) \ln 2$ with each iteration of the map and ultimately diverging.

Iterating the same Baker map "backward" (with the same initial condition, which is unchanged by time-reversal), provides a second fractal structure, the "repeller". Though the details obviously depend on the initial point, the analog of a Baker-map trajectory, generates, over time, exactly the same phase-space structure as does a repeated iteration of the distribution function. All the same features, the deterministic formation of fractal Lyapunov-unstable and time-reversible repeller-attractor pairs, are ubiquitous features of time-reversible nonequilibrium chaotic simulations of dissipative systems. Maps show us that this rich behavior simply represents correlated operations on one-dimensional digit strings.

There *do* appear to be problems simpler than maps, but perhaps conceptually just as fruitful, involving the manipulation of digit strings. Is there an analog of simple shear for digit strings? To what extent does the paradoxical coexistence of reversibility and irreversibility persist in simple models of this kind? It appears that a minimum of four dimensions are required for a simple continuous dissipative and time-reversible ergodic flow. See the paper by Harald Posch and Bill in the 1997 Physical Review E. Are there topological reasons for expecting increased or decreased complexity as the necessary dimensionality increases? Can we claim to "understand" nonequilibrium flows on the basis of two-dimensional maps?

It is evident that the mathematicians still have much more to learn and perhaps more to teach us. Feynman's description in his Princeton story, "A Different Box of Tools" pictures the frustration which can result when physicists request help from mathematicians. The idea of fixed numbers of positive and negative (local) Lyapunov exponents, seemingly essential to a "rigorous" Anosov-based analysis, contradicts results from even the very simplest of realistic models. The four time-reversible differential equations describing an ergodic harmonic oscillator in a temperature gradient, for example, give rise to only a single positive exponent, and two negative ones, even for small fields. Even so, for reasonable gradients, the *local* exponents appear to be able to take on *all possible sixteen combinations* of signs, from $(-,-,-,-)$ to $(+,+,+,+)$. The time-reversed shockwave problem of Chapter 6 provides a similar finding for an interesting Hamiltonian system: the "largest" (in terms of its long-time-averaged value) Lyapunov exponent describing the time-reversed motion can occasionally be *negative*.

Perhaps the paradox linking reversibility and irreversibility is more subtle than the usual distinction between rational and irrational numbers, while including the same fundamental mathematical difficulty, that which underlies the Banach-Tarski paradox. French illustrated that paradox by constructing a one-to-one mapping, carrying a single point set into *two* disjoint sets, congruent with the original. French constructed two oranges from one in this way. Such "paradoxical decompositions" and constructions have the same peculiar nature as do "steady" many-to-one phase-space flows with continuously shrinking distributions. It would be useful for a kindhearted and knowledgable mathematician to express his understanding of such paradoxical behavior in terms more accessible and relevant to physicists. Bridgman was one of those who noticed and criticized the tailoring of Cantor's new clothes.

Because quantum mechanics, like classical mechanics, is a fundamental theory with the same time-reversibility properties, it suffers from the same recurrence and reversibility problems as does the classical approach. But there is an additional fly in the ointment: the Schrödinger Equation contains no forces, only energies, and this complicates the search for a nonequilibrium quantum mechanics. Chirikov has stressed that quantum mechanics, with its discrete spectrum of energy states, is analogous to computer simulations of classical problems in a discrete solution space. This analogy seems quite good. The typical level of precision in classical simulations, sixteen digits, is similar to quantum uncertainties on the scale of Planck's constant. The quantum situation seems more complex, in that adding constraints—through Lagrange multipliers—to the Schrödinger Equation can certainly provide nonequilibrium stationary wave functions, but does not seem to lead to any chaos in the form of exponential divergence.

Does finite-precision arithmetic have any important effect on the properties obtained from classical simulations? The *usual* computer algorithms lead, in principle, from a large but finite basin of attraction to a *very* long periodic orbit. [Periodic orbits from the bit-reversible algorithms *include* their initial points!] Perhaps the distance away from this inevitable periodic orbit can quantify the meaning of "far from equilibrium" by answering the question: "How far?" The difference between having many very long periodic orbits, with fixed precision, and the idealized inaccessible ergodic chaotic orbit implied by classical continuum-based mechanics is reminiscent of the distinctions between quantum chaos (if any) and classical chaos.

How can we better characterize the multifractal distributions which arise in these nonequilibrium flows? The Lyapunov spectra seem to have less structure than do the frequency spectra of solid state physics. Despite this simplicity, approximate representations of fractals as sums of periodic orbits are cumbersome in the extreme. Are there some computational ways, both "interesting" and "practical", to characterize the different types of multifractal attractors? These representations should reflect the dramatic differences in visual appearance of the few simple two- and three-dimensional cases accessible with computer graphics. Despite their similar fractal dimensionality, there are undeniably big differences between puffy cumulus clouds and balled-up wads of paper.

The analysis of Lyapunov-unstable systems is becoming routine. Both the classic Gram-Schmidt analyses and the covariant analyses are being applied to larger systems, both conservative and dissipative. The analysis of the Lyapunov vectors for strongly nonequilibrium situations is bound to lead to new insights and new characterization techniques. The shockwave problems studied in Chapter 6 promise new insight into the irreversibility associated with conservative Newtonian-Hamiltonian dynamics. By cycling repeatedly through a bit-reversible compression-expansion shockwave cycle a symmetry breaking distinguishing the forward and reversed trajectories can be quantified. This parallel to the symmetry breaking found in dissipative systems is an exciting clue on the path toward a future understanding of presentday mysteries.

Computational power continues to grow. In the near future we can look forward to much more detailed insights into time-reversible nonequilibrium systems. It is becoming possible to characterize the large-system limit of the Lyapunov spectrum for nonequilibrium hard-sphere systems and to characterize the corresponding eigenvectors. Likewise, faster machines will be able to determine the small-system fluctuations in Poincaré recurrence times, and improve the characterization of the multifractal phase-space distributions corresponding to nonequilibrium steady states.

10.9 Summary

We have reached the end of our journey. Our goal has been to explain, as simply and clearly as we can, how irreversibility results from time-reversible mechanical laws. To us it is satisfying to understand how velocity differences or temperature differences lead, irreversibly, to dissipation and to entropy increase. In the linear case the explanations of Green, Kubo, and Onsager are undoubtedly correct, and are in full agreement with the predictions of Gibbs' ensemble theory and Boltzmann's kinetic theory. Far from equilibrium, the explanation is even simpler and more compelling, and makes use of concepts from chaos and fractal geometry. Chaos makes it possible to do much more. By now, we have maps, attractors, and typical Lyapunov spectra for a wide variety of simple nonequilibrium problems. These provide understanding bridging the gap between mechanics and thermodynamics.

Consider again the four-chamber heat-flow example from Chapter 5, page 196, in which, for *any* initial condition, the time-averaged flow of heat corresponds to a positive heat conductivity (and so to a flow consistent with the Second Law of Thermodynamics). By accumulating the heat transfers to and from the "hot" and "cold" regions of the system, the entropy change of the hypothetical external heat reservoirs interacting with the two thermostated regions can be recorded as a function of time. See Figure 10.1, which shows a typical time development of the cumulative averages as well as their sum.

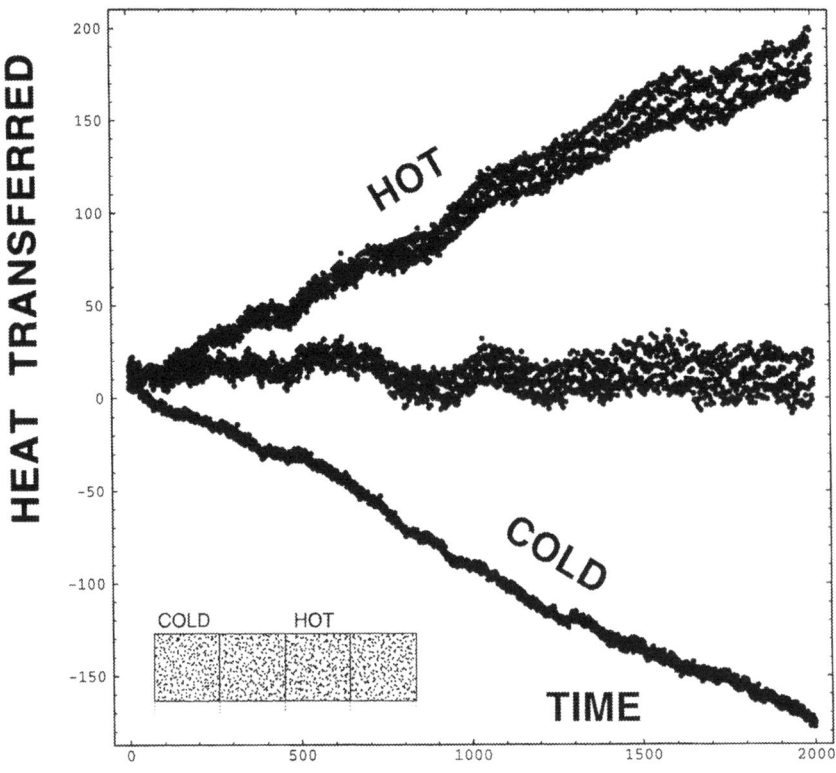

Figure 10.1. Cumulative dissipation in the four-chamber conductivity simulation discussed in Chapter 5, Section 5.9.3. The total heat added to the hot chamber and that extracted from the cold chamber are shown. Their sum fluctuates in the neighborhood of zero, as energy balance implies.

The instantaneous oscillations about the well-characterized mean slopes are typically large and appear without warning. A first step for an interpretive nonequilibrium theory would be an understanding of the frequency and amplitude of these fluctuations. The overall slopes of the integrated heat transmitted to the reservoirs are proportional to the heat conductivity of the fluid confined by the two heat reservoirs. The entire system is fully time-reversible. If we had recorded *all* the particle coordinates and velocities, as functions of time, the entire many-body trajectory could have been played backward, and would have satisfied exactly the same (time-reversible!) motion equations. Thus, on a time-averaged basis, the change in external entropy, due to the fixed temperature difference, is positive in the forward direction of time and would be negative in the backward direction. Like Boltzmann's H Theorem, this is an example of irreversible behavior obtained from time-reversible motion equations. Unlike Boltzmann's, our own results are exact, and with the numerical algorithms applying equally-well to liquids and solids, not just dilute gases.

But now consider the details of the Lyapunov spectra for the forward and backward trajectories. Forward in time the summed spectrum is negative, corresponding to the stable collapse to a multifractal attractor. The diaphanous nature of the attractor fits with our intuition that the number of nonequilibrium states is relatively small. If the trajectory, and its sensitivity to linear perturbations were *both* processed backward in time—a strict reversal of *all* the reference and satellite coordinates back toward their initial values—each of the Lyapunov exponents would change sign, both locally and as an average. Thus the summed spectrum of the reversed stored trajectory would have a *positive* sum and would correspond to a violation of the Second Law, even when time-averaged.

Why is such a reversed trajectory never seen? Precisely because of this instability. Any trajectory in the vicinity of the unstable repeller (corresponding to the reversed trajectory) promptly seeks out the stable attractor. For this reason the formal three-step construction of the past—(i) reversing the velocities, (ii) solving the equations of motion, and (iii) reversing the velocities and the time ordering of the points which result—is ineffective. It leads to an unobservable zero-measure repeller. It is quite clear that the stationary boundary conditions, because they extract information, lead to stationary heat flow which is only *formally* time-reversible. In fact the flow, together with its local Lyapunov spectrum, can *only* be reversed by playing the corresponding stored trajectories backward. This is the main message which the fractals convey in their description of nonequilibrium steady states.

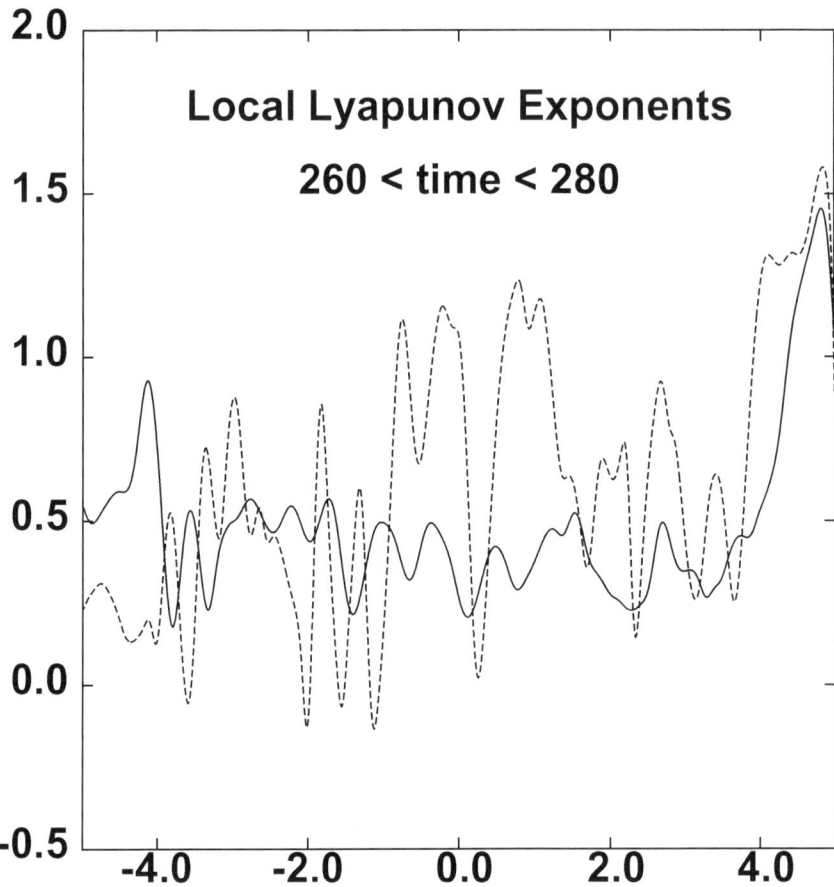

Figure 10.2. Fully-converged Lyapunov exponents for the time-reversed $N = 1600$ shock compression problem of Figure 6.16 (page 236). The motion was first run forward for 30 time units with $\Delta t = 0.01$. The local Lyapunov exponent shown here describes the growth rate of an offset vector linking a satellite trajectory to the bit-reversible reference trajectory. The motion is reversed at times of 30, 70, 110, ... until the satellite trajectory is fully converged. The full line corresponds to the forward trajectory from time 260 to time 270. The dashed line corresponds to exactly the same configurations, from time 270 to 280, but sampled in reversed order. Typically, averaged over the entire period, the Lyapunov exponent is more positive, on average, in the forward direction than in the reverse.

It might be thought that this type of symmetry-breaking is restricted to *dissipative* systems where the attractor has a clear geometrical advantage over the repeller. But the shockwave problem of Chapter 6 shows otherwise. See again Figure 6.16 on page 236. That Figure shows that the phase-space instability associated with the irreversible shockwave compression is much more localized, forward in time, than it is in the reversed motion. This in spite of the perfect time-reversibility of both the underlying equations of motion and the bit-reversible implementation of them.

In retrospect it might seem "obvious" that the instability forward in time is mainly found *at* the shock front, while that in the reversed motion is instead distributed throughout the hotter shocked fluid. But in fact this situation is yet another instance in which computer simulation has shed unexpected light and suggested new avenues to explore. If the bit-reversible forward-backward trajectory is repeated cyclically with a nearby satellite trajectory tethered a fixed distance away, that satellite trajectory converges, to machine accuracy, after just a few cycles. The fixed-distance constraint can be imposed with a Lagrange multiplier or by rescaling the offset vector after every timestep. Figure 10.2 shows the time dependence of the "local" Lyapunov exponent defined by rescaling at every step.
Though all the underlying motion equations are strictly time-reversible the dynamics chooses a broken symmetry. There is no obvious relation between the largest Lyapunov exponent going forward in time and its backward brother. This problem shows there is more to do and much more to learn.

The symmetry-breaking seen here describes phase-space deformation *in the neighborhood* of a trajectory undergoing a highly-irreversible process. Evidently such processes do behave differently, always reflecting the past rather than the future. The forward time direction corresponds to our everyday experiences while the reversed one does not. We hope that further exploration of this problem, as well as the simpler models which it suggests, will lead to quantitative relationships linking phase-space instabilities to the thermodynamics of irrevesible processes. As usual, simulation provides tantalizing suggestions. And it is up to us all to follow them up.

10.10 Acknowledgments

Bill's interest in simulation was awakened during his college days, by the work of Berni Alder, Tom Wainwright, and Bill Wood. Andy De Rocco showed him how to use some of the tools of statistical mechanics, and encouraged his graduate studies. He was also strongly influenced by the clarity, depth, and sense of excitement in the kinetic-theory lectures given by Stuart Rice, at Harvard, and George Uhlenbeck, at Ann Arbor. After graduate school, Berni attracted Bill to the stimulating research atmosphere of the sputnik-era Radiation Laboratory at Livermore. There he met Steve Brush, who soon left for a distinguished career at Maryland, combining history with physics, and Francis Ree and Hugh De Witt, both of whom remain at the Laboratory today, taking advantage of the tremendous computational opportunities there.

A few years later, Bill was given the opportunity to teach statistical mechanics in the now-defunct Livermore campus of the University of California, "Teller Tech". This idyllic University connection eased the programmatic pressures of the Laboratory, and also facilitated the travel so necessary to productive research. Among his students and researchers in the Department of Applied Science at Livermore, Bill Ashurst, Victor Castillo, Errol Craig, Oyeon Kum, Tony Ladd, Bill Moran, and Koichiro Shida contributed the most useful ideas. Carol was one of Bill's best students in his graduate courses in statistical mechanics and kinetic theory, in the late 1970s.

Carol continued at the Livermore Laboratory for the remainder of her working career, serving as Group Leader in the Department of Engineering's Methods Development Group after leading the User Services group at the National Magnetic Fusion Energy Computer Center. In 1987, while at NMFECC, she was visited by a Dr Gupta from Louisiana, and Dr Gupta asked for an introduction to Bill, who hadn't seen much of Carol since her student days. That chance meeting led to a dinner date and proposal, with our marriage, carried out by Bill's son Nathan, just in time for our joint research leaves at Keio University's Hiyoshi campus, in Japan, 1989-1990.

Koichiro Shida, whom we met at Keio, spent several successful months at Livermore, doing research on Maxwell's "Thermal Creep" phenomenon, while preparing the Japanese translation of Bill's *Computational Statistical Mechanics* for publication. Kevin Boercker, Vic, Donovan Jones, and Nancy Owens provided help with the software required to illustrate this book. Nancy Owens is now married, and, as Nancy Fulda, is a well-known author of science fiction. Son Xuan Nguyen carried out the scanning and sizing of the art work. Peter Raboin's Methods Development Group at the Lawrence Livermore Laboratory and the Academy of Applied Science (Concord, New Hampshire) furnished additional welcome support too.

During Bill's sabbatical leaves, in Australia, Austria, and Japan, Carol, Denis Evans, Bill's son Nathan, Toshio Kawai, and Harald Posch were his main collaborators. Harald and Brad Holian have been constant sources of inspiration and new ideas ever since we first met. Other colleagues, including Aurel Bulgac, Philippe Choquard, Christoph Dellago, Bob Dorfman, Marshall Fixman, Joseph Ford, Howard Hanley, Siegfried Hess, Gianni Iacucci, Rainer Klages, Dimitri Kusnezov, Stefano Ruffo, Ian Snook, Karl Travis, Franz Waldner, and Kris Wojciechowski have furnished stimulation and motivation from time to time. During the writing of this book's precursor, *Time Reversibility, Computer Simulation, and Chaos*, Eddy Cohen, Dieter Flamm, Giovanni Gallavotti, and Harald were helpful in locating the relevant portions of Boltzmann's work. Dimitri Kusnezov and John Tully helped Bill to understand Gibbs' views at New Haven.

Brad, Carol, Christoph, Dimitri, and Oyeon kindly provided useful comments on drafts of that earlier book. Carl Dettmann, Pierre Gaspard, David Ruelle, and Tamas Tél all offered generous comprehensive descriptions of their own points of view. Their views were very helpful and stimulating. Giovanni Ciccotti, Jean-Pierre Hansen, Siegfried, Sigeo Ihara, Karl Kratky, Michel Mareschal, Carl Moser, Shuichi Nosé, Bob Watts, and Kris facilitated a number of productive and instructive research tours for which we are very grateful. The present book stimulated a warm rewarding correspondence with an engineer friend from Nathan's childhood, Steve Ramsey.

More recently meetings at Saint Petersburg, arranged by Anton Krivtsov and Vitaly Kuzkin, Cuernavaca, by Thomas Gilbert and Francisco Uribe, and Spain, by Pedro Garrido, Joaquín Marro, and Francisco de los Santos, have helped us to hone our understanding of deterministic, chaotic, and time-reversible systems far from equilibrium. We can't imagine a more interesting field of study or a more supportive group of colleagues!

Last, but not least, we need to mention our Editor at World Scientific Publishers in Singapore, Lakshmi Narayanan. She successfully encouraged our work on this book, despite all our retired opportunities for horseplay and horseback riding in the mountains of Nevada. Lakshmi has been so helpful and supportive over the years, that we are pleased to claim her as a friend and colleague, every bit as important to this work as have been our scientific colleagues all over this rapidly-shrinking interconnected world.

Bibliography

[1] B. J. Alder and T. Wainwright, "Molecular Dynamics by Electronic Computers", 97-131 in *Transport Processes in Statistical Mechanics. Proceedings of the IUPAP Symposium, Brussels, 1956* (Interscience, New York, 1958).

[2] B. J. Alder and T. E. Wainwright, "Phase Transition for a Hard Sphere System", Journal of Chemical Physics **27**, 1208-1209 (1957).

[3] T. A. Bass, *The Eudæmonic Pie* (Houghton Mifflin, New York, 1985).

[4] G. Benettin, L. Galgani, and J. M. Strelcyn, "Kolmogorov Entropy and Numerical Experiments", Physical Review A **14**, 2338-2345 (1976).

[5] G. Benettin, L. Galgani, A. Giorgilli, and J. M. Strelcyn, "Lyapunov Exponents for Smooth Dynamical Systems and for Hamiltonian Systems; a Method for Computing All of Them", Meccanica **15**, 9-30 (1980).

[6] L. Boltzmann, *Lectures on Gas Theory*, S. G. Brush, translator (University of California Press, 1964).

[7] H. Bosetti and H. A. Posch, "Covariant Lyapunov Vectors for Rigid Disk Systems", Chemical Physics **375**, 296-308 (2010).

[8] P. W. Bridgman, "A Physicist's Second Reaction to Mengenlehre", Scripta Mathematica **2**, 101-117 and 224-234 (1934).

[9] P. W. Bridgman, "The Thermodynamics of Plastic Deformation and Generalized Entropy", Reviews of Modern Physics **22**, 56-63 (1950).

[10] J. Bricmont, "Science of Chaos or Chaos in Science?", Physicalia **17**, 159-212 (1995).

[11] V. M. Castillo and Wm. G. Hoover, "Heat Flux at the Transition from Harmonic to Chaotic Flow in Thermal Convection", Physical Review E **58**, 4016-4018 (1997).

[12] V. M. Castillo, Wm. G. Hoover, and C. G. Hoover, "Coexisting Attractors in Compressible Rayleigh-Bénard Flow", Physical Review E **55**, 5546-5550 (1997).

[13] V. M. Castillo, *Cubic Spline Collocation Method for the Simulation of Turbulent Thermal Convection in Compressible Fluids*, (PhD dissertation, University of California, Davis/Livermore, 1998).

[14] C. Cercignani, *Ludwig Boltzmann, The Man Who Trusted Atoms* (Oxford University Press, 1998).

[15] N. I. Chernov, G. L. Eyink, J. L. Lebowitz, and Y. G. Sinai, "Derivation of Ohm's Law in a Deterministic Mechanical Model", Physical Review Letters **70**, 2209-2212 (1993).

[16] B. Chirikov, "Pseudochaos in Statistical Physics", 149-171 in *Nonlinear Dynamics, Chaotic and Complex Systems*, E. Infeld, R. Żelazny, and A. Galkowski, Editors (Cambridge University Press, 1997).

[17] P. Choquard, "Variational Principles for Thermostated Systems", Chaos **8**, 350-356 (1998).

[18] N. Clisby and B. M. McCoy, "Ninth and Tenth Order Virial Coefficients for Hard Spheres in D Dimensions", Journal of Statistical Physics **122**, 15-57 (2006).

[19] E. G. D. Cohen, "Boltzmann and Statistical Mechanics", 9-23 in *Boltzmann's Legacy 150 Years after His Birth* (Lincei Academy, Rome, 1997).

[20] P. Coveney and R. Highfield, *The Arrow of Time* (Fawcett Columbine, New York, 1985).

[21] M. Curd, "Popper on Boltzmann's Theory of the Direction of Time", 263-303 in *Ludwig Boltzmann Gesamtausgabe*, R. Sexl and J. Blackmore, Editors (Akademische Druck, Graz, 1982).

[22] Ch. Dellago, L. Glatz, and H. A. Posch, "Lyapunov Spectrum of the Driven Lorentz Gas", Physical Review E **52**, 4817-4826 (1995).

[23] Ch. Dellago, *Lyapunov Instability of Two-Dimensional Many-Body Systems*, PhD dissertation (University of Vienna, 1996).

[24] C. P. Dettmann and G. P. Morriss, "Hamiltonian Formulation of the Gaussian Isokinetic Thermostat", Physical Review E **54**, 2495-2500 (1996).

[25] C. P. Dettmann, "The Lorentz Gas: A Paradigm for Nonequilibrium Stationary States", in *Hard Ball Systems and the Lorentz Gas*, D. Szasz, Editor. *Encyclopædia of Mathematical Sciences* **101**, 315-365 (2000).

[26] J. R. Dorfman, *An Introduction to Chaos in Nonequilibrium Statistical Mechanics* (Cambridge University Press, 1999).

[27] J. P. Dougherty, "Foundations of Nonequilibrium Statistical Mechanics", 172-178 in *Nonlinear Dynamics, Chaotic and Complex Systems*, E. Infeld, R. Żelazny, and A. Galkowski, Editors (Cambridge University Press, 1997).

[28] J. P. Eckmann and D. Ruelle, "Ergodic Theory of Chaos and Strange Attractors", Reviews of Modern Physics **57**, 617-656 (1985).

[29] P. and T. Ehrenfest, *The Conceptual Foundations of the Statistical Approach in Mechanics*, J. J. Moravcsik, translator (Cornell University Press, 1959).

[30] D. J. Evans and G. P. Morriss, *Statistical Mechanics of Nonequilibrium Liquids* (Academic, New York, 1990).

[31] D. J. Evans, E. G. D. Cohen, and G. P. Morriss, "Probability of Second Law Violations in Shearing Steady States", Physical Review Letters **71**, 2401-2404 (1993).

[32] J. D. Farmer, E. Ott, and J. A. Yorke, "The Dimension of Chaotic Attractors", Physica D **7**, 153-180 (1983).

[33] E. Fermi, J. R. Pasta, and S. M. Ulam, "Studies of Nonlinear Problems", originally a 1955 Los Alamos Laboratory Report LA-1940 and also available

in the *Collected Works of Enrico Fermi* (University of Chicago Press, 1965). The 1972 article by Tuck and Menzel reviews this work and includes some interesting analysis of recurrence in nonlinear chains.

[34] R. P. Feynman, *The Character of Physical Law* (MIT Press, 1967).

[35] R. P. Feynman, *Classic Feynman: All the Adventures of a Curious Character*, R. Leighton, Editor (Norton, New York, 2006).

[36] W. Fleischhacker and T. Schönfeld, Editors, *Pioneering Ideas for the Physical and Chemical Sciences: Josef Loschmidt's Contributions and Modern Developments in Structural Organic Chemistry, Atomistics, and Statistical Mechanics* (Plenum, New York, 1995).

[37] C. Foidl and P. Kasperkowitz, "Systematic Generation of Linear Graphs—Check and Extension of the List of Uhlenbeck and Ford", Journal of Computational Physics **89**, 246-250 (1990).

[38] S. M. Foiles, M. I. Baskes, and M. S. Daw, "Embedded-Atom-Method Functions for the FCC Metals Cu, Ag, Au, Ni, Pd, Pt, and Their Alloys", Physical Review B **33**, 7983-7991 (1986).

[39] J. Ford, "What is Chaos, that We Should be Mindful of It?", 348-372 in *The New Physics* (Cambridge University Press, 1989).

[40] R. M. French, "The Banach-Tarski Theorem", Mathematical Intelligencer **10**(4), 21-28(1988).

[41] D. Frenkel and B. Smit, "Understanding Molecular Simulation; from *Algorithms to Application* (Academic Press, San Diego, 2002).

[42] G. Gallavotti and E. G. D. Cohen, "Dynamical Ensembles in Nonequilibrium Statistical Mechanics", Physical Review Letters **74**, 2694-2697 (1995).

[43] P. Gaspard, *Chaos, Scattering, and Statistical Mechanics* (Cambridge University Press, 1998).

[44] J. W. Gibbs, *Elementary Principles in Statistical Mechanics* (Oxbow Press, Woodbridge, CT, 1991). [First published in 1902.]

[45] T. Gilbert and J. R. Dorfman, "Fluctuation Theorem for Constrained Equilibrium Systems", Physical Review E **73**, 026121 (2006).

[46] T. Gilbert, "Fluctuation Theorem Applied to the Nosé-Hoover Thermostatted Lorentz Gas", Physical Review E **73**, 035102(R) (2006).

[47] F. Ginelli, P. Poggi, A. Turchi, H. Chaté, R. Livi, and A. Politi, "Characterizing Dynamics with Covariant Lyapunov Vectors", Physical Review Letters **99**, 130601 (2007).

[48] J. Gleick, *Chaos, Making a New Science,* (Viking Penguin, New York, 1987).

[49] J. Gleick, *Genius: The Life and Science of Richard Feynman* (Pantheon, New York, 1992).

[50] J. Gleick, *The Information* (Pantheon, New York, 2011).

[51] I. Goldhirsch, P. L. Sulem, and S. A. Orszag, "Stability and Lyapunov Stability of Dynamical Systems: A Differential Approach and a Numerical Method", Physica D **27**, 311-337 (1987).

[52] H. J. M. Hanley, Editor, *Nonlinear Fluid Behavior* (1982 Conference at Boulder, Colorado) published as Physica A **118** (1983).

[53] R. J. Hardy, "Formulas for Determining Local Properties in Molecular-

Dynamics Simulations: Shockwaves", Journal of Chemical Physics **76**, 622-628 (1982).

[54] E. Helfand, "Transport Coefficients from Dissipation in a Canonical Ensemble", Physical Review **119**, 1-9 (1960).

[55] B. L. Holian, Wm. G. Hoover, B. Moran, and G. K. Straub, "Shockwave Structure *via* Nonequilibrium Molecular Dynamics and Navier-Stokes Continuum Mechanics", Physical Review A **22**, 2798-2808 (1987).

[56] B. L. Holian, Wm. G. Hoover, and H. A. Posch, "Resolution of Loschmidt's Paradox: the Origin of Irreversible Behavior in Reversible Atomistic Dynamics", Physical Review Letters **59**, 10-13(1987).

[57] B. L. Holian, "The Character of the Nonequilibrium Steady State: Beautiful Formalism Meets Ugly Reality", 791-822 in *Monte Carlo and Molecular Dynamics of Condensed Matter Systems*, K. Binder and G. P. F. Ciccotti, Editors (Italian Physical Society, Bologna, 1996).

[58] Wm. G. Hoover and F. H. Ree, "Melting Transition and Communal Entropy for Hard Spheres", Journal of Chemical Physics **49**, 3609-3617 (1968).

[59] Wm. G. Hoover, A. C. Hindmarsh, and B. L. Holian, "Number Dependence of Small-Crystal Thermodynamic Properties", Journal of Chemical Physics **57**, 1980-1985 (1972).

[60] Wm. G. Hoover, *Molecular Dynamics*, Lecture Notes in Physics (Springer-Verlag, New York, 1986).

[61] Wm. G. Hoover and H. A. Posch, "Direct Measurement of Equilibrium and Nonequilibrium Lyapunov Spectra", Physics Letters A **123**, 227-230 (1987).

[62] Wm. G. Hoover, H. A. Posch, B. L. Holian, M. J. Gillan, M. Mareschal, and C. Massobrio, "Dissipative Irreversibility from Nosé's Reversible Mechanics", Molecular Simulation **1**, 79-86 (1987).

[63] Wm. G. Hoover, C. G. Hoover, and H. A. Posch, "Lyapunov Instability of Pendulums, Chains, and Strings", Physical Review A **41**, 2999-3004 (1990).

[64] Wm. G. Hoover, *Computational Statistical Mechanics* (Elsevier, New York, 1991). [*Keisan Toukei Rikigaku*, Japanese translation by Koichiro Shida with the supervision of Susumu Kotake (Morikita Shupan, 1999).]

[65] Wm. G. Hoover and B. Moran, "Viscous Attractor for the Galton Board", Chaos **2**, 599-602 (1992).

[66] Wm. G. Hoover, B. L. Holian, and H. A. Posch, "Comment I on 'Possible Experiment to Check the Reality of a Nonequilibrium Temperature' ", Physical Review E **48**, 3196-3198 (1993).

[67] Wm. G. Hoover, T. G. Pierce, C. G. Hoover, J. O. Shugart, C. M. Stein, and A. L. Edwards, "Molecular Dynamics, Smooth Particle Applied Mechanics, and Irreversibility", Computers and Mathematics with Applications **28**, 155-174 (1994).

[68] Wm. G. Hoover, "Temperature, Least Action, and Lagrangian Mechanics", Physics Letters A **204**, 133-135 (1995).

[69] Wm. G. Hoover and H. A. Posch, "Shear Viscosity *via* Global Control of Spatiotemporal Chaos in Two-Dimensional Isoenergetic Dense Fluids", Physical Review E **51**, 273-279 (1995).

[70] Wm. G. Hoover and B. L. Holian, "Kinetic Moments Method for the Canonical Ensemble", Physics Letters A **211**, 253-257 (1996).
[71] Wm. G. Hoover and H. A. Posch, "Numerical Heat Conductivity in Smooth Particle Applied Mechanics", Physical Review E **54**, 5142-5145 (1996).
[72] Wm. G. Hoover, "Mécanique de Nonéquilibre à la Californienne", Physica A **240**, 1-11 (1997).
[73] Wm. G. Hoover, K. Boercker, and H. A. Posch, "Large-System Hydrodynamic Limit for Color Conductivity in Two Dimensions", Physical Review E **57**, 3911-3916 (1998).
[74] Wm. G. Hoover, "Isomorphism Linking Smooth Particles and Embedded Atoms", Physica A **260**, 244-254 (1998).
[75] Wm. G. Hoover, "Time-Reversibility in Nonequilibrium Thermomechanics", Physica D **112**, 225-240 (1998).
[76] Wm. G. Hoover, "Liouville's Theorems, Gibbs' Entropy, and Multifractal Distributions for Nonequilibrium Steady States", Journal of Chemical Physics **109**, 4164-4170 (1998).
[77] Wm. G. Hoover, "The Statistical Thermodynamics of Steady States", Physics Letters A **255**, 37-41 (1999).
[78] Wm. G. Hoover, H. A. Posch, V. M. Castillo, and C. G. Hoover, "Computer Simulation of Irreversible Expansions via Molecular Dynamics, Smooth Particle Applied Mechanics, Eulerian, and Lagrangian Continuum Mechanics", Journal of Statistical Physics **100**, 313-326 (2000).
[79] Wm. G. Hoover, *Time Reversibility, Computer Simulation, and Chaos*, (World Scientific, Singapore, 1999 and 2001).
[80] Wm. G. Hoover, H. A. Posch, V. M. Castillo, and C. G. Hoover, "Computer Simulation of Irreversible Expansions via Molecular Dynamics, Smooth Particle Applied Mechanics, Eulerian, and Lagrangian Continuum Mechanics", Journal of Statistical Physics **100**, 313-326 (2000).
[81] Wm. G. Hoover, *Smooth Particle Applied Mechanics; the State of the Art* (World Scientific, Singapore, 2006).
[82] Wm. G. Hoover and C. G. Hoover, "Hamiltonian Dynamics of Thermostated Systems: Two-Temperature Heat-Conducting ϕ^4 Chains", Journal of Chemical Physics **126**, 164113 (2007).
[83] Wm. G. Hoover, C. G. Hoover, H. A. Posch, and J. A. Codelli, "The Second Law of Thermodynamics and Multifractal Distribution Functions: Bin Counting, Pair Correlations, and the Kaplan-Yorke Conjecture", Communications in Nonlinear Science and Numerical Simulation **12**, 214-231 (2007).
[84] Wm. G. Hoover and C. G. Hoover, "Nonequilibrium Temperature and Thermometry in Heat-Conducting ϕ^4 Models", Physical Review E **77**, 041104 (2008).
[85] Wm. G. Hoover, C. G. Hoover, and J. Petravic, "Simulation of Two- and Three-Dimensional Dense-Fluid Shear Flows via Nonequilibrium Molecular Dynamics: Comparison of Time-and-Space-Averaged Stresses from Homogeneous Doll's and Sllod Shear Algorithms with those from Boundary-Driven Shear", Physical Review E **78**, 046701 (2008).

[86] Wm. G. Hoover, Carol G. Hoover, "Three Lectures: NEMD, SPAM, and Shockwaves", 23-55, Proceedings of the Granada Seminar on the *Foundations of Nonequilibrium Statistical Physics* (La Herradura, Spain, 13-17 September 2010).

[87] Wm. G. Hoover, C. G. Hoover, and F. Uribe, "Flexible Macroscopic Models for Dense-Fluid Shockwaves: Partitioning Heat and Work; Delaying Stress and Heat Flux; Two-Temperature Thermal Relaxation", 261-273, Proceedings of the 38th International Summer School–Conference *Advanced Problems in Mechanics* (Saint Petersburg, Russia, 2010).

[88] E. Hopf, "On Causality, Statistics, and Probability", Journal of Mathematics and Physics **13**, 51-102 (1934).

[89] D. A. Huckaby, "Exact Classical Harmonic Free Energy of the Triangular Lattice", Journal of Chemical Physics **54**, 2910-2911 (1971).

[90] T. J. Hunt and R. S. MacKay, "Anosov Parameter Values for the Triple Linkage and a Physical System with a Uniformly Chaotic Attractor", Nonlinearity **16**, 1499-1510 (2003).

[91] M. Ichiyanagi, "Conceptual Developments of Nonequilibrium Statistical Mechanics in the Early Days of Japan", Physics Reports **262**, 227-310 (1995).

[92] R. Illner and H. Neunzert, "The Concept of Irreversibility in the Kinetic Theory of Gases", Transport Theory and Statistical Physics **16**, 89-112 (1987).

[93] C. Jarzynski, "Equalities and Inequalities: Irreversibility and the Second Law of Thermodynamics at the Nanoscale", Annual Reviews of Condensed Matter Physics **2**, 329-351 (2011).

[94] E. T. Jaynes, "Information Theory and Statistical Mechanics", Physical Review **106**, 620-630 (1957).

[95] E. T. Jaynes, "Violation of Boltzmann's H Theorem in Real Gases", Physical Review A **4**, 747-750 (1971).

[96] D. Jou, J. Casas-Vázquez, and G. Lebon, *Extended Irreversible Thermodynamics*, (Springer, Berlin, 1993).

[97] J. L. Kaplan and J. A. Yorke, "Chaotic Behavior of Multidimensional Difference Equations", in Lecture Notes in Mathematics **730**, 204-227 (Springer-Verlag, Berlin, 1979).

[98] R. Klages, K. Rateitschak, and G. Nicolis, "Thermostating by Deterministic Scattering: Construction of Nonequilibrium Steady States", Physical Review Letters **84**, 4268-4271 (2000).

[99] R. Klages, *Microscopic Chaos, Fractals, and Transport in Nonequilibrium Statistical Mechanics*, Volume 24 in the Advanced Series in Nonlinear Dynamics (World Scientific, Singapore, 2007).

[100] M. J. Klein, "The Physics of J. Willard Gibbs in His Time", 1-21 in *Proceedings of the Gibbs Symposium*, Yale University, May 1989, E. G. Caldi and G. D. Mostow, Editors (American Mathematical Society, 1989).

[101] N. S. Krylov, *Works on the Foundations of Statistical Physics*, A. B. Migdal, Y. G. Sinai, and Y. L. Zeeman, translators (Princeton University Press, 1979).

[102] R. Kubo, *Statistical Mechanics* (North-Holland, Amsterdam, 1965).
[103] O. Kum and Wm. G. Hoover, "Time-Reversible Continuum Mechanics", Journal of Statistical Physics **76**, 1075-1081 (1994).
[104] O. Kum, *Nonequilibrium Flows with Smooth Particle Applied Mechanics*, (PhD dissertation, University of California, Davis/Livermore, 1995).
[105] O. Kum, Wm. G. Hoover, and H. A. Posch, "Viscous Conducting Flows with Smooth-Particle Applied Mechanics", Physical Review E **52**, 4899-4908 (1995).
[106] O. Kum, Wm. G. Hoover, and C. G. Hoover, "Temperature Maxima in Stable Two-Dimensional Shock Waves", Physical Review E **56**, 462-465 (1997).
[107] D. Kusnezov, A. Bulgac, and W. Bauer, "Canonical Ensembles from Chaos", Annals of Physics **204**, 155-185 (1990).
[108] D. Kusnezov, "From Chaos to Nonequilibrium Statistical Mechanics", Czechoslovak Journal of Physics **49**, 35-87 (1999).
[109] D. Kusnezov, "Quantum Lévy Processes and Fractional Kinetics", Physical Review Letters **82**, 1136-1139 (1999).
[110] L. D. Landau and E. M. Lifshitz, *Course of Theoretical Physics* (ten volumes, several publishers and dates).
[111] J. Lebowitz, "Boltzmann's Entropy and Time's Arrow", Physics Today **46**, 32-38 (September, 1993). See also the Letters section in volume **47** (November, 1994).
[112] J. Lebowitz, "Microscopic Origins of Irreversible Behavior", Physica A **263**, 516-527 (1999).
[113] D. Levesque and L. Verlet, "Molecular Dynamics and Time Reversibility", Journal of Statistical Physics **72**, 519-537 (1993).
[114] S. Y. Liem, D. Brown, and J. H. R. Clarke, "Investigation of the Homogeneous-Shear Nonequilibrium-Molecular-Dynamics Method", Physical Review A **45**, 3706-3713 (1992).
[115] E. N. Lorenz, "Deterministic Nonperiodic Flow", Journal of Atmospheric Science **20**, 130-141 (1963).
[116] E. N. Lorenz, *The Essence of Chaos* (University of Washington Press, Seattle, 1993).
[117] L. B. Lucy, "A Numerical Approach to the Testing of the Fission Hypothesis", Astronomical Journal **82**, 1013-1024 (1977).
[118] B. B. Mandelbrot, *The Fractal Geometry of Nature* (Henry Holt, New York, 1982).
[119] M. M. Mansour, F. Baras, and A. Garcia, "On the Validity of Hydrodynamics in Plane Poiseuille Flows", Physica A **240**, 255-267 (1997).
[120] M. Mareschal and E. Kestmont, "Experimental Evidence for Convective Rolls in Finite Two-Dimensional Molecular Models", Nature **329**, 427-428 (1987).
[121] M. Mareschal, Editor, *Microscopic Simulations of Complex Flows* (Plenum, New York, 1989).
[122] M. Mareschal and B. L. Holian, Editors, *Microscopic Simulations of Complex Hydrodynamic Phenomena* (Plenum, New York, 1992).

[123] M. Mareschal, "Microscopic Simulations of Complex Flows", 317-392, Advances in Chemical Physics **100**, I. Prigogine and S. A. Rice, Editors (John Wiley & Sons, New York, 1997).

[124] J. E. Mayer and M. G. Mayer, *Statistical Mechanics* (John Wiley & Sons, New York, 1940).

[125] N. Metropolis, A. W. Rosenbluth, M. N. Rosenbluth, A. H. Teller, and E. Teller, "Equation of State Calculations by Fast Computing Machines", Journal of Chemical Physics **21**, 1087-1092 (1953).

[126] J. J. Monaghan, "Smoothed Particle Hydrodynamics", Annual Review of Astronomy and Astrophysics **30**, 543-574 (1992).

[127] B. Moran, Wm. G. Hoover, and S. Bestiale, "Diffusion in a Periodic Lorentz Gas", Journal of Statistical Physics **48**, 709-726 (1987).

[128] R. Morris, *Time's Arrows* (Simon and Schuster, New York, 1985).

[129] H. M. Mott-Smith, "The Solution of the Boltzmann Equation for a Shockwave", Physical Review **82**, 885-892 (1951).

[130] S. Nosé, "A Unified Formulation of the Constant Temperature Molecular Dynamics Methods", Journal of Chemical Physics **81**, 511-519 (1984).

[131] S. Nosé, "Constant Temperature Molecular Dynamics Methods", Progress of Theoretical Physics Supplement **103**, 1-46 (1991).

[132] L. A. Pars, *A Treatise on Analytical Dynamics* (Oxbow Press, Woodbridge, Connecticut, 1981).

[133] K. Pearson, *The Life, Letters, and Labours of Francis Galton* (Cambridge University Press, London, (1930).

[134] R. Penrose, *The Emperor's New Mind* (Penguin, New York, 1989).

[135] H. A. Posch, Wm. G. Hoover, and F. J. Vesely, "Canonical Dynamics of the Nosé Oscillator; Stability, Order, and Chaos", Physical Review A **33**, 4253-4265 (1986).

[136] H. A. Posch and Wm. G. Hoover, "Equilibrium and Nonequilibrium Lyapunov Spectra for Dense Fluids and Solids", Physical Review A **39**, 2175-2188 (1989).

[137] H. A. Posch and Wm. G. Hoover, "Time-Reversible Dissipative Attractors in Three and Four Phase-Space Dimensions", Physical Review E **55**, 6803-6810 (1997).

[138] H. A. Posch and R. Hirschl, "Simulation of Billiards and Hard Body Fluids", in *Hard Ball Systems and the Lorentz Gas*, D. Szasz, Editor. *Encyclopædia of the Mathematical Sciences* **101**, 279-310 (2000).

[139] H. Price, *Time's Arrow and Archimedes' Point* (Oxford University Press, New York, 1996).

[140] I. Prigogine, E. Kestemont, and M. Mareschal, "The Approach to Equilibrium and Molecular Dynamics", 233-240, in *Microscopic Simulations of Complex Flows*, M. Mareschal, Editor (Plenum, New York, 1990).

[141] A. Puhl, M. Malek Mansour, and M. Mareschal, "Quantitative Comparison of Molecular Dynamics with Hydrodynamics in Rayleigh-Bénard Convection", Physical Review A **40**, 1999-2012 (1989).

[142] D. C. Rapaport, "Time-Dependent Patterns in Atomistically Simulated Convection", Physical Review A **43**, 7046-7048 (1991).

[143] D. C. Rapaport, "Hexagonal Convection Patterns in Atomistically Simulated Fluids", Physical Review E **73**, 025301R (2006).
[144] D. Ruelle, *Chance and Chaos* (Princeton University Press, Princeton, 1991).
[145] D. Ruelle, "Smooth Dynamics and New Theoretical Ideas in Nonequilibrium Statistical Mechanics", Journal of Statistical Physics **95**, 393-468 (1999).
[146] D. Ruelle, "Gaps and New Ideas in our Understanding of Nonequilibrium", Physica A **263**, 540-544 (1999).
[147] J. Schnack, "Molecular Dynamics Investigations on a Quantum System in a Thermostat", Physica A **259**, 49-58 (1998).
[148] M. Schröder, *Fractals, Chaos, Power Laws* (W. H. Freeman, New York, 1991).
[149] E. C. Schrödinger, *Science, Theory, and Man*, Chapter III, "Indeterminism in Physics" (Dover, New York, 1957).
[150] L. Sklar, *Physics and Chance; Philosophical Issues in the Foundations of Statistical Mechanics* (Cambridge University Press, New York, 1993).
[151] S. Smale, "Differentiable Dynamical Systems", Bulletin of the American Mathematical Society **73**, 747-817 (1967).
[152] S. Smale, "On the Problem of Reviving the Ergodic Hypothesis of Boltzmann and Birkhoff", Proceedings of the New York Academy of Sciences **357**, 260-266 (1980).
[153] I. Snook, *The Langevin and Generalised Langevin Approach to the Dynamics of Atomic, Polymeric and Colloidal Systems* (Elsevier, Amsterdam, 2007).
[154] C. Sparrow, *The Lorenz Equations: Bifurcations, Chaos, and Strange Attractors* (Springer, New York, 1982)
[155] J. C. Sprott, *Strange Attractors: Creating Patterns in Chaos* (M & T Books, New York, 1993).
[156] J. C. Sprott, *Chaos and Time-Series Analysis* (Oxford University Press, New York, 2003).
[157] J. C. Sprott, "Simplifications of the Lorenz Attractor", Nonlinear Dynamics, Psychology, and Life Sciences, **13**, 271-278 (2009).
[158] S. D. Stoddard and J. Ford, "Numerical Experiments on the Stochastic Behavior of a Lennard-Jones Gas System", Physical Review A **8**, 1504-1512 (1973).
[159] T. Tél, J. Vollmer, and W. Breymann, "Transient Chaos; the Origin of Transport in Driven Systems", Europhysics Letters **35**, 659-664 (1996).
[160] T. Tél and M. Gruiz, *Chaotic Dynamics; an Introduction Based on Classical Mechanics* (Cambridge University Press, New York, 2006).
[161] J. L. Tuck and M. T. Menzel, "The Superperiod of the Nonlinear Weighted String (Fermi-Pasta-Ulam) Problem", Advances in Mathematics **9**, 399-407 (1972).
[162] G. H. Vineyard, Cover Illustration, Journal of Applied Physics **30**, (August, 1959). The description on page 1322 credits J. G. Gibson, A. N. Goland, M. Milgram, and G. H. Vineyard with the underlying work.

[163] J. Vollmer, T. Tél, and W. Breymann, "Equivalence of Irreversible Entropy Production in Driven Systems: An Elementary Chaotic Map Approach", Physical Review Letters **79**, 2759-2762 (1997).
[164] L. P. Wheeler, *Josiah Willard Gibbs. The History of a Great Mind* (Yale University Press, 1951).
[165] W. W. Wood and J. D. Jacobson, "Preliminary Results from a Recalculation of the Monte Carlo Equation of State of Hard Spheres", Journal of Chemical Physics **27**, 1207-1208 (1957).
[166] W. W. Wood, "Early History of Computer Simulations in Statistical Mechanics", 3-14 in *Proceedings of the International School of Physics "Enrico Fermi"*, Course **97**, *Molecular Dynamics Simulation of Statistical Mechanical Systems*, G. P. F. Ciccotti and Wm. G. Hoover, Editors (North-Holland, 1986).
[167] W. W. Wood, "On Some Additional Recollections, and the Absence Thereof, About the Early History of Computer Simulations in Statistical Mechanics", 908-911 in *Monte Carlo and Molecular Dynamics of Condensed Matter Systems*, K. Binder and C. P. F. Ciccotti, Editors (Italian Physical Society, Bologna, 1996).
[168] K. Zetie, "Time's Quantum Arrow Revisited", Contemporary Physics **39**, 393-395 (1998).
[169] D. N. Zubarev, *Nonequilibrium Statistical Thermodynamics* (New York Consultants, 1974).
[170] R. W. Zwanzig, "Time-Correlation Functions and Transport Coefficients in Statistical Mechanics", Annual Reviews of Physical Chemistry **16**, 67-102 (1965).

Index

algorithm 16, 61, 99
Anosov systems 344, 345
atomistic heat flow 195
atomistic R-Bénard flow 128, 154
atomistic shock 203, 229
attractor 190, 302
attractor-repeller pairs 190, 290

Baker Map 21, 68
ball-plate problem 250
Banach-Tarski paradox 377
Bernoulli map 127
Big Bang 340
birthday problem 11, 350
bit reversibility 44, 236
bit reversible momentum 237
bit-reversible shockwave 62, 236
Block Universe 341
Boltzmann entropy 321
Boltzmann eqn shockwave 353, 358
Boltzmann's equation 15, 320
Boltzmann's equation irreversibility 322
Boltzmann's H Theorem 321, 369
boundary conditions 128, 184, 263
box-counting dimension 284

canonical distribution 92
Cantor set 285
Central Limit Theorem 339
chaos 278, 374
chaotic dbl Hookean pendulum 311

chaotic Rayleigh-Bénard flow 150
coarse-grained entropy 296, 297
coarse-grained Galton Board 192, 312
cold welding 96
color conductivity Lyapunov exp 314
conducting oscillator 192, 298
conducting oscillator chaos 299
conducting oscillator fractal 302
conducting oscillator periodicity 299
confined free expansion 108
conjugate momentum 47
conjugate pairing 283
continuity equation 242–244
continuum energy equation 115, 245
continuum motion equations 115, 245
continuum R-Bénard flow 135, 144
continuum shock 215–229
core potential 266
covariant Lyapunov vectors 309
cubic spline 249

Dettmann Hamiltonian 57, 168
diffusion equation 119, 336
dimensionality loss 315
Direct Simulation Monte Carlo 325, 354
dissipative Baker map 68
dissipative oscillator 192, 299
distributions from trajectories 347, 370
Doll's Tensor Hamiltonian 332
Doll's Tensor shear 184

doubly-thermostated oscillator 193, 300
doubly-thermostated osc fractal 302
dynamical matrix 281

effective viscosity 254
Einstein frequency 105
embedded-atom theory 254
energy flux 203
entropy 91, 93
entropy production 122, 177
equation of motion 47, 115
equilibrium Baker Map 21
equilibrium fluctuations 96, 123
equilibrium Galton Board 25
equilibrium Hookean Pendulum 29
ergodic theory 86
escape-rate theory 343
Eulerian coordinates 243
examples 372
extension in phase 89

fine-grained entropy 297
finite-difference methods 246
finite-element methods 248
finite precision 378
First Law of Thermodynamics 52
fluctuations 96, 123
Fluctuation Theorem 337
Fourier conductivity 117
fractal attractor 190, 290, 302
fractal dimension 284
fractal repeller 190, 290
fractals 68, 189, 361, 374
free expansion 108

Galton Board 25, 73, 189, 373
Galton Board conductivity 191
Gauss' Principle 55
generalized coordinates 47
Gibbs' entropy 91
Gibbs' free energy 82
Gibbs' statistical mechanics 327
Gram-Schmidt procedure 305
Green-Kubo shear 97
Green-Kubo viscosity 180, 332

Grüneisen model 104

Hamiltonian Mechanics 46
hard-disk virial series 106
hard-sphere transition 107
hard-sphere virial series 106
harmonic oscillator 43, 287
Hausdorff dimension 284
heat 51, 115
heat-conducting oscillator 192, 302
heat flow 195
heat flux vector 166
heat theorem 167
Helmholtz' free energy 93
Hookean pendulum 29
H Theorem 321

ideal gas thermometer 54, 58
image particle 265
impact parameter 326
information 10, 114, 352
information dimension 284, 296, 365
information theory 330
initial conditions 340
instability 232, 273
integer space 44, 67
inverted collision 322
irreversibility 6, 58, 60, 113
ireversibility from rev equations 351
irreversible equations 368
irreversible processes 116, 181
isolated system 126

Jaynes' information theory 330

Kaplan-Yorke conjecture 73, 286
Kaplan-Yorke dimension 286, 365
kinetic theory 320

Lagrange multiplier 282, 330
Lagrangian coordinates 243
Lagrangian derivative 244
Lagrangian Mechanics 46
Langevin Equation 60
leapfrog algorithm 44, 61
Levesque-Verlet Algorithm 44, 61
linear response theory 97, 329

Index

linear transport 116, 124
Liouville's Compressible Theorem 294
Liouville's Theorems 49, 172
local Lyapunov exponents 236, 377
local Lyapunov vectors 281
localized chaos 236, 316
Lorenz' Attractor 11, 142, 274, 278
Loschmidt 5
Lucy's weight function 100, 252
Lyapunov exponent 174, 274
Lyapunov instability 76
Lyapunov spectrum 279
Lyapunov spectrum algorithm 303

macroscopic simulation 241
map 21, 68, 349
mass flux 243
mathematics 374, 377
Maxwell-Cat equations 183, 223, 337
Maxwell's relaxation time 336
Mayers' virial series 106, 324
measure 286
mechanical equilibrium 92
microcanonical ensemble 50
microcanonical temperature 92
molecular dynamics 87
momentum flux 166
Monte Carlo algorithm 87
multifractal 378
mu space 93

negative shockwave exponent 377
Newton-Hamilton irreversibility 382
Newtonian mechanics 2
Newtonian pressure tensor 116
Newtonian viscosity 116
nonequilibrium ensembles 367
nonequilibrium entropy 122
nonequilibrium Galton Board 189
noneq. molecular dynamics 334
nonequilibrium steady state 295
nonequilibrium theromostats 376
normal mode 103
Nosé-Hoover mechanics 56, 365
Nosé-Hoover oscillator 32, 88, 276, 298

Nosé 55, 88
number dependence 180
numerical solution 42

Occam's Razor 7
offset vector 281
open system 5, 365
ordinary differential equations 164

partial differential equations 241
particle velocity 202
partition function 84
periodic boundary conditions 184
periodic orbit 338
periodic shear flow 184
phase space 47
Poincaré recurrence 11, 75
Poincaré section 21
pressure tensor 166
Prigogine 348

quantum physics 2, 378
quantum thermostat 89
quasiergodicity 86
quasiharmonic thermodynamics 104

random number generator 63
random walk fractal 288
rarefaction fan 237
R-Bénard problem 128, 135, 140
Rayleigh Number 135, 158
reference trajectory 281
repeller 190, 364, 381
reservoir 178
reversed trajectory 232, 236
reversibility paradox 15, 351
reversible work 51
Reynolds' Number 123
role of examples 372
roundoff error 192
Runge-Kutta algorithm 16, 164
Runge-Kutta reversibility 235

satellite propagation 307
satellite trajectory 281
scaled oscillator exponents 287

scattering angle 326
Second Law 52, 176, 375
Second Law from chaos 292, 296
self similarity 285
Shannon information 365
shape function 249
shear flow 184
shock compression 201
shock density profile 211
shock Lyapunov exponent 232
shock Lyapunov reversibility 236
shock slope 213
shock stability 200, 206
shock temperature 214
shock tensile waves 209
shockwave 181, 199, 209, 240, 373
shock width 200
sinewave profile 207
Sinai 85
Sinai-Ruelle-Bowen measures 344
sinusoidal profile 207
Sklar 340
Smale horseshoe 334
Smooth Galton Board 74
SPAM 251
SPAM averages 99
SPAM core potential 266
SPAM density 253
SPAM gradients 258
SPAM interpolation 168
SPAM kinetic energy 268
SPAM methods 251
SPAM motion equation 254
SPAM R-Bénard algorithm 255
SPAM Rayleigh-Bénard flows 262
stability 206, 279
statistical mechanics 81
steady state 291
stochastic dynamics 58
stochastic reservoir 58
Störmer-Verlet algorithm 75
strange attractor 53
stress 245
surface contributions 96
symmetry breaking 33, 379
symplectic property 50

temperature 54
tent map 274
tensile waves 208
tensor temperature shock 222
thermal boundary 121, 179
thermal diffusivity 120
thermal equilibrium 92
thermodynamic limit 105
thermodynamic reversibility 81
thermodynamics 50
thermomechanics 334
thermometer 54
thermostat 54
thermostated irreversibility 380
thermostated oscillator 373
thermostatics 51
time 2, 13
time ensemble 90
time reversed shock 234
time reversibility 39
time-reversible dissipative maps 348
time-reversible maps 348, 351
Time's Arrow 235
time scaling 56
trajectories 42
turbulence 11
two-grid algorithm 135

Uncertainty Principle 6

variational principle 248
virial series 106
virial theorem 167
viscometer 179
viscous conducting shock 220
viscous shock 217

wave equation 119
wave function 2
weak solution 248
weight function 252
work 51

Zermélo 5
Zeroth Law of Thermodynamics 54
Zubarev 330

The Hoovers moved from California to Ruby Valley Nevada after retiring from the Livermore Laboratory and the University of California. They thoroughly enjoy the mountainous cattle-ranching scene and the wonderful people that are their neighbors. This photograph was taken in early 2012.